101 Spy Gadgets for the Evil Genius

D1257285

Evil Genius™ Series

101 Spy Gadgets for the Evil Genius

Brad Graham

Kathy McGowan

Second Edition

New York   Chicago   San Francisco   Lisbon   London   Madrid
Mexico City   Milan   New Delhi   San Juan   Seoul
Singapore   Sydney   Toronto

The McGraw·Hill Companies

Cataloging-in-Publication Data is on file with the Library of Congress

McGraw-Hill books are available at special quantity discounts to use as premiums and sales promotions, or for use in corporate training programs. To contact a special sales representative, please visit the Contact Us page at www.mhprofessional.com.

101 Spy Gadgets for the Evil Genius, Second Edition

1 2 3 4 5 6 7 8 9 0   QDB/QDB   1 9 8 7 6 5 4 3 2 1

ISBN 978-0-07-177268-6
MHID   0-07-177268-5

The pages within this book were printed on acid-free paper.

Sponsoring Editor Judy Bass	**Copy Editor** Jay Boggis
Acquisitions Coordinator Bridget Thoreson	**Proofreader** Eina Malik, Cenveo Publisher Services
Editorial Supervisor David E. Fogarty	**Indexer** Robert Swanson
Production Supervisor Pamela A. Pelton	**Art Director, Cover** Jeff Weeks
Project Manager Sheena Uprety, Cenveo Publisher Services	**Composition** Cenveo Publisher Services

Contents

PART FOUR   GPS Location Tracking

PART FIVE   Transmission and Interception

PART SIX   Personal Protection

PART SEVEN   Digital Camera Spy Projects

Acknowledgments

Sincerest thanks always to our Evil Genius collaborator, Judy Bass, Senior and Series Editor at McGraw-Hill Professional in New York. Your enthusiasm and support keep us motivated to forge ahead and never give up. You are a very special lady.

Also thanks to many people who take an interest in our projects and encourage us to pursue our dreams. We would especially like to thank Nathan Seidle, CEO of SparkFun Electronics Inc who sent us a wonderful assortment of electronics components and evaluation products to use in our Evil Genius labs! Thanks also go out to all of the people that make SparkFun.com a great resource and community for like minded Evil Geniuses, hackers, and electronics hobbyists.

About the Authors

Brad Graham is an inventor, robotics hobbyist, founder and host of atomiczombie.com Web site, which receives over 2.5 million hits monthly, and a computer professional. He is the coauthor of *101 Spy Gadgets for the Evil Genius, Atomic Zombie's Bicycle Builder's Bonanza*, perhaps the most creative bicycle-building guide ever written, and *Build Your Own All-Terrain Robot*, from McGraw-Hill Professional. Technical manager of a high-tech firm which specializes in computer network setup and maintenance, data storage and recovery, and security services, Mr. Graham is also a Certified Netware Engineer, a Microsoft Certified Professional, and a Certified Electronics and Cabling Technician.

Kathy McGowan provides administrative, logistical and marketing support for Atomic Zombie's™ many robotics, bicycle, technical, and publishing projects. She also manages the daily operations of their high-tech firm and several Web sites, including atomiczombie.com, as well as various Internet-based blogs and forums. Additionally, Ms. McGowan writes articles for ezines, and is collaborating with Mr. Graham on several film and television projects.

PART ZERO
Introduction

Getting Started

Welcome "Noobs"!

If you have been experimenting with electronics for any amount of time, then chances are you can skip right past this chapter and start digging into some of the projects presented in this book. If you are just starting out, then you have a little groundwork to cover before you begin; but don't worry, the electronics hobby is well within reach for anyone with a creative mind and a desire to learn something new.

Like all things new, you have to start from the beginning and expect a few failures along the way. We electronic "nerds" call that "letting out the magic smoke," and you will fully understand this term the first time you connect your power wires in reverse! Don't be intimidated by the huge amount of technical material available on electronic components and devices, because chances are you only need a small amount of what is available to complete a project. All problems can be broken down into smaller parts, and a schematic diagram is a perfect example of this. Once you understand the basic principles, you will be able to look at a huge schematic or circuit and see that it is made up of smaller basic building blocks, just like a brick wall is composed of individual bricks.

Because of limited space in this book, we will cover only the essential basics you need in order to get started in this fun and rewarding hobby, but there are thousands of resources available to research as you move forward one step at a time. You have the Evil Genius itch—now all you need is a good pile of junk and a few basic tools to set your ideas into motion!

The Breadboard

This oddly named tool is probably the most important prototyping device you will ever own, and it is absolutely essential to this hobby. A breadboard or "solderless breadboard" is a device that lets you connect the leads of semiconductors together without wires so you can test and modify your circuit easily without soldering. Essentially, it is nothing more than a board full of small holes that interconnect in rows so you can complete a circuit. In the early days, our Evil Genius forefathers would drive a bunch of nails into an actual board (like a cutting board for bread) and then connect their components to the nails. So you can thank those pioneering Evil Geniuses who once sat in their workshops with a breadboard full of glowing vacuum tubes and wires for the name!

Today's breadboards look nothing like the originals, often containing hundreds of rows to accommodate the increasing complexity and pin count of today's circuitry. It is common to have 50 or more complex integrated circuits (ICs) on a breadboard running at speeds of up to 100 megahertz (MHz), so a lot can be done with breadboards. One of our latest breadboard projects was a fully functional 8-bit computer with a double-buffered video graphics array (VGA) output and complex sound generator. This project ran flawlessly on a breadboard at speeds of 40 MHz, and had an IC count of over 30, so don't let anyone tell you a breadboard is only for simple low-speed prototyping. Let's have a look at a typical solderless breadboard that can be purchased at most electronics supply outlets.

Figure 0-0 A typical solderless breadboard.

A breadboard like the one shown in Figure 0-0 will typically cost you around $30 and will provide years of use. Without a breadboard, you would have to solder your components together and hope your design worked on the first try—something that is only a pipe dream in this hobby. The connections under the plastic holes are designed so that the power strips (marked + and −) are connected horizontally, and the prototyping area holes are connected vertically. The small gutter between the prototyping holes is there so your ICs can press into the board with each row of legs on each side of the gutter. Figure 0-1 shows a close-up of the interconnections underneath the plastic board.

As you can see, the power strip holes connect horizontally, and the prototyping holes connect vertically. This way, you can have power along the entire strip, since ground (GND) and power (VCC)

often have multiple connection points in a circuit. Once you are familiar with a breadboard, it is easy to rig up a test circuit in minutes, even one with a high component count. Once your circuit is tested and working, you can move it to a more permanent home such as a copper-clad board or even a real printed circuit board (PCB).

To make the connections from one row of holes to another, you need wires—many wires. Breadboard wires should be solid, not stranded, have about ¼-inch of bare wire at the end and come in multiple colors and lengths to make tracing your circuit easier. You can purchase various breadboard-ready wiring packs from electronics supply stores, but when you get into larger prototyping, it may get expensive to purchase as many wires as you need. The best solution we have found is to get a good length of "Cat5" wire, which is used for computer networking and then strip the ends for use on the breadboard. The nice thing about Cat5 wire is that it has eight colored wires with a solid copper core that are the perfect size to fit into the breadboard. Figure 0-2 shows some of the Cat5 wiring being cut and stripped for use in our breadboard.

Cat5 wiring comes as four twisted pairs, so you can just cut a bunch of lengths, unwind the wires,

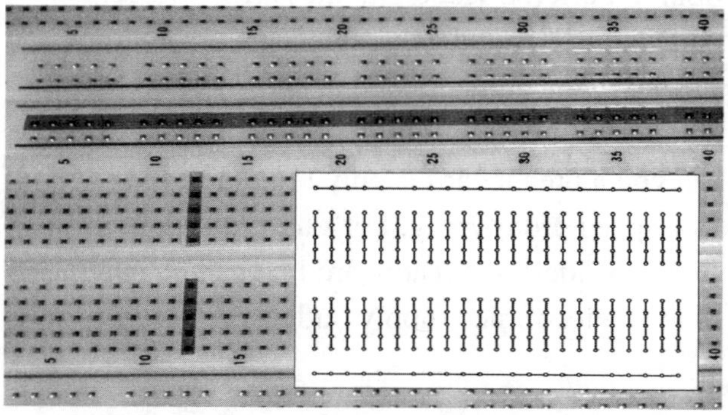

Figure 0-1 Connections between holes on the breadboard.

Figure 0-2 You can never have enough wires.

and then strip the ends of the plastic sheathing using a dull utility knife. Just place the wire over a dull blade and press down with your thumb to score the end so it pulls away from the wire. An actual wire stripper works just as well, but the dull blade seems a lot faster when you want 100 or more wires for your breadboard. For starters, you will need about 20 wires (each) in lengths of 1 inch (in), 2 in, 4 in, 6 in, and a few longer wires for external devices. The 1-in wires should also include red and green (or similar) colors that easily identify your power connections. The power wires will be the most-used wires on your board, so make sure you have enough of them to go around.

Figure 0-3 shows why it is important to have a dull blade for stripping the wires if you choose to do it this way. We purposely sanded the edge of this utility knife so it would be sharp enough to score the wiring shield, yet not cut a thumb after repeated stripping of hundreds of wires in a row. A wire stripping tool also works well, but it is a little slow when we need 128 blue bus wires for a circuit we want to complete before the end of the night. Now that you have a breadboard and an endless supply of wires to press into the holes, you can begin prototyping your first circuit. Learning to look at a schematic, identifying the semiconductors, and then transplanting the connections to your breadboard will open up an entire world of fun, so let's find out how it is done.

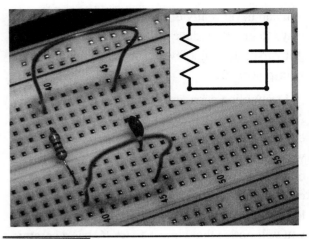

Figure 0-4   Learning to use the breadboard.

Figure 0-4 shows a simple schematic of a capacitor and a resistor wired in parallel. This wiring is transferred to the breadboard by placing the leads of the two components into the holes and then using the wires to join the rows. Because the holes are all connected to each other in vertical rows, the wires can be placed in any one of the five holes that make a row. Although a breadboard circuit may have 500 wires, there is not much more to it than that. Now you can see why prototyping a circuit on a breadboard is easy and lends itself well to modification. There are, however, a few breadboard "gotchas" to keep in mind, and capacitive noise or "crosstalk" is one of them.

Crosstalk or capacitive noise can be a real problem on a breadboard because the metal plates that make up rows of holes are close enough together to act as capacitors. This can induce noise into your circuit, possibly causing it to fail or act differently than expected. Radio frequency (RF) or high-speed digital circuits are prone to noise and crosstalk, and this can create all kinds of Evil Genius headaches. Sometimes you can design a high-speed or RF circuit on a breadboard and have it work perfectly, only to find that it completely fails or acts differently when redone on a permanent circuit board. The capacitance of the plates is actually part of the circuit now! Although

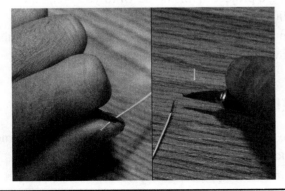

Figure 0-3   Stripping the Cat5 wiring for the breadboard.

you can never really eliminate this error, there are ways to greatly reduce noise on a breadboard, and it involves adding a few decoupling capacitors on your power strips.

Decoupling capacitors act as filters so RF and AC noise don't leak into your power source, causing havoc throughout the entire circuit. When working with high-speed logic, microcontrollers on a clock source, or RF circuits, decoupling capacitors are very important and should not be left out. If you look at an old logic circuit board, you will notice that almost every IC has a ceramic capacitor nearby or directly across the VCC and ground pins. These capacitors are nothing more than 0.01 microfarads (μF) ceramic capacitors placed between VCC and ground on each of your power supply rails as shown in Figure 0-5. Also notice in the figure that the power rails need to be connected to each other, as each strip is independent. If you forget to connect a rail, it will carry neither VCC nor be grounded, so your circuit will fail. Usually decoupling capacitors on each end of the powers strips will be adequate, but in a large high-speed or RF circuit, you may need them closer to the IC power lines or other key components.

When your designs become very large or complex, the standard breadboard may not offer

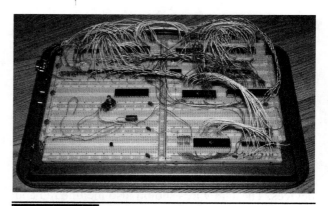

Figure 0-6   Breadboard circuits can get large!

enough real estate. But not to worry—you can purchase individual breadboard sections and snap them together to make a larger prototyping area. Figure 0-6 shows 10 breadboard sections snapped together and then bolted to a steel cookie sheet in order to create a grounding plate. The steel base also helps reduce noise, and all breadboards should have a metal base. This massive cookie sheet breadboard circuit is a fully functional 20-MHz video computer that can display high-resolution graphics to a VGA monitor and generate complex multichannel sound. The entire computer was designed on the breadboard shown in Figure 0-6 and went directly to the final design stage based on the breadboard circuit. We have built high-speed computer systems running at more than 75 MHz on a breadboard, as well as high-power video transmitters, robot motor controllers, and every single project in this book. Until you break the 100-MHz barrier, there is not much you can't do on a breadboard, so become good friends with this powerful prototyping tool!

Electronic Building Blocks

So you've just found a cool schematic on the Internet and you have a brand new breadboard with 100 wires waiting to find a home, but where do you get all of those components? If you have been doing this for a while, then your "junk box" is probably well-

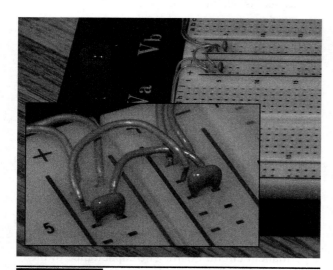

Figure 0-5   Fighting breadboard noise.

equipped, but for those just starting out, you have to be resourceful in order to keep your budget under control. A simple circuit with 10 small components might only cost you $5 at the local electronics shop, but often you may need a lot more than 10 parts or might need some uncommon semiconductors. The best source for free electronic components is from old circuit boards. Dead VCRs, fried TVs, baked radios, even that broken coffeemaker will have a pile of usable components on the circuit board. A small radio PCB might have 200 semiconductors soldered to it and at 50 cents a piece, that adds up fast. You may never use all of the components, but on a dreary day when you are in your mad scientist lab in need of some oddball resistor value, a box of scrap circuit boards is great to have.

We keep several large boxes full of circuit boards that we find, and often discover that they yield most of the common parts we need and often have hard-to-find or discontinued ICs that we need when working on older schematics. Figure 0-7 shows one of the 20 or more large boxes of scrap PCBs we have collected over the years. Removing the parts from an old circuit board is easy, especially the simple 2 or 3 pin parts, like capacitors and transistors. For the larger ICs with many pins, you will need a desoldering tool, also known as a "solder sucker." A low-budget soldering iron (34 watts [W] or higher) with a blunt tip and a solder sucker can make easy work of pulling

Figure 0-8   Removing parts from an old circuit board.

parts from old PCBs. Figure 0-8 shows the hand-operated solder sucker removing an 8-pin IC from an old VCR main board. To operate a solder sucker, you press down on a loading lever, heat up the pad to desolder, and then press a button to suck up the solder away from that pad.

We prefer to desolder using a cheap iron with a higher wattage and fatter tip than used for normal work because it heats up the solder faster, reaching both sides of the board easier. There are other tools that can be used to desolder parts, such as desoldering wicks, spade tips, and even pump vacuums, but the ten-dollar solder sucker has always done the job for us, even on ICs with as many as 40 pins without any problems at all. When you really start collecting parts, you will find that a giant bowl of resistors is more of a pain that having to desolder a new one, so some type of organization will be necessary. You will soon discover that there are many common parts when it comes to resistors, capacitors, transistors, and ICs, so having them sorted will make it very easy to find what you need. Storage bins such as the ones shown in Figure 0-9 are perfect for your electronic components, and you can easily fill 100 small drawers with various parts, so purchase a few of them to get started.

We have an entire closet full of these component drawers and the larger parts or PCBs are stored in

Figure 0-7   Old PCBs are a goldmine of parts.

Figure 0-9   Organization makes finding parts a snap.

plastic tubs. It's a rare day when we can't find the parts we need, even for a retro project that needs some long-discontinued component. Of course, there are always times when you want new parts or need something special, so one of the many online sellers will be glad to take your money and send you the part in a few days.

The Resistor

If you are new to this hobby, then you have probably seen a few schematics and thought that they made about as much sense as cave hieroglyphics. Don't worry—that knowledge will come as you use schematics more and start to decode the electronic component datasheets. If you want to take the fast track, then you might consider getting a book on basic electronics to get you kick-started, but for those who want to learn as you go, here are a few of the basics you will need to complete the projects in this book.

Resistors, like the ones shown in Figure 0-10, are the most basic of the semiconductors you will be using, and they do exactly what their name implies—they resist the flow of current by exchanging some current for heat which is dissipated through the body of the device. On a large circuit board, you could find hundreds

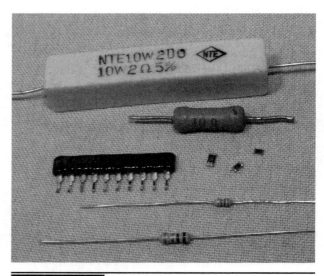

Figure 0-10   Resistors with fixed values.

of resistors populating the board, and even on tiny circuit boards with many surface-mounted components, resistors will usually make up the bulk of the semiconductors. The size of the resistor generally determines how much heat it can dissipate and will be rated in watts, with ¼ and 1/8 W being the most common type you will work with (the two bottom resistors shown in Figure 0-10). Resistors can become very large, and will require ceramic based bodies, especially if they are rated for several watts or more like the 10-W unit shown at the top of Figure 0-10. To save space, some resistors come in packs, like the one in the figure that has multiple legs.

Because of the recent drive to make electronics more "green" and power conservative, large, power-wasting resistors are not common in consumer electronics these days; instead, it is more efficient to convert amperage and voltage using some type of switching power supply or regulator, rather than by letting a fat resistor burn away the energy as heat. On the other hand, small-value resistors are very common, and you will find yourself dealing with them all of the time for simple tasks, such as driving a light-emitting diode (LED) with limited current, pulling up an input pin to a logical "one" state, biasing a simple transistor amplifier, and thousands of other common functions. On most common axial lead resistors, like the ones you will most often use in your projects, the value of the resistor is coded onto the device in the form of four colored bands which tell you the resistance in ohms. Ohms are represented using the Greek omega symbol (Ω), and will often be omitted for values over 999 ohms (Ω), which will be stated as 1 K, 15 K, 47 K, or some other number followed by the letter K, indicating the value is in kilohms (thousands of ohms). Similarly, for values over 999 K, the letter M will be used to show that 1 M is actually 1 megohm, or one million ohms. In a schematic diagram, a resistor is represented by a zigzag line segment, as shown in Figure 0-11, and will either have a letter and a number such as R1 or V3 relating to a parts list, or will simply have the value printed next to it such as 1 M, or 220 Ω. The schematic symbol on the left of Figure 0-11 represents a variable resistor, which

can be set from zero ohms to the full value printed on the body of the variable resistor.

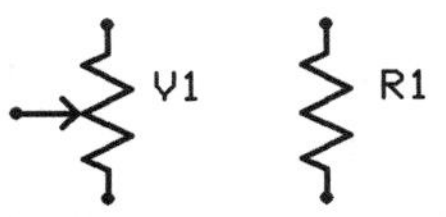

A variable resistor is also known as a "potentiometer," or "pot," and it can take the form of a small circuit board–mounted cylinder with a slot for a screwdriver, or as a cabinet-mounted can with a shaft exiting the can for mating with some type of knob or dial. When you crank up the volume on an amplifier with a knob, you are turning a potentiometer. Variable resistors are great for testing a new design, since you can just turn the dial until the circuit performs as you want it to, and then remove the variable resistor to measure the impedance (resistance) across the leads in order to determine the best value of fixed resistor to install. On a variable resistor, there are usually three leads—the outer two connect to the fixed carbon resistor inside the can, which gives the variable resistor its value, and a center pin connects to a wiper, allowing the selection of resistance from zero to full. Several common variable resistors are shown in Figure 0-12, with the top-left unit dissected to show the resistor band and wiper.

As mentioned earlier, most fixed value resistors will have four color bands painted around their bodies, which can be decoded into a value as shown in Table 0-0. At first, this may seem a bit illogical, but once you get the hang of the color band decoding, you will be able to recognize most

Figure 0-11   Variable (left) and fixed (right) resistor symbols.

Figure 0-12   Common variable resistors.

TABLE 0-0	Resistor Color Codes Chart			
Color	1st band	2nd band	3rd band	Multiplier
Black	0	0	0	1 Ω
Brown	1	1	1	10 Ω
Red	2	2	2	100 Ω
Orange	3	3	3	1 K
Yellow	4	4	4	10 K
Green	5	5	5	100 K
Blue	6	6	6	1 M
Violet	7	7	7	10 M
Grey	8	8	8	
White	9	9	9	0.1
Gold				0.01
Silver				

common values at first glance without having to refer to the chart.

There will almost always be either a silver or gold band included on each resistor, and this will indicate the end of the color sequence, and will not become part of the value. A gold band indicates the resistor has a 5 percent tolerance (margin of error) in the value, so a 10-K resistor could end up being anywhere from 9.5 to 10.5 K in value, although in most cases will be very accurate. A silver band indicates the tolerance is only 10 percent, but we have yet to see a resistor with a silver band that was not on a circuit board that included vacuum tubes, so forget that there is even such a band! Once you ignore the gold band, you are left with three color bands that can be used to determine the exact value as given in Table 0-0.

So, let's assume that we have a resistor with the color bands brown, black, red, and gold. We know that the gold band is the tolerance band and the first three will indicate the values to reference in the chart. Doing so, we get 1 (brown), 0 (black), and 100 Ω (red). The third band is the multiplier, which would indicate that the number of zeroes following the first to values will be 2, or the value is simply multiplied by 100 Ω. This translates to a value of 1000 Ω, or 1 K (10 × 100 Ω). A 370-K

resistor would have the colors orange, violet, and yellow followed by a gold band. You can check the value of the resistor when it is not connected to a circuit by simply placing your millimeter on the appropriate resistance scale and reading back the value. We do not want to get too deep into electronics formulas and theory here, since there are many good books dedicated to the subject, so we will simply leave you with two basic rules regarding the use of resistors: (1) put them in series to add their values together, and (2) put them in parallel to divide them. This second simple rule works great if you are in desperate need of a 20-K resistor for instance, but can only find two 10-K resistors to put in series. In parallel, they will divide down to 5 K. Now you can identify the most common semiconductor that is used in electronics today—the resistor. Now we will move ahead to the next most common semiconductor—the capacitor.

The Capacitor

A capacitor in its most basic form is a small rechargeable battery with a very short charge and discharge cycle. Where a typical AAA battery may be able to power an LED for a full year, a capacitor of similar size will power it for only a few seconds before its energy is fully discharged. Because capacitors can store energy for a predictable duration, they can perform all kinds of useful functions in a circuit, such as filtering AC waves, creating accurate delays, removing impurities from a noise signal, and creating clock and audio oscillators. Because a capacitor is basically a battery, many of the large ones available look much like batteries with two leads connected to one side of a metal can. As shown in Figure 0-13, there are many sizes and shapes of capacitors, some of which look like small batteries.

Just like resistors, capacitors can be as large as a garbage can, or as small as a grain of rice, it really depends on the value. The larger devices can store a lot more energy. Unlike batteries, some capacitors

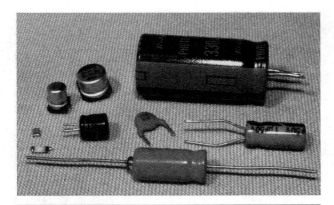

Figure 0-13 **Various common capacitors.**

are nonpolarized, and they can be inserted into a circuit regardless of current flow, while some cannot. The two different types of capacitors are shown by their schematic symbols in Figure 0-14, C1 being a nonpolarized type, and C2 a polarized type. Although there are always exceptions to the rules, generally the disc-style capacitors are nonpolarized, and the larger can-style electrolytic types are polarized. An obvious indicator of a polarized capacitor is the negative markings on the can, which can be clearly seen in the larger capacitor shown at the top right of Figure 0-13.

Another thing that capacitors have in common with batteries is that polarity is very important when inserting polarized capacitors into a circuit. If you install an electrolytic capacitor in reverse and attempt to charge it, the part will likely heat up and release the oil contained inside the case causing a circuit malfunction or dead short. In the past, electrolytic capacitors did not have a pressure release system, and they would explode like firecrackers when overcharged or installed in reverse, leaving behind a huge mess of oily

paper and a smell that was tough to forget. On many capacitors, especially the larger can style, the voltage rating and capacitance value is simply stamped on the case. A capacitor is rated in voltage and in farads, which defines the capacitance of a dielectric for which a potential difference of 1 volt (V) results in a static charge of 1 coulomb (C). This may not make a lot of sense until you start messing around with electronics, but you will soon understand that typically, the larger the capacitor, the larger the farad rating will be, thus the more energy it can store. Since a farad is quite a large value, most capacitors are rated in microfarads (μF), such as the typical value of 4700 μF for a large electrolytic filter capacitor, and 0.1 μF for a small ceramic disc capacitor. Picofarads (pF) are also used to indicate very small values, such as those found in many ceramic capacitors or adjustable capacitors used in radio frequency circuits (a pF is one-millionth of a μF). On most can-style electrolytic capacitors, the value is simply written on the case and will be stated in microfarads and voltage along with a clear indication of which lead is negative. Voltage and polarity are very important in electrolytic capacitors, and they should always be inserted correctly, with a voltage rating higher than necessary for your circuit. Ceramic capacitors will usually only have the value stamped on them if they are in picofarads for some reason, and often no symbol will follow the number, just the value. Normally, ceramic capacitors will have a three-digit number that needs to be decoded into the actual value, and this evil confusing scheme works, as shown in Table 0-1.

Who knows why they just don't write the value on the capacitor? It would have the same amount of digits as the code! Oh well, you get used to seeing these codes, just like resistor color bands, and in no time you will easily recognize the common values such as 104, which would indicate a 0.1 μF value according to the chart. Capacitors behave just like batteries when it comes to parallel and series

C1   C2
         +

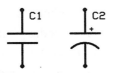

Figure 0-14 **Capacitor symbols.**

TABLE 0-1 Ceramic Capacitor Value Chart

Value, pF	Marking	Value, μF (pF)	Marking	Value, μF	Marking
10	10 or 100	0.001	102	0.10	104
12	12 or 120	0.0012 (1200)	122	0.12	124
15	15 or 150	0.0015	152	0.15	154
18	18 or 180	0.0018 (1800)	182	0.18	184
22	22 or 220	0.0022	222	0.22	224
27	27 or 270	0.0027	272	0.27	274
33	33 or 330	0.0033	332	0.33	334
39	39 or 390	0.0039	392	0.39	394
47	47 or 470	0.0047	472	0.47	474
58	58 or 580	0.0056	562	0.56	564
68	68 or 680	0.0068	682	0.68	684
82	82 or 820	0.0082	822	0.82	824
100	101	0.01	103	1	105 or 1 μf
120	121	0.012	123		
150	151	0.015	153		
180	181	0.018	183		
220	221	0.022	223		
270	271	0.027	273		
330	331	0.033	333		
390	391	0.039	393		
470	471	0.047	473		
560	561	0.056	563		
680	681	0.068	683		
820	821	0.082	823		

connections, so in parallel two identical capacitors will handle the same voltage as a single unit, but double their capacitance rating, and in series, they have the same capacitance rating as a single unit, but can handle twice the voltage. So if you need to filter a very noisy power supply, you might want to install a pair of 4700-μF capacitors in parallel to end up with a capacitance of 9400 μF. When installing parallel capacitors, make sure that the voltage ratings of all the capacitors used are higher than the voltage of that circuit, or there will be a failure—ugly, noisy smelly failure!

The Diode

Diodes allow current to flow through them in one direction only so they can be used to rectify AC into DC, block unwanted current from entering a device, protect a circuit from a power reversal, and even give off light in the case of LEDs. Figure 0-15 shows various sizes and types of diodes, including an easily recognizable LED and the large full-wave rectifier module at the top. A full-wave rectifier is just a block containing four large diodes inside.

Like most other semiconductors, the size of the diode is usually a good indication of how much current it can handle before failure, and this information will be specified by the manufacturer by referencing whatever code is printed on the diode to some datasheet. Unlike resistors and capacitors, there is no common mode of identifying a diode unless you get to know some of the most common manufacturers' codes by memory, so you will be forced to look up the datasheet on the Internet or in a cross reference catalog to determine the exact value and purpose of unknown diodes.

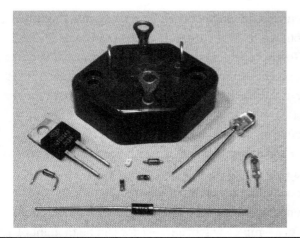

Figure 0-15   Several style of diodes including an LED.

For example the NTE6248 diode shown in Figure 0-15 in the TO220 case (left side of photo) has a datasheet that indicates it is a Schottky barrier rectifier with a peak reverse voltage maximum of 600 V and a maximum forward current rating of 16 amps (A). Datasheets will tell you everything you need to know about a particular device, and you should never exceed any of the recommended values if you want a reliable circuit. The schematic symbol for a diode is shown in Figure 0-16, D1 being a standard diode, and the other a LED (the two arrows represent light leaving the device).

The diode symbol shows an arrow (anode) pointing at a line (cathode), and this indicates which way current flows (from the anode to the cathode, or in the direction of the arrow). On many small diodes, a stripe painted around the case will indicate which end is the cathode, and on LEDs, there will be a flat side on the case nearest the cathode lead. LEDs come in many different sizes, shapes, and wavelengths (colors), and have

ratings that must not be exceeded in order to avoid damaging the device. Reverse voltage and peak forward current are very important values that must not be exceeded when powering LEDs or damage will easily occur, yet at the same time, you will want to get as close as possible to the maximum values if your circuit demands full performance from the LED, so read the datasheets on the device carefully. Larger diodes used to rectify AC or control large current may need to be mounted to the proper heat sink in order to operate at their rated values, and often the case style will be a clear indication due to the metal backing, or mounting hardware that may come with the device. Unless you know how much heat a certain device can dissipate in open air, your best bet is to mount it to a heat sink if it was designed to be installed that way. Like most semiconductors, there are thousands of various sizes and types of diodes, so make sure you are using a part rated for your circuit, and double-check the polarity of the device before you turn on the power for the first time.

The Transistor

A transistor is one of the most useful semiconductors available, and often the building block for many larger integrated circuits and components, such as logic gates, memory, and microprocessors. Before transistors became widely used in electronics, simple devices like radios and amplifiers would need huge wooden cabinets, consume vast amounts of power, and emit large wasteful quantities of heat because of vacuum tubes. The first general-purpose computer, ENIAC (Electronic Numerical Integrator and Computer) was a vacuum tube–based computer that used 17,468 vacuum tubes, 7200 crystal diodes, 1500 relays, 70,000 resistors, 10,000 capacitors and had more than 5 million hand-soldered joints. It weighed 30 tons and was roughly 8 by 3 by 100 feet, and consumed 150 kW of power! A simple computer that would rival the power of

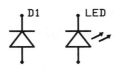

Figure 0-16   Diode schematic symbol (LED on the right).

Figure 0-17   **Various common transistors.**

this power-hungry monster could easily be built on a few square inches of perforated board using a few dollars in parts today by any electronics hobbyist, thanks to the transistor. A transistor is really just a switch that can control a large amount of current by switching a small amount of current, thus creating an amplifier. Several common types and sizes of transistors are shown in Figure 0-17.

Depending on how much current a transistor is designed to switch, it may be as small as a grain of rice or as large as a hockey puck and require a massive steel heat sink or fan to operate correctly. There are thousands of varying transistor types and sizes, but one thing most of them have in common is that they will have three connections that can be called "collector," "emitter," and "base," and they will be represented by one of the two schematic symbols shown in Figure 0-18.

The emitter (E), base (B), and collector (C) on both the NPN and PNP transistors do the same job. The collector/emitter current is controlled by the

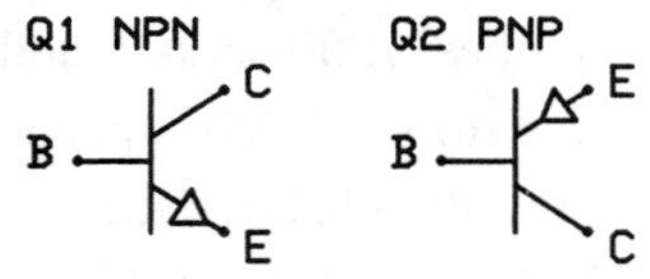

Figure 0-18   **NPN and PNP transistor schematic symbols.**

current flowing between base and emitter terminals, but the flow of current is opposite in each device. Today, most transistors are NPN because it is easier to manufacture a better NPN transistor than a PNP, but there are still occurrences when a circuit may use a PNP transistor because of the direction of current, or in tandem with an NPN transistor to create a matched pair. There is enough transistor theory to cover 10 books of this size, so we will condense that information in order to help you understand the very basics of transistor operation. As a simple switch, a transistor can be thought of as a relay with no mechanical parts. You can turn on a high-current load, such as a light or motor, with a very weak current, such as the output from a logic gate, or light-sensitive photocell. Switching a large load with a small load is very important in electronics, and transistors do this perfectly and at speed that a mechanical switch such as a relay could never come close to achieving. An audio amplifier is nothing more than a very fast switch that takes a very small current, such as the output from a CD player, and uses it as the input into a fast switch that controls a large current, such as the DC power source feeding the speakers. Almost any transistor can easily operate well beyond the frequency of an audio signal, so they are perfectly suited for this job. At much higher frequencies, like those used in radio transmitters, transistors do the same job of amplification, but they are rated for much higher frequencies sometime into the gigahertz range.

Another main difference between the way a mechanical switch and a transistor work is that a transistor is not simply an on/off switch, it can operate as an "analog" switch, varying the amount of current switched by varying the amount of current entering the base of the transistor. A relay can turn on a 100-W light bulb if a 5-V current is applied to the coil, but a transistor could vary the intensity of the same light bulb from zero to full brightness depending on the voltage at the base of the transistor. Like all semiconductors, the transistor must be rated for the job you intend it to do, so maximum current, switching voltage, and speed are things that need to be considered when choosing the correct part. The datasheet for a very common NPN transistor, the 2N2222 (which can be substituted for the 2N3904 often used in this book) is shown in Figure 0-19.

From this datasheet, we can see that this transistor can switch about half a watt (624 mW) with a voltage of 6 V across the base and emitter junction. Of course, these are maximum ratings, so you might decide that the transistor will work safely in a circuit if it had to switch on a 120 mW LED from a 5-V logic level input at the base. As a general rule, look at the maximum switching current of a transistor, and never ask it to handle more than half of the rated maximum value, especially if it was the type of transistor designed

to be mounted to a heat sink. The same principle applies to maximum switching speed—don't expect a 100-MHz transistor to oscillate at 440 MHz in an RF transmitter circuit, since it will have a difficult enough time just reaching the 100-MHz level.

RTFM—Read The Flippin' Manual!

Let's face it, Evil Geniuses don't often read manuals and prefer to learn by trial and error, which really is the best way to learn most of the time. When it comes to determining the specs on an electronic component with more than two pins, you really have no choice but to read the datasheet in order to figure out how it works to avoid letting out the smoke. For transistors and ICs, this is especially

Amplifier Transistors
NPN Silicon

MAXIMUM RATINGS

Rating	Symbol	Value	Unit
Collector–Emitter Voltage	V_{CEO}	40	Vdc
Collector–Base Voltage	V_{CBO}	75	Vdc
Emitter–Base Voltage	V_{EBO}	6.0	Vdc
Collector Current — Continuous	I_C	600	mAdc
Total Device Dissipation @ T_A = 25°C Derate above 25°C	P_D	625 5.0	mW mW/°C
Total Device Dissipation @ T_C = 25°C Derate above 25°C	P_D	1.5 12	Watts mW/°C
Operating and Storage Junction Temperature Range	T_J, T_{stg}	−55 to +150	°C

THERMAL CHARACTERISTICS

Characteristic	Symbol	Max	Unit
Thermal Resistance, Junction to Ambient	R_{0JA}	200	°C/W
Thermal Resistance, Junction to Case	R_{0JC}	83.3	°C/W

P2N2222A

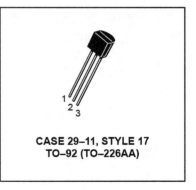

CASE 29–11, STYLE 17
TO–92 (TO–226AA)

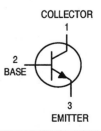

COLLECTOR
1

2
BASE

3
EMITTER

Figure 0-19   Datasheet for the common 2N2222 NPN transistor.

true as you really have no idea what is inside the "black box" without the datasheet. That 8-pin IC may be just a simple timer or it could be a state-of-the-art 100-Mhz microprocessor with a built-in USB port and a video output. Without the datasheet, you would never know. There are also times when you find some cool schematic on the Internet and it has a parts list that you determine contains mostly discontinued transistors, so you will have to compare the datasheets to find suitable replacements. Once you learn the basics, you will be able to make smart substitutions for almost any part.

You can dig up just about any datasheet on any device, even those that have been off the market for decades, so learn how to find them and you will never be in the dark when it comes to component specs. The datasheet for the very common 2N2222 transistor was found by typing "2N2222 datasheet" into Google (Figure 0-20) and looking for the PDF file. Although there are a few bogus datasheet servers out there that try to suck you into joining so they can spam your e-mail address for life, most of the time you will find a datasheet with little or no fuss. If you know the manufacturer of the device, try

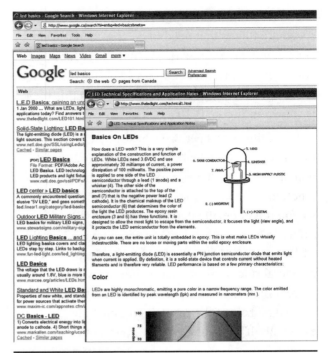

Figure 0-21 Finding the basics online.

their Web site first, or try a large online electronic supply site like Digikey.com as they will have online datasheets for thousands of components. The Internet is also a great source for general information on electronics, so if you find yourself needing to know the basics, consult your favorite search engine.

The wealth of information found by searching "LED basics" was almost endless. Even the first site (Figure 0-21) has more than enough information on the LEDs presented in a very easy to understand format. The fact is that almost all of the information you will need when learning electronics can be found on the Internet with a little patience.

Asking for Help

When the Internet fails to provide you with answers or you really feel you need guidance from those who may know the answers, there are countless forums that you can join and look for help or discuss your projects with other Evil Geniuses. Like all things in life, forums have their own

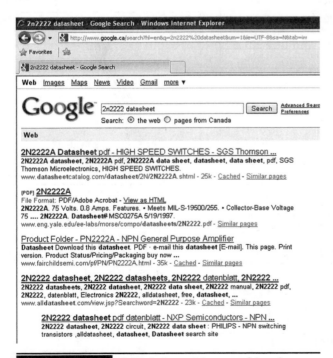

Figure 0-20 The Internet is your friend.

special rules of etiquette, so before you jump in and scream "help me!" please take the time to read their posting rules, and consider the following.

Most of the knowledgeable people on a forum who would consider answering your question are doing so on their own time just to be nice. They do this because they remember what it felt like to be starting out and in need of a little guidance to make that project a success. If you have not bothered to cover the basics or given much effort to solve your own problems first, you will get zero respect in a forum and will either be ignored or "spanked" publicly. Here is how *not* to ask for help in a forum (we see this a lot):

Urgent: can't program the microcontroller. Pls help me, urgent! i am a noob and don't know how 2 get the code in2 the avr. What do I need 2 buy? Can some1 pls send me the file.

This type of post may never get a response for many reasons. First of all, why would your problem be of a concern to anyone else? Most people on a forum also have a life and a job, so their first thought will be, "If your problem is so urgent, pay someone to solve it." Be patient, and remember that if someone is kind enough to help you in the next few days, then you are very lucky. Also, making multiple posts of the same or similar request in other threads hoping that more posts from you will encourage responses will likely get you into trouble with moderators and regular members. It may take days for a response; but remember that everyone else has urgent things in their lives to look after before taking the time to help you and others.

The next reason why most forum members will ignore this kind of post is because of the sloppy use of language. Posting like that would be the same as showing up for a formal job interview wearing your garage clothes and is a clear indication of laziness. The message also indicates that the poster has done nothing to help him or herself and expects to be "spoon-fed" by those who have done the groundwork or have invested in many years of formal and informal education. The question also

has little information, so the forum members are left to guess. Here is a much better way to ask the same question and likely get a response:

Project X—Question regarding the AVR selection—Hello everyone. I'm working on the Mind Reader project and was wondering if the AVR644p could be used to replace the AVR324p? I have read the datasheets on both chips but just want to make sure that there aren't some issues with this chip that I should know about. Has anyone out there compiled a HEX file for this device yet? If not, I'm probably going to give it a try and upload the file if it works. Thanks for your help.

Now the poster has asked a clear question that indicates both the project and the part numbers, as well as what he or she has done to help himself or herself first. People who are willing to try to solve their own problems first and then ask politely for additional help are welcomed in a forum because they are likely the type of person to come back later and help others when they make the transition from newcomer to experienced hobbyist. Consider these points before asking for help online.

Tools of the Trade

Once you have a good supply of junk to mess around with and a basic understanding of the parts and schematic symbols, you can begin to turn your ideas into reality. Of course, you will need a few basic tools, which can be purchased at most electronics suppliers. Much like the breadboard, the soldering iron is the workhorse in the electronics industry. You can create and test your circuit on a breadboard, but to move it to a more permanent home, you will have to solder those components down. We like to have two soldering irons for this hobby—one is a cheap unit to use when unsoldering parts from scrap boards and the other is a better heat-controlled unit with interchangeable tips for fine work.

The ability to control the heat and replace the tips on a soldering iron like the one shown in

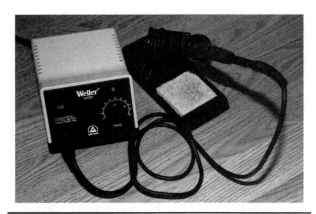

Figure 0-22 A midrange heat-controlled soldering iron.

Figure 0-22 makes it easy to work on all types of projects. For delicate surface-mounted components, you need a sharp point and low heat, but for soldering down a large heat sink, you need a blunt tip and lots of heat, so the ability to turn a dial and switch the tip is very handy. Also shown in Figure 0-22 is the "holster" and sponge bath for cleaning the tip. You can certainly get away with using a ten-dollar soldering iron for most of the projects in this book, but if you plan to dive deep into this hobby, then consider investing in a quality soldering station.

Unless everything you build is going to run from a 9-V battery, you will need some type of adjustable power supply like the one shown in Figure 0-23. Often, a circuit will need two different voltages, so a dual supply makes it easy to just dial up the voltage you need. These test-bench power supplies also allow you to set the current, so you can help eliminate fried components by slowly adjusting the voltage while you watch for spikes in the current. If you accidentally reverse the power pins on your last microcontroller and drop on a 1-A "wall wart," you will most certainly see some fireworks. If you limit the current on an adjustable supply and slowly turn up the voltage, chances are very high that you will catch the error before any damage is done to the chip. Having an amp meter also lets you figure out how much power your circuit needs when it comes time to choose a regulator. An adjustable power supply is another one of those "must have" devices for the electronics hobby.

A multimeter like the one shown in Figure 0-24 will also be needed in order to measure component values, check voltages, and test your circuits. A very basic multimeter will measure voltage, current, and resistance, and a more advanced model might also include a frequency counter, capacitance tester, transistor checker, and even some basic graphing functions. You will use your multimeter to check voltage and measure resistance most of the time, so even a basic model will be fine for most work.

Figure 0-23 A variable dual power supply.

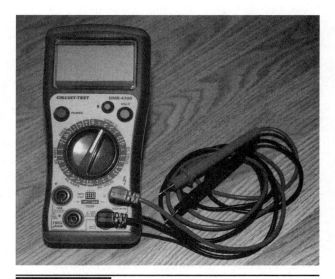

Figure 0-24 A basic digital multimeter.

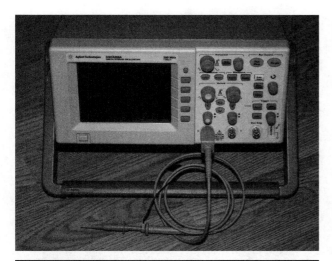

Figure 0-25   An oscilloscope for more advanced work.

If you really want to run with this hobby or learn electronics on a professional level, you will eventually want an oscilloscope (Figure 0-25) for your bench. A modern oscilloscope can measure just about anything and then display it on a high-resolution color screen or download the results to your computer for analysis. Even an older tube-based scope can be very useful when you need to see instant feedback from a circuit on a screen. As your circuits become larger, faster, and more complex, a simple multimeter may not be enough to debug problems as it cannot respond in real-time or detect such high-speed voltage changes. This becomes important when designing high-speed digital devices that expect a certain sequence of digital information to be transmitter or received. An oscilloscope lets you "dig deep" into a signal and look at it one bit at a time until you find the bugs. You can also display analog signals in real-time, compare multiple signals, and even save the results to a PC for further analysis. Although you will not need an oscilloscope to build the projects in this book, we could not have done many of them without one, so when you start designing your own circuits, a scope is invaluable.

In the time before millions of transistors were crammed onto a single IC, you had a few basic logic gates, a handful of transistors, and your usual resistors and capacitors. Now the selection of components is so vast that printing a supplier's catalog is almost a dream. If you can dream up a use for ones and zeros, there are probably 10 companies making a chip for it, so the selection is massive; often, mass-scale production, finding a single IC may be very difficult. Maybe you just tore into a VCR and isolated the on-screen display chip, found the datasheet, and thought "Cool, we want that for my robot." After a bit of search you find that the suppliers minimum-order quantity is 100,000 units and the programmer needed to make the chip work is the same price as a small car. The simple fact is that as electronics become more complex and the faster the industry moves, the less chance us hobbyists have of sourcing the latest parts. Sure, transistors, resistors, and the basic components will always be available, but complex ICs designed for consumer devices will either not be available in small quantities, or will have packages so small you need a microscope just to see all of the pins. Enter the microcontroller.

A microcontroller is like a blank IC ready to be made into whatever you want it to be. A five-dollar microcontroller can take on the function of a 16-input logic gate or become something as complex as a fully functional Pong game that will display video on your television. A microcontroller is a chip that contains a processor, some memory, and a few peripherals, such as serial ports or even a high-speed USB port. There are even microcontrollers that have MP3 decoders and high-resolution video generators onboard. Although learning to program a microcontroller is almost a hobby in itself, it does open up a door that allows you to create just about any function on a single IC, so you don't have to worry about what the market is doing, or if a chip will be available. Sure, there are limits, but not that many, and microcontrollers are king in the hobby world of electronics. Figure 0-26 shows a few microcontrollers with varying pin counts and a programmer from Microchip, as well as Parallax.

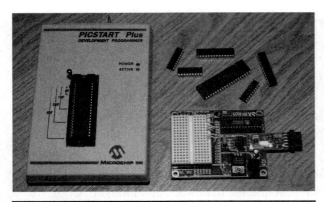

Figure 0-26   Microcontrollers and programmers.

Figure 0-27   Perf board becomes a permanent home.

To make a microcontroller do your bidding, you first write code on your PC using C, Assembly, or Basic, and then you compile and send your program to the programmer, which fits it into the microcontroller onboard flash memory. If your code works properly, your microcontroller now becomes your own custom IC. Microcontrollers typically cost between $5 and $20, depending on flash memory size and onboard peripherals, and programmers cost between $50 and $200, depending on chip support and debug functionality. There are even Web sites that show you how to build your own programmers for a few bucks, so microcontrollers are certainly well within anyone's budget.

Beyond the Breadboard

Once you have your prototype up and running on your breadboard, you will need to move the components to a more permanent home so your device can fit into a cabinet and become real hardware. This migration usually involves moving everything to a fiberglass board with holes with or without copper traces and then adding the wires. This perforated board is usually called "perf board," and is available in many sizes with or without copper-plated holes or strips that copy the connections on a solderless breadboard. Figure 0-27 shows an empty and populated copper-hole perf board with a dual-stepper motor drive prototype soldered and tested.

For small circuits, we usually use the non-copper-plated perf board and just solder wires on the bottom of the board where the component leads stick through, as this is easy, quick, and inexpensive. Larger circuits become a mess quickly, so having copper-plated holes or even traces like a solderless breadboard can make migration a lot easier. Perforated board is available as a large sheet, and it is easy to just cut off a piece the size you need using a utility knife to score the surface and then break it apart. A perf board circuit made correctly is just as good as a manufactured PCB, so don't underestimate what can be done with a handful of wires, a few dollars worth of perf board, and some hard work. Have a look at the complex VGA computer prototype on the huge breadboard in the top half of Figure 0-28. The prototype now lives on an 8- by 5-in perf board (lower half of Figure 0-28) and functions perfectly at 20 MHz.

The next step after hardwiring your project to a perf board will be an actual printed circuit board like the one shown in Figure 0-29. When you are using surface-mounted components with very small pitch leads, you don't have many options besides a real PCB. A PCB also makes more sense when

Figure 0-28 Massive high-speed circuits can be built on a breadboard.

Figure 0-29 A professionally manufactured printed circuit board.

you need to duplicate a project or want to get into selling your designs. Bringing a professional PCB to market is a costly and time-consuming process often involving many failed attempts and board revisions, but for us hobbyists, there are alternatives that may be viable if a real PCB is needed. If you search the Internet for "PCB production," you will find several companies that will allow you to download their free software to design your PCB and then order a few boards for around $100. This is actually a fair deal considering the amount of investment and time it would take to produce your own PCB at home using a chemical etching or photo-based system. Of course, you will have to decide if the cost is worth it.

Now let's dig into some of the projects and have some fun. Remember that we all had to start from the beginning, and if something does not make sense at first, turn off the power supply, take a break, and search the Internet for more information. Electronics is a great hobby, and with a little investment in time and money, you will be able to create just about anything. When you start to get a good handle on things, feel free to try out your own modifications to the projects presented in this book, or mix and match them together to create completely new devices. Oh, and don't feel too bad when you accidentally let the smoke out of your semiconductors, they won't feel a thing!

PART ONE
Spy Technologies 101

Desoldering Basics

IN ORDER TO build an electronic circuit from plans or from scratch, you will need a number of semiconductors and components. Often, these individual components cost only pennies now, but in order to purchase from a large supplier a minimum order number or price level has to be met. It seems crazy to pay $25 in shipping to receive a few 10-cent capacitors or resistors, especially when you can salvage them from just about any old circuit board. Almost all of the components we use are salvaged from old boards. We never had to order an odd value resistor or capacitor as we have always found what we needed by scavenging our huge junk piles.

A few dead TV or VCR circuit boards will net you enough raw components to fill an entire electronics parts bin and allow you to experiment with varying values as you breadboard a new or published circuit from a schematic. Almost any discarded appliance that includes a power cord will become a great source of raw semiconductors, so tell everyone you know not to throw out their broken appliances and instead send them to your mad scientist lab for dissection!

A decent soldering iron will make your work so much easier, allowing the heat to be adjusted to the job at hand. Sure, you could do a lot with a ten-dollar soldering gun, but when it comes to removing components or soldering small pin devices, the budget soldering iron may not be of much use. A fine point and a lot of heat will make component extraction very easy, especially on multilayered boards, which require the heat to reach the inner copper layers in order to melt the solder. A soldering station with a heat control and interchangeable tips is one of those must-have items if you plan to invest any time in the electronics hobby (Figure 1-1). A decent soldering station can be purchased at most electronics suppliers, and will range in price from $50 to $200,

Figure 1-0   Most of your parts inventory can be salvaged from old circuit boards.

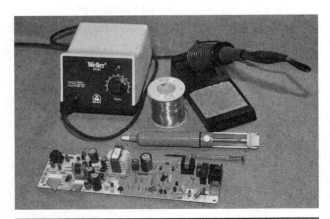

Figure 1-1   The basic tools you will require for any soldering or unsoldering work.

depending on what features and accessories are included. If your soldering system does not include a sponge bath, then just wet an ordinary kitchen sponge with some water, as this will be needed in order to clean off the tip as it gums up with flux and old solder.

Large through-hole semiconductors can often be ripped up from old circuit boards using a soldering iron and your fingers to lift the parts, but there will be limits to this technique, especially on smaller components or those with more than two or three pins. A desoldering tool, often called a "solder sucker" is essential for removing short-lead semiconductors, as well as those with multiple leads, such as integrated circuits (ICs). This simple tool contains a spring-loaded plunger that basically sucks up the heated solder, clearing the area around the lead and hole so that the part can be removed from the board. On most boards, it is possible to cleanly lift up a 40-pin IC in a few minutes so that it can be reused or even placed into a solderless breadboard. A solder sucker is also good for removing large components that have been placed down with a lot of solder to secure them to the board or transfer heat.

You will also require a spool of solder, even when unsoldering components, as there will be times when you have to melt a bit of solder into a hole in order to use the solder sucker to clean up the hole. Molten solder is a liquid, so the transfer of heat between the newly added solder and the hard-to-reach solder stuck in the center of a multilayer board will allow the solder sucker to pull up every bit in one shot. For electronics work, you will need flux core (rosin core) solder, which is a thin, hollow solder with special antioxidizing chemicals in the center that clean and protect the copper when soldering. There are various diameters of solder, with 0.032 inch being a good choice for all-around electronics work. The type of solid heavy solder used for plumbing is of no use for electronics work!

If your soldering system includes a dial to set the heat (amperage), then this will come in handy when

Figure 1-2 A higher heat setting is best for unsoldering larger components.

working with the larger through-hole components and board-mounted hardware, such as switches, heat sinks, connectors, and large-pin devices. You don't have to worry too much about damaging a component with a soldering iron, as it has been designed to withstand a lot more heat during the manufacturing process than you will be subjecting it to here. When we unsolder, we always crank up the station to the highest setting, as this releases the part faster, subjecting it to less time under the gun (Figure 1-2).

Once you become proficient at arming, aiming, and snapping the desoldering tool, you will be able to lift a 40-pin IC from a circuit board in about 60 seconds flat. This tool will allow you to cleanly rip up just about any component from a board in such a way that the pins look almost brand new again, ready to be inserted into a solderless breadboard. It does take some practice to get the rhythm down though, and there is a split-second where the solder sucker must be positioned right over the pin just as the solder pools in order to completely evacuate a hole (Figure 1-3). If you are too slow or fail to aim the top over the hole, you will only pull up part of the solder; this will require you to actually use new solder to fill the hole back in again so you can try a second time. Proficiency will come with practice, and after your tenth TV circuit board, you will be able to wield the soldering gun and desoldering

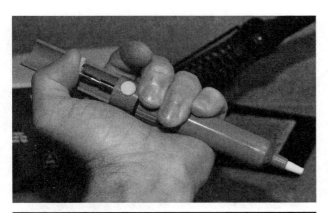

Figure 1-3 Get used to arming and aiming the solder sucker with accuracy.

tool with Jedi reflexes. Oh, but your thumb muscles will ache for a while as you get used to pressing down that button.

Like all products, the desoldering tool will range in quality depending on the manufacturer and price, and it is worth investing in one that will last for some time. Each time you evacuate a hole, the molten solder will solidify in the tube and will drop back out on the next press of the plunger. Of course, a small amount of solder will stick to the internal spring or casing, eventually clogging the device. To clean the solder sucker, unscrew the two halves and then clean the debris out as much as you can, ensuring the small rubber ring is also clean. A bit of vegetable oil can be used to lubricate the unit for continued efficiency. This cleaning process should be done after a full day of use or when the loading system seems to stick (Figure 1-4).

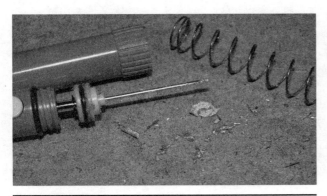

Figure 1-4 The desoldering tool will need to be cleaned every so often.

Although the current trend is to reduce the size and number of components on a printed circuit board to reduce cost, there will always be something to salvage, and older boards will be a regular gold mine of usable parts. Large capacitors, resistors, transistors, light-emitting diodes (LEDs), diodes, and any other two- or three-pin through-hole component can easily be removed using only your fingers and a soldering iron by simply heating and lifting them free (Figure 1-5). If the component is small enough that you can heat all of the pads at the same time, then the process is very simple—just heat and remove. Larger devices require a bit of working to release them from the circuit board.

Heat one of the pads as you rock the component to one side in order to slide the lead out of the hole as much as you can before the part hits the one next to it. If the device has more than two pins, then you can't bend it as far, as you may bend one of the other pins. Transistors and other large devices need to be worked out a small bit at a time, heating each pad one at a time.

Once you have melted the solder on one of the pads, bend the component as far as you can before it hits the one next to it, and then withdraw the heat to let the lead settle back in the hole. The process is then repeated on the other lead, working

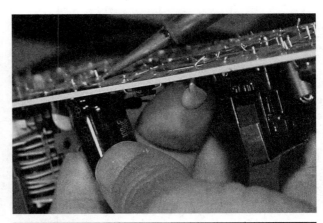

Figure 1-5 Larger components with two or three pins are easy to remove from a board.

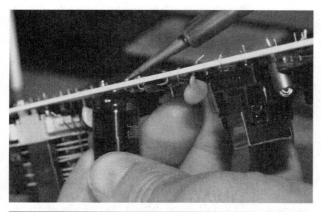

Figure 1-6   Slide the lead out of the hole a bit at a time and then let the solder cool.

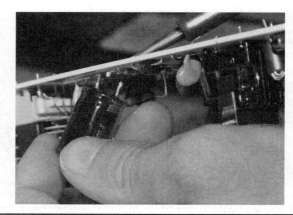

Figure 1-8   After heating the pads several times, the part is finally released.

the component out of the board a bit at a time (Figures 1-6 and 1-7).

When you are working a component off the board using the heat and rock technique, be careful not to bend the leads so far that they break right off. Some capacitors have such short leads that there is no space between the underside and the board, so too much movement may rip the lead right out of the casing. Also, some devices, such as large resistors, transmit heat to their outer bodies, so you will need to work fast or use tweezers with this technique in order to avoid burning your fingers.

This lead rocking process only takes a few seconds if you have your soldering iron set on high, and will not damage the part if done carefully (Figure 1-8). Some cleanup of the leads might be necessary after, though, as the molten solder may stick to the lead as you release the component from the board. This lead cleanup is only necessary if you plan to place the part in a solderless breadboard, as the solder blob may be too wide for the slot on the breadboard.

Some devices are inserted into a circuit board with very short leads or in such a way that makes it impossible to manipulate the part with your fingers (Figures 1-9 and 1-10). Resistors that are

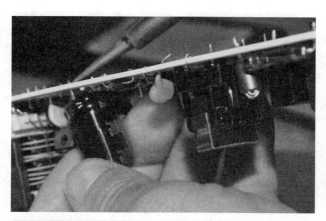

Figure 1-7   Keep working each lead until the component is freed from the board.

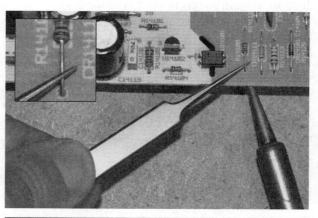

Figure 1-9   Smaller components with short leads can be heated and pried up.

Figure 1-10   Delicate components need to be removed carefully.

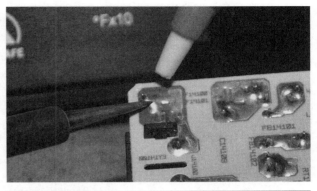

Figure 1-11   Timing and aim are the key to using the desoldering tool.

laid down horizontally are often difficult to lift up by hand as there is not much to hold on to, and they become extremely hot in a hurry. If you can get a hold of the device and pull it quickly, you may be able to lift it by hand, but after a while, your fingers will start to develop heat blisters. Resistors and diodes can be pulled up from a board by heating the lead while you pry them up with tweezers or a small screwdriver blade. Actually, you only need to pry up one end and then you can easily pull the component from the board by melting the pad on the other lead and pulling it away from the hole.

When there is no way to safely grip or pry up a component, you will need to evacuate the hole by using the desoldering tool in order to remove the part. If the lead has been bent outwards during the manufacturing process, try to straighten it as much as you can so that the solder sucker can pull out the solder evenly around the lead and the hole. Many ICs and components are inserted by a machine that forces the pins to fall against the edge of the holes. This can make solder evacuation more difficult.

Depending on the amount of solder and size of the pad, you may need to make multiple strikes with the desoldering tool in order to free the part (Figure 1-11). Heat up the pad until the color of the solder changes (molten solder becomes very shiny), and then quickly place the loaded solder-sucker tip right over the pin and press the release button. If you managed to make a seal between the work and

the tip while the solder was still molten, then most of the hole will be cleared out. Larger pads will require a few attempts to clear out all of the solder (Figure 1-12).

On larger devices with thick leads, the solder sucker may not evacuate the entire hole, leaving a small connection between the edge of the solder pad and the lead. Often, you can just grip the lead with pliers or tweezers to move it away from the edge of the hole (Figure 1-13). This is the reason that leads should be a straight as possible before using the desoldering tool. On smaller components, a quick push with your fingernail will easily release the lead from the edge of the hole. If you cannot free the lead, then it may be necessary to fill the hole with fresh solder and then use the desoldering tool again.

Figure 1-12   Sometimes the lead will be stuck to the side of the hole.

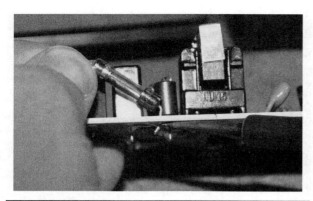

Figure 1-13   Once one lead has been freed, the fuse can be pulled from the board.

Figure 1-15   Clean the leads with the desoldering tool after removal.

The fuse shown in Figure 1-13 couldn't be pried up until one end was already removed from the board as the glass was too smooth to get a grip on, and any prying might have broken the glass. Large ceramic resistors are also similar and made of very fragile material that cannot by pried easily.

To prepare a part for reinsertion into a board, use the desoldering tool to remove any solder blobs from the leads and then press them with flat pliers to straighten the leads (Figures 1-14 and 1-15). This becomes important if you plan to install larger pin devices into a solderless breadboard as there is a limit to how large a pin will fit before causing stress on the internal plates that connect the rows.

When cleaning up the leftover solder from a larger part like a capacitor, be careful not to

overheat the leads as it may be possible to damage the internals now that there are no copper solder pads to help dissipate the heat. Only heat up the part as much as necessary in order to melt the solder for extraction. Another trick that can be used to remove solder blobs from the pin is the heat-and-shake method, which involves melting the solder and then flicking the part to send the molten solder flying off. Do this into a garbage pail or onto a surface that will not be tarnished by the molten solder.

On larger pin devices, a twisting motion with ridged pliers can also help straighten up and remove leftover solder from the leads. Since the solder is much softer than the lead, it can be removed by the grinding action of the pliers. Only use enough pressure to remove the solder without bending or twisting the lead (Figure 1-16).

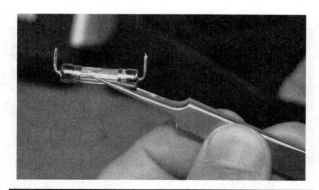

Figure 1-14   Leads are cleaned and straightened after removing a part.

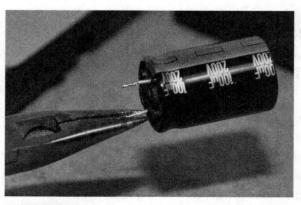

Figure 1-16   A twisting motion can also remove the excess solder.

Figure 1-17   Trim both leads to an equal length for breadboarding.

For use in a solderless breadboard, leads should be cut to a length of no more than half an inch. It is also important that leads be straightened and positioned so that they fit into the holes without stressing the part. For use on a perforated board, it may be best to keep leads as long as possible as they could be used to help form solder points and traces on the underside of the board. Trim both leads to an equal length for breadboarding (Figure 1-17).

When you are first building up your semiconductor inventory, there are obvious common values that you will need with capacitors, resistors, transistors, and logic ICs (Figure 1-18). You will find more

than enough common components on just about any salvage board such as 1-, 10-, and 100-µF capacitors, 1-, 10-, and 10-K resistors, and small NPN signal transistors. As you build projects from plans and schematics, common values will become obvious, and even resistor color codes will become easily recognizable right away. These common values will be the starting point for building a well-organized inventory that will probably end up spanning several multidrawer component bins.

As you desolder and solder, the tip will need constant cleaning to allow for a proper transfer of heat to the work (Figure 1-19). The wet sponge is used to clean the tip by dragging it across the surface to remove the old solder buildup. Do not let your soldering iron sit in the holster while a blob of old solder is stuck to the end as this will slowly degrade and pit the tip, making it eventually break or deform. Before putting your soldering iron down, wipe the tip first.

Most soldering stations will include a holster and a sponge, but if you only have a basic system, any kitchen or bath sponge will work. Just wet the sponge until it is saturated (but not dripping) and then it will be good for a few days. To clean the solder blobs from the tip, drag the hot soldering iron across the sponge and spin it as you go (Figure 1-20). The solder blobs will be stuck on the surface of the sponge and when it finally dries, they can be shaken off into a garbage can. Remember to always clean the tip when you are

Figure 1-18   A pair of solderless breadboard compatible salvaged capacitors.

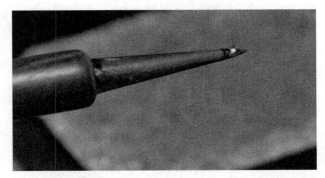

Figure 1-19   Old solder will build up on the tip of the soldering iron as it is used.

Figure 1-20   Drag the tip with a turning motion
to clean off the old solder blobs.

Figure 1-22   Surface-mounted components add
a level of difficulty to soldering.

going to shut off or holster the iron for any amount
of time to avoid ruining the fine point in the tip.

The clean soldering iron tip will have a difficult
time melting a large blob of solder unless it has a
bit of solder on it already to help transfer the heat
to the target. This wetting of the tip or "tinning"
is important when you are soldering large pads or
bare copper pads and new leads (Figure 1-21). To
tin the tip, just melt a small amount of new solder
to the tip and then run it across the sponge to get
rid of any excess solder. This process places a
small coating of solder on the tip, which allows
the transfer of heat to the work. As you solder, this
difference will become obvious.

Surface-mounted devices (SMDs) can be
quite a task to solder by hand because of the fact
that the capillary action of the molten solder is
enough to trap them or make them stick to the

tip (Figure 1-22). To work with surface mounted
semiconductors, you will definitely need a good
set of tweezers, a magnifier, and a steady hand.
SMD work is not all that bad once you get past the
pitfalls, and desoldering components can be done
with a sharp tip and a pair of good tweezers. The
desoldering tool is almost useless for SMD work
as the parts are so small they will end up lost inside
the vacuum chamber or broken by the force.

SMD components are so small that even static
electricity can cause havoc as you handle them.
There are also different types of marking and coding
used because of the small amount of space available,
so you will need to learn another facet to this hobby
when working with SMDs (Figure 1-23). Oh, and
don't bother looking around on the floor if you drop
one of these parts...the chance of recovering one is
not worth the effort!

Figure 1-21   Tinning the tip will help the transfer
of heat to the work.

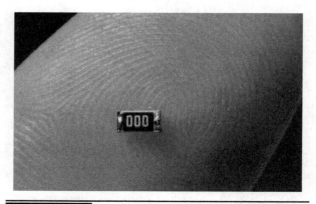

Figure 1-23   A surface-mounted resistor is
smaller than a grain of rice.

Figure 1-24   Lifting a surface-mounted IC using the pry method.

Figure 1-26   Another successful surface-mounted IC extraction.

Surface-mounted ICs have many sizes of packages and pins, and many are well within the realm of hand soldering, so don't be afraid to try to save these parts from a circuit board. Many of the common logic ICs or op amps have few enough pins that they can be easily lifted from a board using only a small screwdriver and a soldering iron (Figure 1-24). There are specialty tools available for SMD work, but a lot can be done with just the basic tools as well.

One easy method of lifting a surface-mounted IC is to gently pry up on one end as you drag the hot soldering iron tip back and forth across the leads on that side. You will need to tin the tip and slide it back and forth in order to evenly heat all of the pins at once so that end will lift form the pads. Don't pry too hard or push too much on the part or it will be damaged (Figure 1-25). When the pins on one side become free from the pads, you will hear a snap sound and the part will be lifted slightly on that side.

Once the side you are working on has popped from the board, stop prying so the pins on the other side don't become bent or broken. The heating and sliding process can then be done on the other side using tweezers or your fingers to lift the IC completely from the board (Figure 1-26).

Once the surface-mounted IC has been fully removed from the pads, straighten and clean any pins that have been bent or may still have solder balls or bridges on them. This method of extracting IC packages works well for most ICs that have pins on both sides. For ICs that have pins on all four sides, this method will probably be of no use, and will require a heat gun to be used to evenly heat the entire part.

We used to spend hours unsoldering components from boards, throwing them all in a huge bin for later sorting, but found this to be a huge process (Figure 1-27). Looking up datasheets and

Figure 1-25   One side of the surface-mounted IC has lifted from the pads.

Figure 1-27   Sorting and organizing a huge inventory can take some time

Figure 1-28   Parts cabinets with small drawers are perfect for staying organized.

cross-referencing all of these parts took a lot longer than the removal process. Instead, now we go hunting for certain parts, like a 0.01-μF disc capacitor and sort them in groups as they are extracted. Having a massive tangle of resistors is completely useless when you need to find a specific value, so sorting after extraction is important. Sometimes we just leave resistors on the boards, as it would be impossible to sort all but the most common values into groups. ICs, capacitors, transistors, and larger components are usually easy to sort, so they are done as they are extracted.

We have 10 or more large parts cabinets completely full of salvaged semiconductors, all sorted by value, type, or use (Figure 1-28). Although 90 percent of the inventory came from free scrap boards, there must be thousands of dollars in usable components in there now. Every resistor costs something if you have to order it from a supplier, so don't pass up any scrap boards for salvage. Tell your friends and family that all nonfunctioning and functioning appliances are welcome in your lab for dissection, and build up that junk pile as much as you can!

Spy Camera Basics

At one time, a decent low-light video camera was the size of soda can and cost several hundred dollars. Now, you can purchase a decent micro spy camera that will fit into a marker cap for well under a hundred dollars. Tiny security cameras also offer a huge variety of lenses to allow for practically any angle of view and covert installation. These cameras are also very easy to connect because they have used the same basic video standard since the invention of the television. This introduction to security cameras will give you the required basic information needed to install and power up practically any type of security camera.

We use these small video cameras for security work, machine vision experiments, and robotic projects. They range in size from half an inch square to quite large depending on the features, imaging system, and type of lens installed. For most security installations, a small board camera

with a fixed medium- to wide-angle lens will be perfect, but there are times when you need to see a much larger area or over great distances, so these cameras have several body types that allow the use of multiple lens styles. Extremely tiny spy cameras will not have this option, as they use a very tiny glass or plastic lens built right into the housing. There will be a bit of a tradeoff between image quality and size.

For most covert work, size will be the key factor in camera selection, so you probably won't be dealing with multiple lenses or even have a choice. In Figure 2-1, you can see the huge difference in size between the long-range motorized lens telephoto camera on the left and the tiny spy camera at the front of the photo. The tiny spy camera has a simple fixed medium-angle lens and will run from a 6-volt (V) power source for many hours. The larger camera uses a computer to control the motorized zoom lens and requires multiple power sources and electronic control

Figure 2-0   Micro spy cameras are inexpensive and easy to install in covert locations.

Figure 2-1   A few of the basic security cameras in an ever-growing collection.

systems in order to operate. Of course, the zoom lens can see the color of your eyes from 500 feet away, but the tiny spy cam can be hidden just about anywhere. There will always be a camera available to suit your needs as long as you have the budget to afford it.

Security cameras and micro spy cameras will have a connection point for DC (direct current) power and a video output, and some will have an audio output or several control lines for onboard features such as gain, color, text overlay, and lens control. The most basic cameras you will probably work with most often will have only power and video connectors, or simply include three wires coming out of the tiny enclosure. The tiny board cameras with three wires will usually follow a simply color scheme of black or green = ground, red = positive power, and white, yellow, or brown = composite video output. Of course, it is always a good idea to check the manual for polarity and voltage before making any guesses.

The camera in the left of Figure 2-2 is quite advanced, having a built-in on-screen display system, many internal function parameters, lens control, and audio. This camera is somewhat large, so it has many buttons on the back side, audio and video jacks, DC power jack, and a special connector to control a motorized lens. The basic low-lux micro camera on the right of Figure 2-2 only has a tiny connector with a

Figure 2-3 Large cameras with removable lenses have C or CS mounts.

three-wire cable to allow power, ground, and video output connections. We will discuss lux later in this section. The large camera is great for daytime long-range imaging where color quality is important, and the tiny camera is perfect for extremely low light or infrared night vision imaging in a stealthy location.

Larger security cameras designed to be used with higher-quality interchangeable lenses like the one shown in Figure 2-3 will usually have a C- or CS-type mount. C-mount lenses provide a male thread, which mates with a female thread on the camera. The thread is nominally 1 inch (25 mm) in diameter, with 32 threads per inch. A CS lens has a flange focal distance of 12.526 millimeters (0.4931 in), and is otherwise identical to the C-mount. Some camera bodies have a smaller lens mount and will often come with an adapter to allow the attachment of C or CS lenses, although some focal range may be lost. The camera shown in Figure 2-3 has been fitted with a C-mount adapter.

With the lens unscrewed form the camera body, you can see the charge-coupled device (CCD) mounted to the circuit board (Figure 2-4). This IC has a top glass window to allow the lens to focus the image on the extremely tiny, high-resolution array, which forms the image to be displayed. When removing a line, be careful not to touch the top surface of the CCD sensor or it will become smudged and require a proper cleaning.

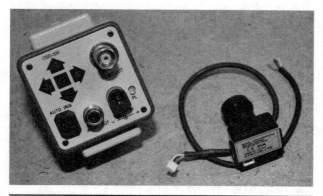

Figure 2-2 Camera connections for power, ground, and audio.

Figure 2-4 The CCD or CMOS imager shown with the lens removed.

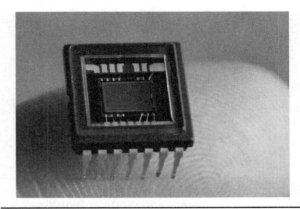

Figure 2-5 The CCD imaging chip taken from a high-quality camcorder.

The CCD determines the type of image (color or monochrome), as well as the resolution in horizontal and vertical pixels. Most color cameras will also include a special set of filters to remove infrared light, which is why monochrome cameras are much better for night vision operations. CCD imagers are available in extremely high resolutions, but for composite video application, the video bandwidth limits the usable resolution to about 720 by 480 pixels.

When digging into the technical details of a security camera, one of the most important aspects will be the type of imaging device used, which will either be a CCD imager or complementary metal-oxide semiconductor (CMOS) imager. Both types of imagers convert light into voltage and process it into electronic signals. In a CCD sensor, every pixel is transferred through a very limited number of output nodes (often just one) to be converted to voltage, buffered, and sent off-chip as an analog signal. All of the pixels can be devoted to light capture, and the image quality is high (Figure 2-5). In a CMOS sensor, each pixel has its own charge-to-voltage conversion, and the sensor often also includes amplifiers, noise-correction, and digitization circuits, so that the chip outputs digital bits. These other functions increase the design complexity and reduce the area available for light capture.

These various qualities means that for low-resolution and low-light imaging, CCD cameras

are currently the best choice. CMOS imagers are used in very high-resolution imaging systems, such as digital cameras and scanners, as proper lighting is usually not a problem in those situations. If you have the choice between a hundred-dollar CCD security camera with a 0.5 lux rating and a fifty-dollar CMOS camera with a 1.5 lux rating, the CCD camera will offer a far superior image and work well in a low-light or night vision application. Resolution is not much concern in a security camera, as most are already beyond the actual capabilities of the NTSC (National Television System Committee) or PAL (Phase Alternate Line) composite video standard anyhow. The lux rating will probably be the most important specification on a security camera next to the type of lens and field of view.

Lux is the measure of light or luminous power per area. It is used in photography as a measure of the intensity, as perceived by the human eye. The lower the lux rating on a camera, the better it will see in the dark. For instance, a monochrome camera with a lux rating of only 0.5 lux will see much better in the dark than you could with your naked eye, and with the help of an infrared illuminator will be able to see into the darkness and display it on a monitor as if it were midday. Color cameras require much more light to achieve a decent image, and since they usually include infrared filters, they are not the best choice for night vision applications.

Figure 2-6 C-type lenses offer the highest quality and range of features.

Figure 2-7 Fixed-lens micro cameras are very small and easy to conceal.

Although security cameras with removable C-type lens mounts are much more expensive and bulky than fixed-lens micro cameras, the availability of different lens types is vast. If your security application requires control over field of view, zoom, or iris, then you will likely be looking for a camera body with a C- or CS-type mount to mate to the lens. A long-range zoom lens like the large ones shown in Figure 2-6 may cost upwards of a $1000, but the ability to pinpoint the tiniest detail at long distances may be the goal. Smaller focusable C-mount lenses range in price from $20 to $500 depending on the size and style of the optics, but they are always top quality, comparable to the optics in high-grade camcorders.

The large motorized zoom lenses shown in Figure 2-6 are expensive and much larger than the camera body, but they allow for fine digital control over focus, zoom, and iris through a set of analog control lines. It would be next to impossible to hide this lens, but since it can see so far away, stealth installation is not really necessary. C-mount lenses are also available in special pinhole and snake configurations, but these are incredibly expensive and bulky compared to the lower-end micro camera lenses of this type.

Micro video cameras or board cameras like the ones shown in Figure 2-7 are very inexpensive, offer good quality, and normally only need a power and video connection to operate. These cameras

are perfect for covert operations as they range in size from about an inch square to the size of a pencil eraser. These cameras will run for hours from a single 9-V battery and have many fixed-lens options ranging from the small glass medium-wide-angle type to the ultra-tiny pinhole type. The only downside to these smaller lens types is that they are not really adjustable—they are designed to have only a single field of view and focal range.

With the small lens unscrewed from the mounting hardware, the CCD can be seen installed on the camera circuit board. This 1/3-inch color CCD is no different from the ones found in lower-end color camcorders. The quality of the image far exceeds what is necessary for basic surveillance operations. Most micro camera lenses use the same thread type and are often interchangeable, although they are not easy to source by themselves.

The focus can be set on these micro video lenses by loosening the tiny set screw, as shown in Figure 2-8, in order to turn the lens clockwise or counterclockwise to gain the best focus for a certain installation. Most of these lenses will see perfectly from about 10 feet to infinity, but for closer focus, it will probably be necessary to make initial adjustments. Although the optics on these tiny lenses range from good to poor, the image quality is always decent because of the low

Figure 2-8 Setting the initial focus on a simple micro-camera fixed lens.

resolution of the composite video signal along with the high quality of the CCD imaging device used. We found that a $30 color board camera will exceed the quality of a $1500 camcorder that was made five years earlier. Technology is always improving and becoming less expensive, so high-quality surveillance equipment is becoming more and more affordable for the "average" person.

With the price of security gear at an all-time low, you can often purchase a complete night vision camera like the one shown in Figure 2-9 for well under $50. Inside the weatherproof aluminum shell, you will find a high-quality board camera, power supply, and another board with an array of infrared

light-emitting diodes (LEDs). By itself, this inexpensive security camera is a high-performing night vision camera with a range of about 50 feet, and has a very good color image for daylight operations. As for parts, the camera module is the same one as shown in Figure 2-8, and the cost of all of the components separately would be a lot more. Sadly, we have seen many online spy stores sell the bare camera for three or four times the price of this entire unit, so shop around and be wary of any site claiming that they sell mainly to law enforcement agencies.

There are a few different types of connectors used on security cameras, but the most common type is the RCA jack shown in Figure 2-10. This connector is the same one used on the back of a television set to connect older video players and camcorders. On the back of a TV or monitor, this jack will often be labeled "Video Input," "Line Input," or "External Input," and will use the same connector for audio and video. Most security cameras will output composite video in either NTSC or PAL format to be connected to any TV or monitor that can accept this input. Some of the smaller spy cameras may not have any connector, just bare wires, so it makes sense to adapt them to the RCA jack if you plan to use them with standard TVs or monitors.

On the back of the monitor, video recorder, or television, the RCA connectors will look like

Figure 2-9 An outdoor security camera with built-in night vision LEDs.

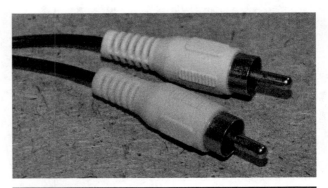

Figure 2-10 This is the most common type of composite video connector.

Figure 2-11   The back of a composite video monitor showing the RCA inputs.

Figure 2-13   Adapting the BNC connector to the RCA-style connector.

the ones shown in Figure 2-11, often having two inputs for stereo audio along with the video input. Although there is no real standard, the color coding for the inputs will usually be yellow for video and red or white for audio.

On the larger, more expensive types of security cameras, the type of connector may be the BNC style as shown in Figure 2-12. Except for the way the connector locks together, there is no difference between the RCA connector and BNC connector for the transmission of the composite video signal. It is best to simply add an adapter to the BNC connector in order to convert it back to the much more commonly available RCA connector.

The inexpensive adapters shown in Figure 2-13 can be used to adapt the BNC-style connectors

back to the more commonly used RCA-style connectors. Although the BNC-type connector is better for high-frequency signal transmission, this means nothing on a composite video camera, which will send a low-bandwidth signal. BNC connectors are not found on TVs or video recorders, so it is pointless to mess around with this type of connection unless you already have a complete system, including cables.

The BNC-to-RCA adapter is shown hooked to the back of the high-quality machine vision camera in Figure 2-14. Oddly, the audio output was already using an RCA connector, so it seemed like a strange move to sell the camera with a BNC-style connector. The camera can now feed its video output into any video recording system or TV with an RCA-style composite input connector.

Figure 2-12   Some security cameras use a BNC-style video connector.

Figure 2-14   The camera is now adapted for use on any composite monitor.

Figure 2-15   The board camera– and micro camera–style connector.

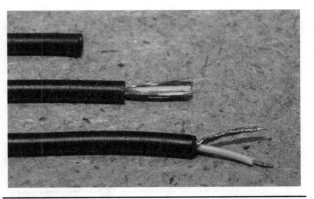

Figure 2-16   Coaxial cable is needed to carry the audio and video signals.

On the smaller micro spy cams and board cameras, the connector will often be like the one shown in Figure 2-15, just a small plastic plug with as many connection points as needed. The majority of small cameras in our collection have a three-pin connector with pins for ground, power, and video output. The cameras that include audio also have an audio output pin. Since these connectors are often custom-made or hard to source, it is important to ensure that your camera includes the cable and connector when purchased. The other end of the cable will often break out into a set of female RCA connectors for audio and video, as well as a coaxial plug for the DC power source. Also make sure that you have the correct power supply voltage and polarity for your camera.

Because the audio and video signals are prone to interference over longer distances, shielded coaxial cable is necessary in order to make the connection between the camera and the monitor or recording device. If your camera has RCA connectors at both ends, then you can use any standard patch cable to make the connection to your TV or monitor, but often you will need longer cables or have to install your own connectors. The standard coaxial cable type used for security cameras is the RG-59 type. This cable has a single insulated signal wire surrounded by a conductive shield. The "Siamese RG-59" version also includes a second shielded pair to allow camera power supplies to be located

indoors at the location of the video recorder or monitor. For short connections of only a few feet, just about any coaxial wire will work just fine.

Figure 2-16 shows the various stages of stripping down a coax cable to reveal the grounded outer shielding, as well as the protected inner signal wire. When connecting coaxial cable to an audio or video source, the inner wire carried the signal and the outer shield is connected to a common ground. When using coaxial cable to carry power, the positive supply is connected to the inner wire and the common ground becomes the outer shield. In almost all instances, the ground is common to all signals, including the power supply.

A lot of electronics prototyping will be done using a solderless breadboard, so it will make your job a lot easier if you create a set of breadboard-pluggable cables for your camera. This way you can test various power supplies, audio preamplifiers, and video systems without having to cut and strip the wires every time. We have several sets of breadboard cables for the various camera models in our collection, as well as some standard DC connectors and the RCA-style coaxial connectors. Figure 2-17 shows how to make a solderless breadboard-compatible cable by soldering the ends of the cables to a set of header pins. You could also just solder the ends of the wires to a set of solid copper wires that will push into the breadboard sockets.

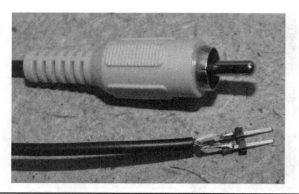

Figure 2-17 Making a set of breadboard-compatible connector cables.

We often work with machine vision, robotics, and video generation, so it is useful to have the ability to drop an NTSC camera right onto our solderless breadboard. For this task, we have made the necessary power connector for a 9-V battery, as well as the necessary three-pin camera connector and another RCA cable to feed back to the video monitor. With these breadboard-compatible cables, it takes only seconds to add a video signal to a machine vision project or connect a monitor to our video generator (Figure 2-18).

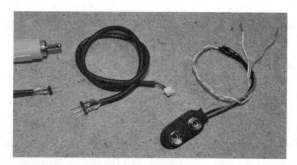

Figure 2-18 A complete set of breadboard-compatible cables for a micro camera.

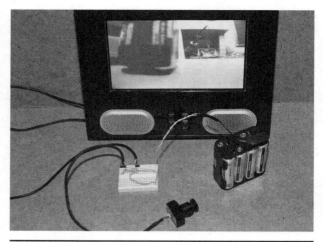

Figure 2-19 A micro video camera image displayed on a small LCD monitor.

If you are planning to do a lot of work with video cameras, then a small liquid crystal display (LCD) monitor with a composite input is a must-have tool for your kit (Figure 2-19). You can often find these small video monitors at any gaming supply outlet or even in the auto department of a hardware store. These small monitors are used as portable gaming screens or as video displays in recreation vehicles and passenger vehicles for rearview cameras. Many small televisions also include RCA jacks on the back to input composite video sources, and these are normally fully compatible with most security cameras, too.

Now that you have the basic understanding of how to get the video signal out of your security camera, you can begin to design your covert spy system, machine vision robot, or practically any project that requires a video source or video screen.

Invisible Light Basics

NIGHT VISION IS one of the most important factors when considering any kind of video-operated spy gadget as this technology allows the viewer to see in complete darkness while the subject is completely unaware. Because infrared light (radiation) falls just below red on the visible light spectrum, making up the wavelengths from about 750 to 1500 nanometers (nm), this light cannot be seen by human eyes but it can easily be seen by many video cameras, making it useful as a covert lighting method in night vision systems. A common example of infrared light is the medium for communication between your remote control and television set. The LED on the end of your remote sends out pulses of infrared light, which is received by the infrared detector on the TV and demodulated back into data. Of course, you cannot see the pulses because they are out of our visual range, but the infrared receiver in the television can see the pulses perfectly.

Security cameras and mini spy cams can also see infrared radiation very well. They are easy to connect, inexpensive, and can be easily hidden. There are many good-quality security cameras available on the market that includes a low-lux video camera in a weatherproof housing along with an array of infrared LEDs for night vision applications. Black-and-white security cameras and small board cameras are particularly sensitive to infrared light. These ultra low lux cameras can usually be purchased for about $100 or less, especially from online sellers. Add 10 or more infrared LEDs, and you now have a night vision system that is better than those that were sold for thousands of dollars in the 1980s.

The light spectrum shown in the lower half of Figure 3-1 covers light from ultraviolet right to infrared and shows the small portion of the light spectrum that can be seen by human eyes. Light is visible to our eyes from approximately 400 nm (violet) to approximately 700 nm (red); 550 nm (green) is at the midpoint. What is interesting is that the imaging system in a video camera can see the light extending past both ends of the visible light scale, which includes both infrared and ultraviolet. Since the infrared portion of the light spectrum is not visible to our eyes, the correct term for this portion of the spectrum is *infrared radiation*, not infrared light, but often both terms are used.

The infrared portion of the light spectrum that can be seen by most video cameras covers the range of 700 nm well into 1500 nm, with exceptional sensitivity around 800 nm. Video cameras can also see ultraviolet radiation, but

Figure 3-0 Infrared radiation can illuminate a scene for night vision operations.

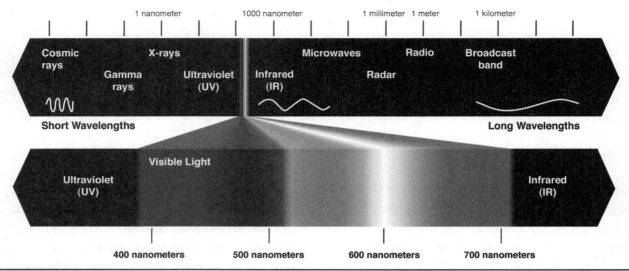

| 1 nanometer | 1000 nanometer | 1 millimeter | 1 meter | 1 kilometer |

Cosmic rays | X-rays | Microwaves | Radio | Broadcast band
Gamma rays | Ultraviolet (UV) | Infrared (IR) | Radar

Short Wavelengths

Long Wavelengths

Visible Light

Ultraviolet (UV)

Infrared (IR)

400 nanometers | 500 nanometers | 600 nanometers | 700 nanometers

Figure 3-1 The light spectrum, showing the small segment we call visible light.

infrared radiation is easier to generate and at higher output levels will not cause any biological dangers as ultraviolet will. Not all video cameras will be able to see infrared light, though, especially those designed for high-quality color imaging. Camcorders and digital still cameras contain a glass filter that essentially blocks out all infrared light, leaving only the visible portion of the light spectrum so that the image quality is maintained. For this reason, camcorders will not be able to see the infrared light from an illumination system unless you are willing to open up the case and remove the tiny glass filter that has been affixed to the CCD imager.

The LEDs shown in Figure 3-2 are all infrared types that are commonly used to send pulses in

Figure 3-2 Infrared LEDs are commonly used in remote control applications.

remote control applications. The perfect example of such an application is your television remote control. A television remote control will use one or more LEDs to send 40-Kilohertz (KHz) modulated bursts of light to the infrared receiver module in the TV in order to control the device. There are many other uses for infrared LEDs, but because of the vast market for consumer remote controls, the availability of these LEDs makes them the perfect choice for night vision illuminator projects. Extremely high-power infrared LEDs are also available, but the cost of these devices makes them less attractive than simply using an array of inexpensive infrared LEDs.

The peak wavelength of an infrared LED will determine the wavelength of the radiation. This data is important when matching the LEDs to any device that expects a narrow bandwidth. Television remote controls, for instance, use 940-nm infrared LEDs, so the receiver module is matched to that region of the light spectrum, often including a dark plastic filter lens that removes all but the required light radiation. Because our eyes cannot see into the infrared spectrum, some LEDs that include a plastic filter appear completely black, like the one shown in the right of Figure 3-2. To the imaging system, this dark plastic will appear completely

Figure 3-3 These 5-mm axial lead infrared LEDs are inexpensive and easy to work with.

transparent as the infrared radiation is picked up as normal white light.

The infrared LEDs shown in Figure 3-3 can be purchased from many electronics suppliers for just pennies each, especially when ordering them in bulk quantities. A night vision illuminator made with 32 infrared LEDs will be quite powerful, and may only cost $20 to build if you order your LEDs from an online electronics distributor. When you compare the cost of a 1-watt (W) high-power infrared emitter to the cost of 50 common infrared LEDs, the common LEDs will be less expensive and not require any special power supply, so they are currently the best choice. For extremely size-sensitive applications, the high-power emitters are better, but expect to pay a premium for both the LED base and the special driver circuit needed to power it properly.

The 5-mm-diameter plastic lead package shown in Figures 3-2, 3-3, and 3-4 are the most

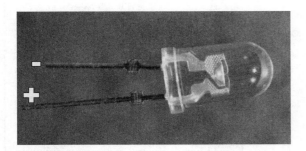

Figure 3-4 An LED must be connected using the proper polarity.

commonly used infrared LED, but there are also smaller packages available, including surface mount devices (SMDs). Since the goal is to output infrared radiation in a controlled manner, the 5-mm body is ideal as it allows the manufacturer to include a beam-shaping lens at the top of the body. SMD LEDs are usually nonfocused and for this reason not often used in infrared applications. Being able to control the focal range of the emitted radiation allows infrared LEDs to be tailored to the application. Remote control applications use a medium-angle focus to allow the remote control to operate from various angles, yet output a fairly focused and intense beam to the receiver. Infrared LEDs designed for use in night vision systems may have a wider focal range in order to illuminate more of the area being seen by the camera. The focal range of the LED will be shown in the manufacturer's datasheet along with all of the other important specifications.

To make an LED emit light, it must be connected using the correct polarity. Unlike a flashlight bulb, which will function from both AC and DC current, LED requires the correct DC current in order to function. A new LED that has not had its leads trimmed will have one lead longer than the other, and this lead may be the positive lead. If you are salvaging your parts from an old circuit board, then lead identification will be impossible, but the good news is that there are two other ways to identify the polarity of the LED. If you look at the underside of the LED, one side of the round base will have a flat side near one of the leads to indicate that it is the negative lead. Also, the negative lead will be connected to the larger cup-shaped carrier inside the lead as shown in Figure 3-4, although you may not be able to see through all of the various plastic colors.

You can make a very simple LED tester that will check the polarity and functioning of all visible color LEDs by simply keeping a 3-volt (V) coin battery (such as a CR2450) on hand to drop across the leads (Figure 3-5). If you connect an LED

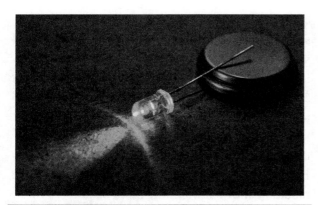

Figure 3-5 A 3-V coin battery becomes a universal LED tester.

backwards, it will not give off any light. If you connect it properly, it will visibly light up as long as the wavelength falls in our human range. The 3-V lithium coin battery has such low internal resistance, that it cannot damage an LED, even one rated for less than 3 V, so it makes the perfect tester for all visible LEDs. To test an infrared LED, you will need to view the output from a security camera on a video monitor, as will be

discussed later. Since most LEDs will show some output between 1.5 and 3 V, this simple LED tester becomes very handy, especially when dealing with salvaged LEDs that you have no datasheet for.

When hunting for new LEDs from any manufacturer, you will need to refer to the datasheet in order to choose the appropriate LED for the job. The most important specifications will be peak wavelength in nanometers, field of view in degrees, output power in millicandela (mcd), and forward voltage and current. There are many other specifications that you may also need to know such as peak pulsed current limitations, package size, and lead type. Also note that the brightness of an infrared LED is usually not rated in millicandela, as that is a rating for visible light. Infrared LED brightness is specified as optical power output in milliwatts per steradian (mW/sr). The datasheet segment shown in Figure 3-6 indicates that this LED is a common remote control type LED, operating at 940 nm and requiring a forward voltage of 1.2 V.

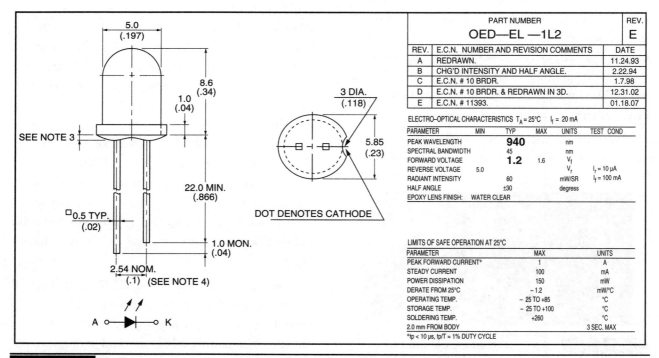

Figure 3-6 The datasheet is the ultimate source of information for an LED.

LEDs are not the only source of infrared radiation. Just like LEDs, laser diodes are available in many different wavelengths, visible and nonvisible. Laser pointers that are considered "high power" can be defocused in order to spread out the beam, making them suitable for certain night vision illumination systems, especially when long-range operation is important. You can easily purchase a laser pointer for under $100 from many online distributors with enough power to pop a balloon at 20 feet and burn electrical tape, but make no mistake—these lasers are not toys. Your first mistake will also be your last as a single shot into your eyes will likely permanently damage your vision. If you intend to work with high-power lasers, purchase a set of laser safety goggles and wear them in your lab when experimenting with these lasers.

The two laser pointers shown in Figure 3-7 are both extremely high-power Class IIIb lasers, capable of outputting 50 mW and 250 mW of infrared laser radiation. Calling these lasers "pointers" is really a bad term to use as they are completely invisible and not safe for pointing at objects. The laser shown in the top of Figure 3-1 will typically burn a black object from up to 20 feet away, and has a rated output power of over 250 mW. The laser shown at the bottom of Figure 3-7 will output about 50-mW of power and has been converted for infrared operation by changing the original visible red laser diode with one from a DVD burner. Both of these lasers work very

well as long-range night vision illuminators once their beam has been spread by some type of lens.

If you are looking to purchase a pointer-style laser for night vision experimentation, then choose one that allows the collimating lens to be adjusted or fully removed. The goal will be to spread the beam out to an area of several feet in diameter at a distance of several hundred feet.

Laser pointers and laser modules are basically the same thing—a cylindrical enclosure containing a laser diode, front-end optics, and some kind of current limiting circuit. In the case of laser pointers, the enclosure also makes room for a battery pack, which means that the laser driver circuitry can be made simpler, or often just run the diode directly off of the battery. Laser modules are designed with much more robust driver electronics and also include power regulation, so they can be run from a range of voltages. Laser modules also include much higher quality optics, and are often fully adjustable so that the beam can be focused or spread for a given distance.

This ability to defocus the beam will be the key to making a working laser illumination system, so the laser module may be the best choice for this project if your budget permits. The small laser modules shown in Figure 3-8 have various wavelengths, and output low-power (5 mW) laser radiation in the red, green, and infrared bands.

Figure 3-7 Laser pointers are also available with infrared output.

Figure 3-8 Laser modules are small and often include quality adjustable optics.

We also have several high-power infrared laser modules that we use in night vision systems, and they range in power from 50 mW to almost 1 W in power! A 1-W laser diode is more like a weapon than an illuminator, and special care must be taken in order to avoid serious eye damage or spontaneous fires in the lab!

If you are looking for a less expensive source for high-power lasers, then you may consider using a bare laser diode along with a home brew driver circuit or a simple battery power pack. Laser diodes are usually less expensive than laser modules, and available at much higher power levels than most laser pointers. The downside to using bare laser diodes is that some type of current limiting circuit needs to be made, and there are no included optics. The good news is that for illumination use, optics can be found from many salvaged sources and often laser diodes can be powered directly from a 3-V battery without any driver circuitry at all.

You can purchase a used or brand new DVD/CD burner combo pretty cheap these days. Inside you will find a pair of very powerful laser diodes. The DVD burner will include a high-power 150-mW or better visible red laser diode working in the 650-nm bandwidth, and the CD burner will include a 60-mW or better infrared laser diode running in the 780-nm (infrared) bandwidth. For night vision illumination, the infrared laser diode will be the one you want, so look for a DVD/CD writer combo, or just a CD writer.

The bare laser diode is shown in Figure 3-9. Most of them will look exactly the same, regardless of wavelength and output power. The small opening at the top will usually include a protective glass cover, but this is not a focusing lens, just a way to protect the actual diode from the elements. There will be as few as two connecting leads and as many as four, depending on the internal wiring of the laser diode, but all of them will function the same way. The small can will contain both a laser diode, as well as an optical sensor that can be used to monitor the output in a closed loop driver circuit.

Figure 3-9 This is the bare 60-mW infrared laser diode from a CD burner.

When there are only two connecting leads, that means the steel can is tied to ground and the leads will be the positive side of the laser diode and the optical sensor. With three leads, the wiring is often the same as with two leads, but will include a lead tied directly to the can. Laser diodes with four leads have separate ground connections for both the laser diode and the optical sensor. We have not yet seen a laser diode with five leads. The three-lead package is by far the most commonly used.

A filter that blocks out all light except for the small portion of the spectrum that falls between 800 and 1000 nm is called an infrared pass filter. This effect is exactly the same effect seen by placing a colored lens over your eyes to see the world in a different color tone. If you place a green piece of translucent plastic over your eyes, the world will look green because only the light from the 490- to 560-nm wavelengths will reach your eyes. An infrared filter will do the exact thing, but since you cannot see infrared light, the filter material will seem completely dark to your eyes. When you place an object made of translucent infrared material in front of a video camera, it will look completely clear, as if the camera has some special ability to see through a solid object.

This infrared pass filter effect can be exploited to create a very powerful infrared illuminator by

Figure 3-10 Infrared passing filters can be made from many different materials.

Figure 3-11 This small black and white spy cam will be used to view the infrared light.

passing white light through the filter to extract and send out only the infrared light that the video camera can see. White light encompasses a broad spectral range, including the infrared portion of the light spectrum. The benefit to this approach over using infrared LEDs is that a very small and powerful illuminator can be made, as well as a very large and extremely bright illuminator. The objects shown in Figure 3-10 all exhibit some infrared passing abilities, which will be explored using a small black and white security camera and some white light from an incandescent flashlight bulb as shown later.

To view the infrared radiation, which is completely invisible to the human eyes, a security camera must be used so that the image can be seen on a video monitor. To the video camera, infrared radiation appears as white light. This effect forms the basis of most night vision systems. These inexpensive camera modules run from 9 to 12 V, and output a standard composite color or black and white video signal. There are many suppliers on the Internet. You can expect to pay between $10 and $100, depending on the quality of the camera. Standard camcorders will not work for viewing infrared light, though, as they contain infrared blocking filters in their optics in order to achieve a higher quality visible light reception. You can try a camcorder, but chances are that it will see only

a tiny fraction of the infrared or ultraviolet light spectrum.

The small spy camera shown in Figure 3-11 is a model KPC-EX20H from the company KT&C in Japan. It's a very low-lux and high-resolution camera that has worked extremely well in our night vision experiments. This camera is also distributed by SuperCircuits.com under the model name of PC182XS, selling for around $100. Decent low-lux monochrome cameras can often be purchased for much less, but this one has a very low-lux rating and higher resolution CCD imager.

Depending on your night vision application, your video monitor may be as small as a camera viewfinder or as large as a television set. Since the idea is to operate in complete darkness in a way that does not compromise your stealth, a head-worn screen or camera viewfinder will work well, but for these simple tests we are using the small liquid crystal display (LCD) video screen shown in Figure 3-12 so that we can photograph the results. Typically, security cameras and small spy cams output a composite video signal, which will usually be labeled as "line input" or "video input" on the back of most TV sets and video monitors.

If you have not yet built any of the infrared illuminator projects presented here, then you can use a television remote control to test your camera's

Figure 3-12   The camera is connected to a portable composite video LCD monitor.

response to the infrared radiation (Figure 3-13). You won't get enough infrared power from a TV remote to light up a room at night, but you can certainly see the bursts on the screen as the pulses are sent to the one or more infrared LEDs at the end of the remote control. Try the tests with the room completely dark and you should be able to light up your face using the remote held at a distance of about 2 feet in front. You will notice that the infrared radiation is sent in bursts from the TV remote, as this is how it communicates with the receiving module in the set.

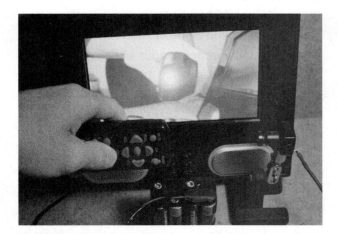

Figure 3-13   Testing the camera response using a television remote control.

Figure 3-14   A simple infrared illuminator in a ring configuration.

Figure 3-14 shows one of the infrared illuminations projects using a series-connected array of infrared LEDs to form a ring light that will slide over the lens on a small spy camera. By using 10 LEDs rated at 1.2 V each, the series-connected chain is able to be powered directly from a 12-V battery pack, which is also used to power the camera. This illuminator has a decent range of about 20 feet indoors, and will run for many hours from a single 12-V battery pack made up of 8-AA batteries.

To make a simple LED infrared illuminator output more radiation, the LEDs can actually be run beyond their continuous current rating by using pulses that have a short duty cycle (on time). This operation is called "pulse mode," and is often rated in the datasheet along with the other specifications. Continuous current for a common infrared LED may be rated at 100 mW, but the pulsed current may be 10 times this value, often reaching a watt or more!

The simple illuminator shown in Figure 3-14 is rewired to run in pulsed current mode, and is shown in Figure 3-15, running from the small circuit board that includes a short pulse oscillator feeding a power transistor. Running LEDs in pulsed mode requires understanding of the LED's datasheet and some type of oscillator that will send short duty cycles pulses to the transistor driving the LEDs. Substantial gains can be made in the output power using pulsed mode operation, but at a trade-off of simplicity and power consumption.

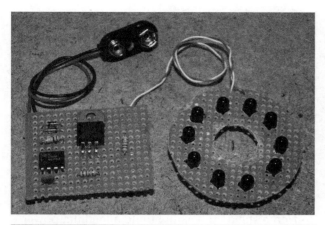

Figure 3-15   Pulsing LEDs with higher current will increase the infrared output.

To illuminate a larger area for an outdoor security camera, more LEDs will be needed in order to expand the field of view. The illuminator shown in Figure 3-16 uses 266 LEDs wired in a series and parallel configuration in order to create a large bright array of LEDs. This illuminator is designed to fit into the shell of an outdoor weatherproof security light and will give an outdoor security camera the ability to see an entire yard in complete darkness. The power supply is a 12-V DC power pack that has been installed in the light case so that the unit can be plugged directly into any AC outlet.

Figure 3-16   Using a large array of infrared LEDs to create a high-power illuminator.

Figure 3-17   This massive infrared LED array illuminator contains over 1500 LEDs.

There are times when you may need an extremely intense infrared light source, and the use of an incandescent-based system may be prohibitive due to the massive heat generated. The huge infrared array shown in Figure 3-17 is made by soldering over 1500 LEDs onto a printed circuit board that has been wired to give the LEDs their correct forward voltage in a series parallel configuration. This massive panel was used for several large indoor security jobs where an intense amount of infrared radiation was needed along with stealth operation. To conceal such a massive array, it was placed inside a gutted 20-inch LCD monitor, which had a one-way mirror installed to make it look like a regular monitor. It was no easy task hand soldering all of these LEDs, but when your spy mission must succeed, you do whatever it takes to acquire the information!

To create a portable night vision viewer that has the imaging system and illuminator combined in one small package, you will need some kind of tiny video output device such as a camcorder viewfinder. The night vision viewer shown in Figure 3-18 combines a hacked camcorder viewfinder along with a small infrared LED array and a low-lux video camera to create a handheld night vision device. This project is fairly simple to build if you can find an older camcorder to salvage the CRT based viewfinder, which is basically a tiny composite video monitor. This night vision device has an illumination range of about 50 feet, and works as

Figure 3-18 A portable night vision viewer made from an old camera viewfinder.

well as many of the military-grade night visions systems that were available only 10 years ago.

Since a color camcorder cannot see infrared light due to the inclusion of an infrared blocking filter over the CCD sensor, it cannot be used as the basis for an infrared viewer. Of course, there are always ways around every problem, and the converted camcorder shown in Figure 3-19 uses a small

low-lux spy camera to bypass the camcorder's internal video system, giving the camcorder the ability to record infrared illuminated scenes as well as displaying them live on the viewfinder. This project also includes a small infrared LED illuminator to create an all-in-one portable night vision recorder and viewing system.

Since many materials can be used to pass only infrared light while blocking out most of the visible light, a simple illuminator can be made to convert a standard flashlight into a medium power night vision illuminator. The flashlights shown in Figure 3-20 both have filters placed over the lenses that only pass infrared light and block out all of the visible light from the incandescent light bulb. Materials such as floppy disk surfaces, 35-mm film, and even tinted glass bottles can be made into infrared pass filters. By using a powerful white light source, a fairly intense source of infrared radiation can be made that will form the basis for many night vision security projects.

Lasers that output infrared radiation can also be used as an illumination source. Lasers are exceptionally effective at sending light over long distances, and infrared lasers operate

Figure 3-19 A camcorder converted for night vision operation using an external camera.

Figure 3-20 This infrared illuminator converts white light into infrared light.

exactly the same way as visible light lasers. The rig shown in Figure 3-21 uses a powerful 250-mW infrared laser with a beam spreader to cast an intense infrared beam several hundred feet away so that a camera focused through an optical magnification system can view the scene in complete darkness. This ultra covert night vision system is able to view a distant scene as though it was lit by a high-power flood light, yet the only limitation is the optical magnification power of the camera lens.

The solutions to your night vision applications are almost infinite in number, so with a little creativity and a good supply of junk, you can hack together your own stealthy spy gear.

Figure 3-21   A long-range night vision system using a laser and optical magnification.

PART TWO
Peering into the Night

Infrared Light Converter

ALTHOUGH INFRARED LIGHT-EMITTING DIODES (LEDs) are the most common source of invisible light for a night vision device, they are certainly not the only option available, nor are they always the best. Depending on your camera type and setup, you may need a hand-held source of infrared light that can be rapidly moved around the scene, or possibly an infrared light source that differs in wavelength from the standard 800 to 950 nanometer (nm) wavelength of the standard infrared LEDs.

PARTS LIST

- Filters: Undeveloped 35-mm film, floppy disc surface, UV light bulb, infrared photo filter

- Light source: Incandescent flashlight bulb, black light

- Camera: Security camera and monitor to view infrared

Figure 4-0 This project will explore several ways to convert visible light into infrared light.

Infrared light falls just below red on the light spectrum, making up the wavelengths from about 750 to about 1500 nm. This light cannot be seen by human eyes, but it can easily be seen by most video cameras, making it useful as a covert lighting method in night vision systems. Some video cameras can even see part of the ultraviolet light spectrum from 200 to 400 nm. That will be covered here as well. The goal will be to pass white light through various materials that will attempt to block out all of the visible light and only pass the light that is invisible to the human eyes, yet visible to most security cameras and spy cameras.

A filter that blocks out all light except for the small portion of the spectrum that falls between 800 and 1000 nm is called an infrared pass filter. This effect is exactly the same effect seen by placing a colored lens over your eyes to see the world in a different color tone. If you place a green piece of translucent plastic over your eyes, the world will look green because only the light from the 490 to 560 nm wavelength will reach your eyes. An infrared filter will do exactly the same thing, but since you cannot see infrared light, the filter material will seem completely dark to your eyes. When you place an object made of translucent infrared material in front of a video camera, it will look completely clear, as if the camera had some special ability to see through a solid object.

This infrared pass filter effect can be exploited to create a very powerful infrared illuminator by passing white light through the filter to extract and send out only the infrared light that the video camera can see. The benefit of this approach over

Figure 4-1   Infrared pass filters can be made from many different materials.

the use of infrared LEDs is that a very small and powerful as well as a very large and extremely bright illuminator can be made. The objects shown in Figure 4-1 all exhibit some infrared passing abilities, which will be explored using a small black-and-white security camera and some white light from an incandescent flashlight bulb.

To view the invisible (to humans) infrared radiation, an ultra low-light (lux) security camera will be used so that the infrared light can be viewed on a monitor as though the camera was looking at white light (Figure 4-2). These inexpensive camera modules run from 9 to 12 volts (V), and output a standard composite color or black-and-white video signal. There are many suppliers on the Internet. You can expect to pay between $10 and $100, depending on the quality of the camera. Standard

camcorders will not work for viewing infrared light, though, as they contain infrared blocking filters in their optics in order to achieve a higher quality visible-light reception. You can try a camcorder, but chances are that it will see only a tiny fraction of the infrared or ultraviolet light spectrum.

The camera we used is a model KPC-EX20H from the company KT&C in Japan. It's a very low-lux and high-resolution camera that has worked extremely well in our night vision experiments. This camera is also distributed by SuperCircuits.com under the model name of PC182XS, selling for around $100. Decent low-lux monochrome cameras can often be purchased for much less, but this one has a very low-lux rating and high-resolution CCD imager. The liquid crystal display (LCD) monitor is one of those portable gaming screens, and can be connected to any composite video input.

Some of the more obvious materials that can be used to pass only infrared light (radiation) can be found in remote control devices that use infrared LEDs for communication. Older TV remotes often place the LEDs behind a small infrared filter window; the same is true for the receiver in the actual appliance. This plastic will look perfectly black to human eyes, but as you can see in Figure 4-3, the video camera sees right through it as if made of clear plastic!

The long plastic bezel that is shown perfectly clear to the camera in Figure 4-3 is the face plate taken from an older infrared wireless headphone

Figure 4-2   The test rig will consist of a low-lux video camera and small LCD monitor.

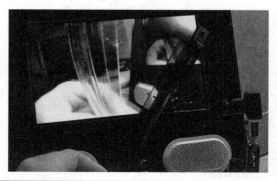

Figure 4-3   The camera can see right through any translucent infrared materials.

set. To our eyes, the part looks perfectly black. We can only see a tiny amount of light through it if we hold it up to our room lighting. But to the camera, it looks like clear plastic, able to pass all of the infrared light to the charge-coupled device (CCD) imager, which is then displayed on the LCD monitor as white light. To a color security camera, the infrared light would look slightly reddish, yet still appear to be lit from a visible light source. Interestingly, the Nikon camera we used to take this photo could see through the plastic just enough to make it look semitranslucent. To our own eyes, though, the plastic is almost completely black.

The bulb shown in Figure 4-4 is called a "black light" bulb and will emit mostly ultraviolet radiation in the 340 to 400 nm range. These bulbs are often used for special-effect lighting, where a certain phosphorescent color is made to look as though it is glowing in the dark. These bulbs are also used for medical purposes, as well as to identify the anticounterfeiting security strip on money. In this project, the ultraviolet light will be used as an illumination source, giving the security camera the ability to see the entire room with light that human eyes cannot perceive.

As you probably already know, ultraviolet radiation in high doses is not healthy for the skin or eyes, and although a black light is designed for "human safe" operation, it is certainly not a good idea to expose your eyes to the light for extended periods of time. If you are going to purchase

a black light bulb, follow the warnings on the package, and remember that just because you can't see much light coming from the bulb, it doesn't mean that there isn't light. In reality, the most dangerous effect of the black light might be the intense heat that is generated by the bulb due to having most of the visible light blocked by the glass. These things get extremely hot in seconds, unlike "standard" white light bulbs.

The black light (shown in the monitor in Figure 4-4) becomes so transparent that you can even see the tungsten filament and connecting wires inside the bulb. To our eyes and the digital camera used to take this photo, the black light bulb looks completely dark. It is strange to see the camera see right through it without any effort at all. This means that the black light bulb will be an amazing source of illumination for the video camera, which is obviously sensitive to the ultraviolet light. The downside is that the bulb also passes some light in the visible violet end of the spectrum, so the area being illuminated will have a dull purple glow that can be seen by human eyes. Of course, there are ways to block the violet light, as well other techniques shown in the next few steps.

Once you have a low-lux black-and-white security camera connected, you can test all kinds of dark-looking materials for their ability to perform as infrared pass filters. Some of the materials that pass infrared light are shown in Figure 4-5—exposed

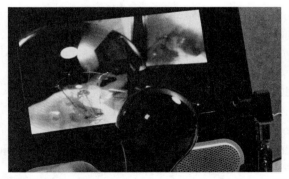

Figure 4-4   An ultraviolet light bulb also looks like clear glass to the video camera.

Figure 4-5   Developed film and floppy disk material can be made into infrared pass filters.

camera film and floppy disk material. Having a massive collection of retro junk on hand, we were able to find some old 5.25-inch (in) floppy discs to hack up, but the smaller 3.5-in types will also give the same results. Oh, and don't worry, we didn't hack up the Commodore Dos 3.3 disks—those are classics!

Exposed film is also a good material that will block most visible light and pass infrared radiation, but we found that the film would also pass some deep red light as well. The film used is just common 35-mm film that has been exposed and then developed. In other words, you load a new roll of film into your camera, point it at an evenly lit surface, and then snap away until you have as many frames as you need to cut up and make your filter. Even the film from a disposable camera will work fine, and remember to tell film processors that you purposely made blank photos when you bring it in for developing, or they may toss out your negatives.

As for the light source, a standard halogen bulb hand-held flashlight will be used, as these light sources also include a large infrared component. Because the film and the floppy disk material cannot withstand a great deal of heat, the flashlight is a better choice than a larger incandescent light source. A 60-watt (W) white light bulb might offer more illumination, but the thin film would certainly melt within seconds of being exposed to such intense heat. Remember that for a filter to block light, it must also absorb the light, which means a great deal of heat must be contained as well.

To test the material for its ability to work as a night vision illuminator, it is placed in front of a small halogen bulb flashlight and viewed on the monitor. As shown in Figure 4-6, the exposed camera film passes a great deal of infrared light, creating a huge bloom on the camera as it shines onto the CCD sensor. The camera film was so good at passing infrared light that it almost looked the same is if the flashlight was shone directly at the camera. The downside to the camera film is that it also passes some amount of deep red visible

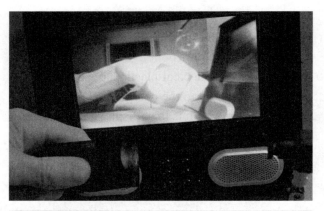

Figure 4-6 The camera film allows a lot of infrared light to pass through to the camera.

light, so it wouldn't be the optimal solution for a purely covert night vision illuminator. But if we placed four strips of the film together to form a thicker filter, then most of the visible red light was absorbed at the expense of about half of the infrared brightness. This makes the camera film a good candidate for use as an infrared pass filter, as long as several layers are used along with a fairly bright white light source. When dealing with thin film that could be easily melted, care must be taken not to use a light source that will radiate a lot of heat.

Since the camera film seemed to be a good choice of material to make a cheap infrared pass filter, we wanted to see if we could find a strong light source that would not melt the film. Ultrabright white LEDs seemed like a logical choice, and they were magnitudes brighter than any of the infrared LEDs we had on hand. Unfortunately, the light given off by a white LED is not the same full spectrum of radiation given off by a halogen flashlight bulb, as we found out. In fact, a white LED is manufactured by placing a special florescent coating over an ultrabright blue LED, which makes the coating give off white light that has very little infrared component. As you can see in Figure 4-7, the ultrabright 3 LED closet light is only slightly visible to the camera after it passes through the infrared pass filter made by the camera film. Results of using white LEDs as a possible light source = total fail.

Figure 4-7   White LEDs were tested but failed to produce enough infrared light.

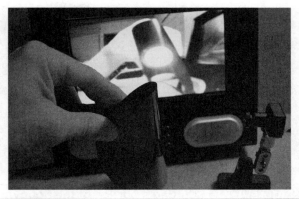

Figure 4-9   The disk material passes a good amount of infrared and blocks most visible light.

Incandescent flashlights use incandescent lamps like the ones shown in Figure 4-8, which consist of a glass bulb and a tungsten filament. A tungsten filament gives off radiant energy in the visible light spectrum as well as the nonvisible infrared spectrum. The amount of infrared radiation given off by these bulbs is high enough that after blocking out all of the visible light, a decent amount of illumination can be generated for a night vision illuminator. The good news is that the small battery-operated bulbs do not give off enough heat to melt thin plastic filters. The bad news is that the illumination is not strong enough for outdoor use. If your goals are to illuminate a small room for covert surveillance, then the flashlight bulb may be the perfect source for your illuminator. Larger

halogen lights such as those used in auto headlights will also offer the same wide spectrum of light, but with much increased heat radiation.

The next material tested was the inner film surface taken from one of the old 5.25-in floppy disks shown in Figure 4-5. As can be seen in Figure 4-9, the disk material passes a good amount of the visible light from the flashlight, but passes almost no visible light that can be seen by human eyes or the camera that took the photo. The floppy disk material was actually far superior as an infrared pass filter than the exposed camera film, as it did not require multiple layers to block out visible light. Of course, it is not easy to dig up a pile of old floppy disks unless you are an intense junk collector, so the exposed camera film may be the better choice. To make the camera film work as well as the floppy disk material, four layers of film had to be stacked to block the same amount of visible light.

The first illuminator we made was using the floppy disk material and the high brightness pocket flashlight shown in Figure 4-10. The beam from this flashlight is highly focused, but very bright so the illuminator would have the same characteristics when viewed using the spy camera. Because we do a lot of infrared experimentation, it was nice to have a small and highly focused source of infrared light that could be added to our tool kit.

Figure 4-8   The light given off by an incandescent flashlight bulb is high in infrared.

Figure 4-10   Cutting a portion of the floppy disc to fit over the flashlight lens.

Figure 4-11   Making a larger more powerful illuminator using a flashlight.

The 5.25-in floppy disc had enough area to allow a 2-in-diameter circle to be cut, which was much more than necessary for the small pocket flashlight. Even the disc from a 2.5-in floppy disc would be plenty of material for a small flashlight lens. The disc was traced for cutting as shown in Figure 4-10.

The small pocket light worked very well and was able to illuminate an area about as large as the original flashlight would as viewed by the naked eye. The floppy disc material blocked enough of the visible light that you could barely see the dull red glow when looking at the beam from a few feet away. Knowing that the disc material worked so well as an infrared pass filter, we decided to make another portable illuminator, but this time using a much larger and brighter flashlight.

The flashlight shown in Figure 4-11 has a lens diameter of about 4 in, and a much stronger beam powered by a 6-V lantern battery. The heat produced by the small bulb is still low enough that the film would not melt, even during extended use. Because of the increased diameter of the larger lens, we had to cut two half-circles from the floppy disc surface to cover the entire lens, but this wasn't a problem as long as there was a slight overlap to block any visible light from escaping. The infrared pass filter was placed between the lens and the reflector and held in place by friction.

Both infrared beam flashlights shown in Figure 4-12 work very well in conjunction with any security camera or low-lux spy camera. The small light is good for close-up work that requires a sharp focused beam, and the larger flashlight can illuminate an entire room for covert security operations. The only drawback to using this type of illumination source for stealthy operations is that there is a very slight noticeable glow visible at close range, so you couldn't use these illuminators in a head-worn night vision illuminator and expect to remain completely invisible. For use as

Figure 4-12   The two infrared beam flashlights made using the disc material.

Figure 4-13 The large infrared flashlight illuminates the ceiling in our lab with ease.

Figure 4-14 Creating a similar infrared beam flashlight using the camera film.

security camera night vision illuminators, the two flashlights will work very well.

Figure 4-13 shows what the security camera sees when we turn off all of the room lights and point the larger infrared beam flashlight up at the ceiling. The beam is so bright to the camera that the automatic iris ends up closing to block out the overabundance of light. To our eyes and the camera taking the photo, there is only a dull red glow coming from the center of the flashlight lens, but it is not very bright at all. After a few minutes of operation, the lens and infrared pass filter material remain cold.

Since we had plenty of film on the roll we had developed, we decided to make a third infrared flashlight using four layers of the exposed camera film as well. The floppy disc material worked a bit better when it came to blocking visible light, but the camera film would probably offer a slightly brighter beam and shift into the nearer infrared spectrum of between 750 and 850 nm. Actually, the exact bandwidth of the infrared light depends on the type of film and how it was developed, so your results will probably vary.

A typical D-Cell battery flashlight was used for this version of the handheld illuminator, and eight slices of film were cut up as shown in Figure 4-14 in order to create the two semicircles that would be

needed to cover the lens area on the flashlight. To avoid any visible light leaks, the two half-circles are made to slightly overlap. Also notice that in Figure 4-14 the camera used to take this photo can see through the developed film somewhat, as compared to not being able to see through the floppy disc material at all. This would indicate that the color of the film has some overlap between the end of the infrared spectrum and the start of the visible red spectrum.

With four layers of camera film covering the flashlight lens shown in Figure 4-15, not much of the visible red light escapes, but the deep red glow is still more noticeable than when using the floppy disc material as an infrared pass filter. The illumination from this version of the handheld infrared illuminator is very bright, but the noticeable deep red glow detracts from any

Figure 4-15 The completed near-infrared beam flashlight using the camera film.

possible stealth night vision use. For infrared photography, this illuminator would be ideal, and as many have discovered, placing the exposed film over the lens on your digital camera gives the same type of infrared filtering as an expensive infrared filter from a camera shop. By using both the infrared illuminator and the infrared pass filter on a digital camera, you can experiment with some very interesting infrared photography.

If you have the budget, then you can also purchase high-quality infrared pass filters that are designed to attach to various camera lenses. The two filters shown in Figure 4-16 have threads on the outside, are made from glass, and have very sharp infrared band pass abilities, making them ideal for infrared photography as well as illumination filters. The downside to using these types of filters for illuminations is that they are expensive, and may cost you between $100 and $300, depending on the size and type of material used. The benefit to using a glass filter is that it can withstand a lot more heat, so you can send a much brighter beam into the filter to extract much more infrared radiation. Most camera shops will be able to order infrared filters in various sizes, and some industrial supply centers can even cut custom-sized infrared filters or source them in sheets.

The infrared illuminator shown in Figure 4-17 was made by fitting a glass infrared pass filter

Figure 4-17   A glass infrared filter is installed in front of a 60-W light source.

over the end of a steel can that contains a 60-W incandescent light source. This infrared illuminator throws out a lot of infrared radiation, but also gets extremely hot in a short time. For this reason, such a system is really only usable outdoors or for short controlled intervals. The efficiency of such a system is also dependant on the infrared capabilities of the original light source, with halogen being a good choice, followed by a full spectrum incandescent bulb. High-efficiency florescent lights are not suitable for infrared illumination as they do not output much infrared radiation.

The completed glass filter illuminator is shown in Figure 4-18, and can plug directly into a standard AC outlet. The illumination from this version is very strong and does block most of the

Figure 4-16   These are manufactured infrared pass filters designed to fit camera lenses.

Figure 4-18   This illuminator is quite powerful, but runs extremely hot.

visible light, but this comes at the cost of running extremely hot. This infrared pass filter will also work well with the low-power flashlight bulbs, but is a little too large and thick to fit into the flashlight lens cover. Another decent light source that was tested on this filter was a round halogen motorcycle headlight running from a 12-V DC power source. The resulting heat was slightly less than the incandescent bulb, but required a less portable power supply.

If you don't mind having a deep purple glow around the room, then the black light shown in Figure 4-19 will make a ready-to-use night vision illumination system right out of the box. Because the CCD imager in most security cameras can see ultraviolet light in the same way it can see infrared light, the black light makes it easy to illuminate a scene with light that human eyes cannot see. The downside to the ultraviolet light is that it also casts a low level of visible violet light and will run as hot as a toaster within seconds of use. Interesting side effects of the ultraviolet illumination include phosphorescence of certain materials and the ability for many digital cameras to pick up the ultraviolet light, which is strong, around 250 nm.

The interesting shot shown in Figure 4-20 was taken by running the 60-W incandescent black

Figure 4-20   An interesting night shot using the 60-W ultraviolet light bulb.

light in our desk lamp as we photographed both the light and the scene as viewed on the monitor by the low-lux black-and-white security camera. The digital camera used to take the photo picked up the dull violet glow from the ultraviolet light and was able to see right through the bulb to capture the lit tungsten filament. On the monitor, the camera easily sees the nearby objects in the room as well as the super bright phosphorescence of our black sweater. Another interesting effect of using the ultraviolet illumination is the ability to see right through certain materials, revealing what is underneath somewhat. This effect has been dubbed "X-ray vision," and the effect can be seen from illumination by both infrared and ultraviolet light sources. The X-ray effect can almost see through certain types of clothing, but the effect really depends on materials, lighting, and infrared heat.

The image shown in Figure 4-21 was captured by feeding the video input on a standard camcorder with the output from our small spy cam while illuminating the scene with the small handheld infrared flashlight shown in Figure 4-12. By looking through the camcorder's viewfinder, we essentially have a portable night vision viewer that can record everything it sees in pure darkness, allowing me to navigate by looking at the scene in the viewfinder. The handheld

Figure 4-19   This black light is a plug-and-play illumination source for video cameras.

Figure 4-21 **Using the small infrared flashlight with a portable night vision viewer.**

flashlight makes it easy to illuminate the part of the room we need to see while we walk around in complete darkness.

Some of the other materials that we found had some infrared passing abilities were dark glass bottles, camera lens caps, thin black plastics, some black paints, and even some cloth materials. Have fun creating light that only your spy cameras can see. If you find a new material that has great filtering capabilities, stop by the LucidScience.com forum and tell us about it.

Simple Infrared Illuminator

INFRARED LIGHT FALLS just below red on the light spectrum, making up the wavelengths from about 750 to about 1500 nanometers (nm). This light cannot be seen by human eyes, but it can easily be seen by many video cameras, making it useful as a covert lighting method in night vision systems. A common example of infrared light is the medium of communication between your remote control and television set. The light-emitting diode (LED) at the end of your remote sends out pulses of infrared light, which are received by the infrared detector on the TV and demodulated back into data. Of course, you cannot see the pulses because they are out of our visual range, but any video camera that is not equipped with an infrared filter can see this light easily.

Figure 5-0	Infrared LEDs are invisible to humans but visible to security cameras.

and small-board cameras are particularly sensitive to infrared light. These ultra low-lux cameras can usually be purchased for about $100 or less, especially from online sellers. Add 10 or more infrared LEDs, and you now have a night vision system that is better than those that were selling for thousands of dollars in the 1980s.

This project represents the most basic LED illuminator possible, and is nothing more than a series string of LEDs running from a DC power source or battery pack. You can build this infrared illuminator from a single LED and coin battery, or add as many LEDs as your power pack can handle. With 10 LEDs, you can easily light up a room for a video camera, and with 100 LEDs, you could light up your entire yard to make it seem like midday to a security camera. Our Night Vision Viewer projects also use infrared LEDs as an invisible light source.

The most basic example of infrared illumination can be found at the end of your TV remote control. This 940 nm infrared LED is pulsed on and off at

---

PARTS LIST

- LEDs: 800- to 940-nm, 5-mm body infrared LEDs
- Resistors: Optional current limiter based on LED ratings
- Battery: 9- to 12-V battery or pack

There are many good-quality security cameras available in the market that include a low-lux video camera in a weatherproof housing along with an array of infrared LEDs for night vision applications. Of course, you may want to just make your own simple infrared illuminator for projects that you need to add night vision to. This can be done in a few hours with a few dollars worth of infrared LEDs. Black-and-white security cameras

40 Kilohertz (KHz) to transmit the control codes to the infrared light receiver in the appliance. Take a look at the LED and press the button in the remote. You will see nothing, even though the light from the LED is probably as bright as the light from a white LED keychain flashlight. Now, if you view this LED light through a monitor connected to a video camera, it will look very bright on the screen. You might also be able to see the light through the viewfinder on a camcorder, but it will become a very dull purple glow due to the infrared filters being used to correct the color balance.

There are several varieties of infrared LEDs, ranging in size, field of view, output power, and effective light color. The most commonly used infrared LED's output is 940-nm infrared light which is far beyond the human visual range, and fairly detectable by any nonfiltered video camera. There are also infrared LEDs available for the 800 to 900 nm range, and these are even better for use in night vision applications, but there will be an ever so slightly detectable red glow as the human eye can faintly detect this band of light. If you have seen an outdoor night vision security camera after dark, then you are probably familiar with this dull red glow. The TV remotes shown in Figure 5-1 all have 940 nm LEDs, whereas our huge collection of night vision LEDs are designed to output 850 nm light.

At the top of the LED, the thickness and shape of the plastic casing forms the optical properties of the output lens. A wide angle lens allows the light to cover a larger area over a given distance, whereas a narrow angle lens allows a brighter beam to reach a further point. For TV remote controls, a medium angle is a good trade-off between distance and viewing angle, but for night vision, a wider angle is usually better. In this project, we are using typical 940-nm TV remote type LEDs with a narrow field of view. You can see in the images that the resulting beam is sharp and focused much like a small flashlight.

If you are planning to use your night vision system indoors in a small area such as a room, then try to match the field of view to both the camera and the LEDs so that you can illuminate the entire scene. For a portable or head mounted night vision system, a narrower beam may work better as this will give you more light as you look around the darkness. The three LEDs shown in Figure 5-2 have different colored bodies, but all of them are basically the same as far as specifications. The black body on the LED shown in the right of Figure 5-2 is actually completely transparent when viewed with a video camera as the plastic has been manufactured to pass only infrared light, light that we cannot see. Field of view, power output, voltage, and wavelength are all specifications stated on the datasheet, so consider these when sourcing new infrared or visible light LEDs for any project.

A new LED that has not had its leads trimmed will have one lead longer than the other, and this lead will be the positive lead. If you are salvaging

Figure 5-1   Your TV remote control communicates over an infrared light beam.

Figure 5-2   Infrared LEDs are available with various lens types and wavelengths.

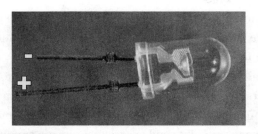

Figure 5-3 To make an LED emit light, it must be connected with proper polarity.

your parts from an old circuit board, then lead identification will be impossible, but the good news is that there are two other ways to identify the polarity of the LED. If you look at the underside of the LED, one side of the round base will have a flat side near one of the leads to indicate that it is the negative lead. Also, the negative lead will be connected to the larger cup-shaped carrier inside the lead as shown in Figure 5-3, although you may not be able to see through all of the various plastic colors.

You can make a very easy LED tester that will check the polarity and functioning of all visible color LEDs by simply keeping a 3-volt (V) coin battery (such as a CR2450) on hand to drop across the leads. If you connect an LED backwards, it will not give off any light. If you connect it properly, it will visibly light up as long as the wavelength falls in our human range. The 3-V lithium coin battery has such low internal resistance, that it cannot damage an LED, even one rated for less than 3 V, so it makes the perfect tester for all visible LEDs. To test an infrared LED, you will need to view the output from a security camera on a video monitor.

When hunting for new LEDs from any manufacturer, you will need to refer to the datasheet in order to choose the appropriate LED for the job (Figure 5-4). The most important specifications will be peak wavelength in nanometers, field of view in degrees, output power in millicandela (mcd), and forward voltage and current. There are many other specifications that you may also need to know such as peak pulsed current limitations, package size, and lead type. Also note that the brightness of an infrared LED is usually not rated in millicandela as that is a rating for visible light. Infrared LED brightness is

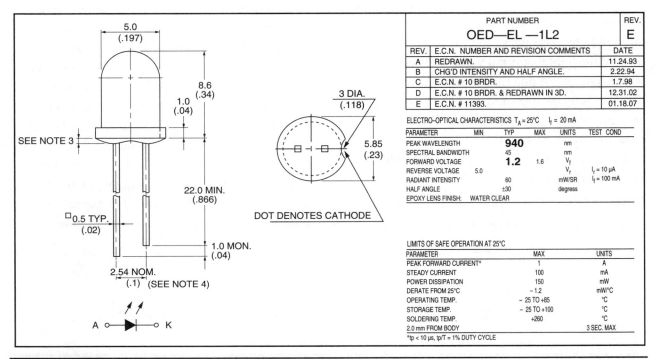

Figure 5-4 A typical LED datasheet showing the important parameters.

Figure 5-5   Some infrared LEDs include a band pass filter in their casings.

specified as optical power output in milliwatts per steradian (mW/sr). The LEDs we are using in this project are very common TV remote types, and have the following ratings: 950-nm wavelength, 1.2-V forward voltage, 100-milliamps (mA) forward current, 250-mW/sr output power.

To the human eye, the 950-nm infrared LEDs shown in Figure 5-5 look completely black, even when held up to a bright light source, but when viewed by a video camera, they appear perfectly clear, even when not powered. The reason the camera can see right through the LEDs is because the plastic is made up of materials that only pass infrared light, creating a band pass filter that blocks most of the light that is not within the specified wavelength. So, if this LED was outputting light between 800 and 1000 nm, the band pass filter may help cut all unwanted light except for the output close to 950 nm as specified in the datasheet. Other LEDs are perfectly clear, or have slightly blue tinted plastic bodies.

Most LEDs will require between 1.2 and 1.6 V to operate at their full capacity. Exceeding their maximum voltage is a very risky game because some LEDs will fry if you exceed the rated voltage by a few percent. Slightly lower voltages are certainly safe, but the output power will begin to roll off rapidly once you start lowering the voltage. The goal is to come as close to the rated maximum voltage as possible and ensure that you are supplying the required current. Our LEDs require 1.2 V for maximum operation, so by placing 10 of them in series, we can power all 10 from a single 12-V battery pack. A series connection means that each LED will receive the sum of the battery pack voltage divided by the number of LEDs in the chain.

Figure 5-6 shows the simple series wiring diagram that is used to divide the 12-V battery pack down to 1.2 V per LED. In a series circuit, voltage is divided equally by all elements in the chain as long as they are drawing the same amount of current. For this reason, it is important to ensure that all of the LEDs are either exactly the same make and model or have identical specifications. LEDs can be chained together like this to run from just about any voltage source. For LEDs that require 1.5 V, the 12-V battery would be able to drive 8 LEDs safely in series, since (12 V/8 = 1.5 V). If your division does not work out perfectly, then you need to round down to the next value to ensure that you do not overpower the LEDs. For instance, 1.2-V LEDs running from a 9-V battery would require a series chain of eight LEDs, but only provide each LED with 1.125 V. This is less than the rated 1.2 V, but with seven LEDs, the maximum voltage rating would be exceeded at 1.285 V. You might get away with slightly more voltage, though, but expect some failures when exceeding device limitations.

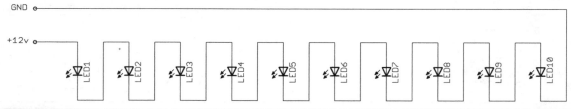

Figure 5-6   Placing LEDs in series allows a higher voltage to be used.

The goal of this small infrared illuminator is to give a micro security camera the ability to view a room in complete darkness. One simple solution to the placement of LEDs is to form a ring around the camera lens so that the light appears uniform and directly along the line of sight of the camera. Of course, you can arrange the LEDs in any pattern or number you need in order to satisfy the requirements of your night vision illuminator, but keep in mind that there are limits to both the field of view as well as the ultimate distance that can be achieved with LEDs.

Perforated board is a good medium to use as a base to support the LEDs as you can either drill holes to mount the LEDs or simply stand them right off their own leads. Perforated board that includes solder pads on one side would be best if planning to stand the LEDs on their own leads as this will ensure that they are all placed flat on the surface. Practically any kind of material from cardboard to thin steel can be used for this project. The goal is to simply secure the LEDs in a circular pattern around your camera lens. Our camera had a 1-inch (in) square body and a ½-in wide lens, so we traced a 2-in circle out on the perforated board shown in Figure 5-7, with a ½-in circle in the center for the camera lens. A Dremel tool with a small disc cutter will be used to cut around the circular markings.

The board was rough cut around the marked line using the cutting wheel and then ground

Figure 5-8 The perforated board is cut into a donut shape to fit around the camera lens.

smooth to remove sharp edges using the grinder attachment. When designing the LED carrier, keep in mind the space needed between the camera body and LED leads if you plan to place the disc right up against the camera body as we are doing in our version of this project. Figure 5-8 shows the disc resting on the camera body, and there is plenty of clearance between the edge of the camera body and the outer edge of the circle in order to make room for the LEDs.

Perforated board can be quite brittle when drilling or cutting, so you will need to start with a small drill bit and then work up to the larger one to prevent the edges from chipping or from snapping the board right in half. Our LEDs required a 3/16-in hole to fit snugly, so we started by drilling the holes with a 1/16 bit, followed by a 1/8 bit. As you can see in Figure 5-9, the edges

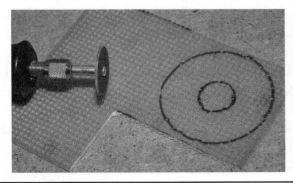

Figure 5-7 Making a camera ring to hold the 10 infrared LEDs.

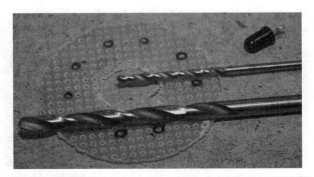

Figure 5-9 Using a pilot drill bit to stop the perf board from chipping.

Figure 5-10   Checking the camera body and
LED lead clearances.

Figure 5-12   The underside of the perf board
with the completed series wiring.

and center of the board have been slightly chipped
from the cutting process, but the board is still
usable. A square board would have been much
easier to make, but that would require a much
larger enclosure to be used if we wanted to use the
system outdoors.

If you are planning to place the ring light
directly against the camera body as shown in
Figure 5-10, then there needs to be enough
clearance between the leads and the edge of the
camera casing. Another option would be to bend
the leads flat and then place an insulating disc over
the back of the board so that the leads can rest on
the camera body.

Connecting up 10 LEDs in series is certainly
not a lot of work; any kind of wire will work. If
your LEDs are new, then the long leads can be bent
to join with each other, and you will only need to
add the power wires to the circuit. We always use
the wire shown in Figure 5-11 for our perf board

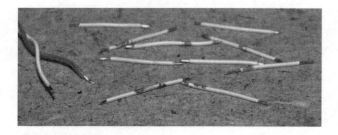

Figure 5-11   Preparing the connecting wires
to make the series LED chain.

circuits and for breadboarding, which is cut and
stripped Cat5 network wiring.

Following the circuit shown in Figure 5-6, we
soldered the connecting wires from the positive
lead on one LED to the negative lead on another
until every LED minus one was chained to the
next. The last two unsoldered leads (one positive
and one negative) were soldered to a long
connecting wire that will carry the 12-V DC power
to the series circuit. (See Figure 5-12.) To avoid
wiring mistakes, we always use a color scheme
that is obvious such as black or green for ground
and red or white for power. The good news is
that if an LED is soldered in backwards or if you
reverse the power, your illuminator will just fail to
light, but there will be no damage. The same thing
is certainly not true for the camera, this we know
from our painful experience!

To attach the ring light illuminator to our tiny
camera in a way that would allow it to be removed
easily, we opted to use a few bits of double-sided
tape between the top face of the camera body and
the back of the perforated board. (See Figure 5-13.)
Double-sided tape is great for holding down circuit
boards or batteries in project boxes, and is easy
enough to remove without damaging anything. Hot
glue will work too, but it is a permanent solution
that can be difficult to pry apart if needed. Zip ties
will also work depending on the shape of your
camera body.

Figure 5-13 | Fastening the illuminator to the camera with two-sided tape.

Figure 5-14 | The completed infrared illuminator mounted to the small video camera.

In case you are interested in the camera we are using here, it is a model KPC-EX20H from the company KT&C in Japan, and is a very low-lux and high-resolution camera that has work extremely well in all of our night vision experiments. This camera is also distributed by SuperCircuits.com under the model name of PC182XS, and sells for around $100. Decent low-lux monochrome cameras can often be purchased for far less, but this one has a very low-lux rating and a high-resolution CCD imager.

Once the wiring was completed, we dropped the power line onto the 12-V battery and had a look on the monitor to ensure that the ring light was outputting and that all 10 LEDs were brightly lit. If you have a bad LED, either none of the LEDs will be lit or the bad one will be out, causing the others to receive a bit too much voltage. If this happens, remove the power and try feeding each LED with the required voltage or with 3 V form a lithium coin cell until you find the problem. Figure 5-14 shows the completed ring light infrared illuminator attached to the small video camera.

We use the small liquid crystal display (LCD) video screen shown in Figure 5-15 to test all of our projects as it is portable and can run off the same 12-V DC power supply that is being used for the camera and illuminator. These small video monitors can be purchased at video gaming stores

or at any electronics store. Basically, any video system that offers a composite video input jack will work with a security camera or small board camera. In Figure 5-15, the camera is powered up and looking at the illuminator, which is shown in the screen. The small switch on the breadboard is connected to the illuminator power line, so we can turn it on and off for testing purposes.

When the room lights are on, the infrared illuminator makes only a slight difference in the well-lit scene, but it is still easy to see the output on the video monitor. Figure 5-16 shows the ring of LEDs glowing white on the monitor, yet we can

Figure 5-15 | To test the infrared illuminator, you will need a video source and a monitor.

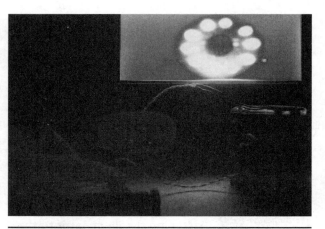

Figure 5-16   Looking at the infrared output on the monitor in a brightly-lit room.

Figure 17   With the lights off, the illuminator becomes a very bright source of light

only see pure black when looking at them. The illuminator is certainly working, so the next tests will be done in the darkness.

Figure 5-17 shows the intense glow from the 10 infrared LEDs as they shine directly into the camera lens while the room is dark. The LEDs seem so bright to the camera in the dark that they swamp out everything else completely. This camera almost sees in the dark by itself, but with the help of the infrared illuminator, it can see an entire room in compete darkness. Interestingly, the Nikon D90

that we are using to take these photos only sees a very slight blue glow from the LEDs because it contains an infrared filter over its image sensor that blocks out most of the infrared light. We have not been brave enough to dig into the camera to see if we can remove the filter to make a night vision camera, but one day that will make for another interesting hack.

Figure 5-18 shows the minimal amount of infrared light that passes through the infrared filter on the Nikon D90, and shows how well the video

Figure 5-18   The infrared illuminator lights up our camera as we take this photo.

camera is able to see us as we take this photo. Although it is difficult to tell from the photo of the monitor, the image on the screen is almost as bright as if the room lights were on. Because the infrared light is detected by the video camera as white light, the images look no different than what would be seen if the room lights were turned on. At 10 feet (ft) away from the illuminator, the image is almost too bright, reflecting off the camera like the light from a powerful flashlight beam. The back wall of the room 20 ft behind me is clearly visible.

Further tests revealed that the 10 LED infrared illuminator worked well in any room up to about 30 ft before the light started to fade. The beam angle on these LEDs was slightly too narrow, so a second row of 10 more LEDs pointing slightly outward would have been better. This small illuminator is probably not powerful enough for much outdoor spy use, but it was certainly good enough for use in a portable indoor night vision system or spy camera.

Figure 5-19   The completed Simple Infrared Illuminator.

The completed Simple Infrared Illuminator shown in Figure 5-19 was well worth the minimal amount invested in parts and about an hour of time it took to build. You can now give your small spy camera night vision, test infrared viewers, or even give your TV remote super extended range by feeding its output through an amplifier that feeds the illuminator. Because this illuminator is so simple, it can form the output for a number of infrared experiments and give practically any security camera a decent boost in the darkness.

PROJECT 6

LED Array Illuminator

THERE ARE TIMES when the small infrared light-emitting diode (LED) ring built into a security camera will not cover the range or field of view you require, so you will need to find another invisible light source. Some large infrared illuminators use powerful incandescent light sources that are passed through an infrared pass filter, causing only the infrared component of the light to come through the filter. These types of infrared illuminators create intense heat due to the fact that the white light source must be fully enclosed and burn the unwanted light energy off as radiated heat. Because of this intense heat, incandescent filtered illuminators cannot be used indoors and may not be suitable for many outdoor installations.

The good news is that LEDs can be used to create a very powerful infrared illumination system if you use enough of them. Okay, you need a lot

of them, but these days they can be purchased for only pennies a piece if ordered in quantities of hundreds or more. The bad news is that you will need to do a lot of soldering, even on a small array of 16 × 16 LEDs, which will have more than 512 connection points. Of course, many circuit board houses offer proto service, and you could have a very large LED array circuit board made for under $100 if you shop around. If you are patient and like to solder, then any size array can be made on some perforated circuit board, resulting in a very high-power illumination system that will only cost you one-tenth of what a manufactured unit would cost.

Before you decide to make a huge array that will light up an entire city block, do a little research on bulk LED prices and power requirements because an array will become hungry on both counts. We built two version of the LED array—one using a hand-wired perforated board having 13 × 19 LEDs and a much larger printed circuit board (PCB) version having 32 × 48 LEDs. So the smaller LED array has 247 LEDs, and the larger array has a whopping 1526 LEDs! Make no mistake—it takes a good chunk of power to crank up 1526 LEDs to their maximum potential, and even at 10 cents per LED, that adds up to $154 just for the LEDs.

Start by calculating how much infrared radiation you will need in order to light your scene. Limitations will likely be the focal range of your camera since details are lost on most security cameras after about 50 ft. This distance is also about as far as an LED can reach, no matter how many you add to the array, so the equation then becomes how wide and how bright do you need

Figure 6-0 A long-range infrared illuminator can be made using many LEDs.

the scene? A 20 × 20 ft interior room will shine like midday with an array of 16 × 16 LEDs at each corner of the room, but the massive array we built is almost too bright to be used indoors. If your camera will stay in a fixed position, then a single array is best, but for general room illumination, it is better to divide up your LEDs into two or more arrays for even lighting. Think of a 16 × 16 LED array to be about the same as a typical hand held flashlight for both output power and field of view. Our 32 × 48 array acts more like a 500-watt (W) halogen light source when placed in a small room.

There are several varieties of infrared LEDs, ranging in size, field of view, output power, and effective light color. The most commonly used infrared LEDs output 940-nanometer (nm) infrared light, which is far beyond the human visual range, and fairly detectable by any nonfiltered video camera. There are also infrared LEDs available for the 800 to 900 nm range. These are even better for use in night vision applications, but there will be slightly detectable red glow as the human eye can faintly detect this band of light. If you have seen an outdoor night vision security camera after dark, then you are probably familiar with this dull red glow. The LEDs shown in Figure 6-1 are commonly available in 940-nm types purchased in bulk from an Internet-based supplier.

Small-to-medium-sized LED arrays can be made by placing the leads into the holes and then soldering them on the underside using copper strips and wires to create the series parallel connections. Most infrared LEDs will be available as 5-mm wide plastic through hole devices like the ones shown in this project, so they can be packed onto a perforated board at about 16 LEDs per square inch. Yes, we mixed up our units of measure, but that is a common thing to do in the electronics industry. If you plan on making your own circuit board, then your only limitations on the number of LEDs in a given area will be the width of the LED bodies— 16 LEDs per square inch is actually very good for most illuminators, as it gives a decent range and field of view.

If you really have a lot of spare time and own a Dremel tool or small bench top grinder, you could actually grind down one side of the LEDs slightly and then pack them down to a smaller area. Figure 6-2 shows the LEDs trying to fit into the perforated board without having an empty hole between them, but without a little grinding, they just won't live together happily. A little buzz with a grinder and the LED would have a flat side, allowing them to be packed more tightly, but this is probably not necessary in this project as the idea is to use more LEDs to cover a wider area.

Figure 6-1 LEDs purchased in large quantities can often be found at bargain prices.

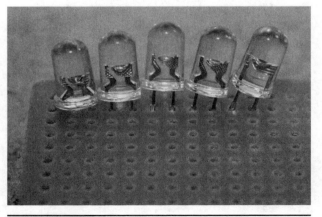

Figure 6-2 When using perforated board, the shape of the LED will dictate placement.

Because the spacing on most prototyping boards is the same as the spacing on standard dual inline chip packages, you will be able to insert the 5-mm LEDs to create arrays with 16 LEDs per square inch, skipping one hole between each LED. Using 3-mm LEDs would allow insertion without skipping a hole, but this size of LED is almost impossible to find with an infrared output. Your array can also have any number of rows and columns, but usually a square is best for the most even field of view. Another idea is to base the size of the array on the wiring requirement, using the number of LEDs in series as the width or depth of the array. For instance, if you have 1.2-volt (V) LEDs, and plan to run 10 of them in series to make an even 12-V power supply, then you could make your array with 10 or 20 LEDs across and as many parallel rows as you have power for. A 10 × 10 array will be large enough to light up a small room for practically any video camera (Figure 6-3).

For our smaller version of the LED array illuminator, we decided to use 19 × 13 LEDs for a total of 247 LEDs. This choice was completely based on the size of the perf board and the enclosure that will be used, so the array does not have even series rows. Not having an even

Figure 6-3 Using 5-mm LEDs, you can make an array with 16 LEDs per square inch.

number of LEDs to create a series row or column means that the wiring will be a bit more complex (and ugly) in the backside of the board, but that really is not important. The array has to fit into the inside of a halogen work light, so size is the only real concern. We were able to find LEDs in bulk 100 pieces packs, so the cost of 250 LEDs was about $30, with 50 to spare. Figure 6-4 shows the populated array of 247 LEDs after an hour of carefully observing the hole spacing and polarity of each LED. On new LEDs, the longer lead is the positive lead, and the flat side of the round case is the negative side.

Figure 6-4 Making a 19 × 13 LED array on perforated prototyping board.

When you are connecting large numbers of LEDs that could reach into the many hundreds or even thousands, you will have to consider the power requirements, which will certainly be quite large as well. Since Watts (W) = Amps (A) × Volts (V), powering up 900 LEDs that require 100 milliamps (mA) each in a 30 × 30 array wired to use 12 V would require over 100 W! Let us explain how this power requirement is calculated for a 30 × 30 LED array based on the series and parallel wiring configuration of the LED array.

Let's start with the forward voltage requirement of each LED. A common forward voltage value for an infrared LED would be 1.2 V. Since it is unlikely that you will find a powerful 1.2-V DC power supply, the idea is to run as many LEDs in series as needed in order to allow the use of a much more common voltage source, such as 12 V. Since 12 V divided by 10 equals 1.2 V, which means that you can safely run 10 LEDs in series from a 12-V DC power supply or battery. When working out this voltage calculation for an uneven number such as 12-V driving LEDs with a forward voltage of 1.4 V, always work it out so that the LEDs are slightly underpowered rather than slightly overpowered. So for 1.4-V LEDs, you can safely power 9 LEDs, giving each LED 1.33 V (12/9 = 1.33).

When working with series-connected LEDs, the amp rating of a single LED is the same requirement as the entire chain. So 10 LEDs that consume 100 mA each in a series chain wired for 12 V will only consume 100 mA at 12 V. Since the array has 900 LEDs, and 10 of them can be run in a series chain, this means that there will be a total of 90 series wired chains that contain 10 LEDs each. Each of these 90 chains is now connected in parallel, since parallel wiring does not alter the voltage requirement. Now there are 90 series chains each consuming 100 mA each, for a total power consumption of 9000 mA, or 9 A, and that's some serious juice! Since Watts = Amps × Volts, the 9 A becomes 108 W (9 × 12 = 108). So, consider carefully the number of LEDs you intend to run, especially if battery power will be your primary power source. A large 12-V lead acid security battery with a 15 A hour rating would only power a 30 × 30 LED array for about an hour before it started to fade. DC power supplies with this kind of output power are not difficult to find, though.

The schematic shown in Figure 6-5 shows an example of 40 LEDs wired in a series parallel combination of four chains of 10 series connected LEDs. This would be an ideal wiring diagram for 1.2-V LEDs running from a 12-V DC power

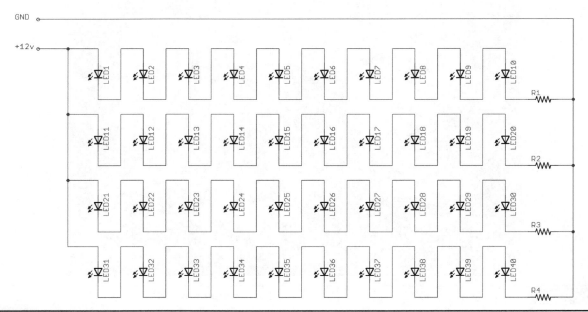

Figure 6-5 Series and parallel wiring makes power supply selection easier.

source. If each LED needed 90 mA, then the total power requirement for the entire array would be 4.32 W—not much power at all. On the other hand, our massive 48 × 32 LED array requires 180 W just to start throwing any decent amount of infrared radiation! Current limiting resistors (R1–R4) should be added to each series chain to avoid damage that could occur if one LED in a chain failed to a short circuit. The low-impedance current limiting resistors (10 to 50 ohms) should be of a size that can handle the current in the chain. A 1-W resistor would be a good choice for a chain of 10 LEDs running from a 12-V DC power source.

NOTE Thanks to Swink and the HackaDay viewers for pointing out an error in the previous version of this wiring diagram. Sometimes a fresh perspective makes all the difference!

Figure 6-6 shows an example of wiring one chain of 10 LEDs on the underside of a perforated board using the leads as traces. Leads are simply cut short and then bent towards each other to form traces. This system of wiring works out nicely if your series chains are dividable by one dimension of the array. If you cannot make even series chains, then you will have to use a wire to carry over to the next row in order to complete the chain. Having even series rows also means that the parallel wiring can be easily done by using one continuous copper strip or wire to connect them all as shown in the next photo.

When an LED array is wired so that the series chains divide evenly by a row, you can simply

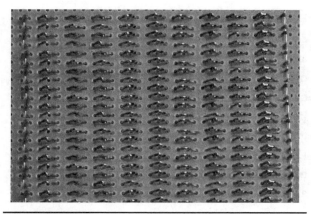

Figure 6-7 Parallel connections are made along the ends of each series chain.

run copper wire along the columns to create the parallel wiring connection for each series chain (Figure 6-7). Remember that the parallel connecting wires need to carry the entire amperage of the array, so use a copper wire or strip that is appropriate for the task. Not sure what wiring size to use? Look at the connecting wires from your power pack, or just strip down some 30-A copper house wiring and use that to solder the connection points.

The plan was to install the 18 × 13 LED array into the weatherproof cabinet of the halogen light shown in Figure 6-8. Using the light enclosure will allow the night vision illuminator to live outdoors since the cabinet is sealed from the elements.

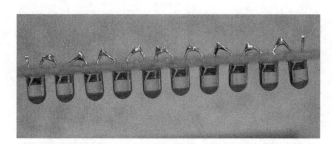

Figure 6-6 A series chain of 10 LEDs wired on the underside of the perforated board.

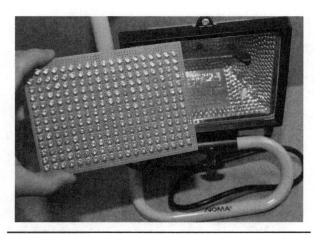

Figure 6-8 The completed 18 × 13 LED array will be installed into the light cabinet.

Being a long-range illuminator for outdoor use, concealment of the LEDs was not such a concern, and the light enclosure will not look out of place mounted to a wall or pole in a yard. With the original AC-lighting components removed from the light housing, there will be just enough room for the LED array circuit board, as well as an AC-to-DC power pack. Having the power adapter means that the light can be plugged into an AC socket just as it was before modification.

Once we found a suitable DC adapter to power up the LED array, we simply cut the AC wiring inside of the light enclosure and affixed it directly to the AC power adapter as shown in Figure 6-9. The AC wires were connected to the prongs on the AC adapter by soldering them directly and then sealing the connection using some heat-shrink tubing. Since the goal was to mount the array on a garage wall, no switch will be added to the enclosure as it would be too high to reach. To turn the LED array off in the daytime, a standard switched outlet will be used to control power to the DC power supply inside the light. For automatic use, a simple relay controlled by a light sensor could switch on the LEDs at dusk, similar to how a streetlight control operates.

The nice thing about the LED array illuminator shown in Figure 6-10 is that it can plug directly into an AC power outlet and live outdoors in most weather conditions. Because the LEDs do not give off any real heat and are completely invisible to the human eyes, indoor use is also possible, although

Figure 6-10   The completed LED array ready to withstand the elements.

the light may look a little out of place if stealth installation is necessary. For a security camera covering a large area, two of these units would be best for maximum illumination of the scene, but depending on the level of ambient light and the distance from the lens to the subject, the single array may be plenty.

Despite extreme weather where we live, our illuminator has lived happily outdoors through rain, hail, snow, and temperature ranges that would seem normal for the North Pole and some deserts. The freezing cold winter scene shown in Figure 6-11 is lit by only the single illuminator and the small amount of ambient light from a distant streetlight. Because the LED array is plugged into a switched AC outlet, we can turn on the night vision from inside if necessary. The low-lux security camera

Figure 6-9   The AC-to-DC power pack was installed in the light enclosure.

Figure 6-11   The all weather illuminator survives through hail, rain, snow, and the deep freeze.

Figure 6-12   This massive infrared LED array contains 1536 LEDs!

does a decent job of seeing at night all on its own, but with the addition of the LED illuminator, details such as faces or license plates can be captured at night in perfect clarity.

There are times when you need to illuminate a much larger outdoor area using a single infrared light source, or when you want to light up an interior scene to look like midday. This kind of illuminating power calls for a very large array of LEDs, like the massive 48 × 32 LED array shown in Figure 6-12. This beast contains 1536 infrared LEDs and will gobble up more than 180 W when running at full intensity! The infrared radiation that comes from this massive array can actually be felt like sunlight hitting your face, and will make any size interior room look as though a window was wide open in a nice sunny day. Outdoors, this massive array is capable of lighting up an entire back yard like a spotlight, allowing any security camera to capture perfect detail in pitch darkness. An array like this is probably overkill for most stealthy security needs, but we like to take things to the limit, so we made a few of these units to use in our various spy missions.

So how long do you think it would take to solder down the approximately 3000 connecting wires for an array of this size? Too long! When working with LED arrays spanning more than 32 × 32 LEDs, it

makes more sense to create a circuit board rather than trying to wire them all by hand. Since this many LEDs will probably cost you at least $150, the added cost of having a real printed circuit board made is certainly not going to seem high. Many of the prototyping board companies will make a board like this for under $100 in small quantities, and you will only have to spend the time soldering the LEDs to the board, which will probably take a few days. Hand wiring a board with 1500 LEDs would take weeks and the chance of making a mistake will be extremely high. Yes, we did it once…only once!

Figure 6-13 shows the backside of the massive LED array where the LEDs are wired into series

Figure 6-13   Using a printed circuit board is a much better solution for an array of this size.

string each containing 10 LEDs for basic 12-V DC operation. The series connection traces do not carry much current (50 to 100 mA), so they can be made small and bounce around the board to make the circuit easy to layout. It is the parallel traces that will have to endure the massive amounts of current to power the entire board, so these must be designed using much larger and heavier traces or even made using copper strips.

Figure 6-14 shows the topside of the massive LED array where the parallel connecting traces join up all 153 of the 10 LED series chains together. This LED array can consume up to 180 W at 12 V, so the traces will need to carry 15 A of current! Of course, we usually do not run the array at full power as it is actually too bright for most indoor use, but for some of the jobs we do, having more light available is better than falling short. Although this massive LED array is 6 times larger than the last one, the series and parallel wiring is exactly the same, based on the schematic shown in Figure 6-5.

When using an infrared illuminator as a stealthy indoor security device, it will be necessary to conceal the LEDs somehow, as they will certainly look suspicious out in the open. In one of our many undercover operations, we had to use the massive 1536 LED array illuminator in an interior room that needed to be extremely well lit in pitch darkness, yet hide the illuminator for total stealth operation. To conceal a 2-foot square array of LEDs, some

Figure 6-15 | This one-way mirror glass can be used to conceal an infrared illuminator.

type of lens will be needed that will hide the LEDs while at the same time passing the infrared light. There are several solutions to this problem.

One-way mirror glass, like the sheet shown in Figure 6-15 is inexpensive and available to be cut into any size or shape. This glass seems to pass most of the infrared light, yet completely hides the LEDs from human view. There are several types of one-way mirror glass, some that look exactly like a standard mirror and others that have various degrees of shading and opacity. We chose a darker shaded one-way mirror glass so that we could install the LEDs in an old liquid crystal display (LCD) monitor case to make it look "normal." The glass was cut by the supplier into a piece exactly the same size as the LCD panel that was removed from the monitor.

Figure 6-16 shows how the glass completely blocks anything behind it, yet allows the light from a visible green LED to shine through it.

Figure 6-14 | The parallel connecting traces must be large enough to carry high current.

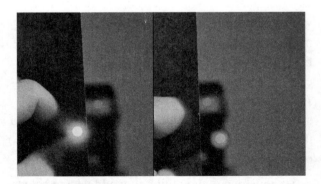

Figure 6-16 | The visible green LED is hidden by the glass, but the light can still pass.

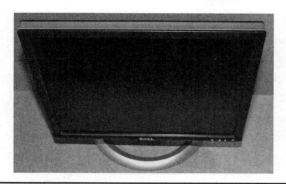

Figure 6-17 This broken 19-in LCD monitor will soon become a giant stealth illuminator.

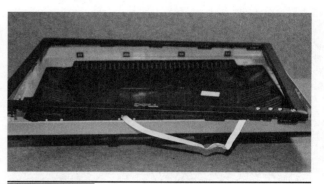

Figure 6-18 Removing the front bezel and the broken LCD panel from the monitor.

This one-way glass mirror only blocks about 10 percent of the LED light, and seems to block even less of the infrared light. Besides the reflection of the camera taking the photo, nothing from the other side of the glass can be seen, so the LEDs are completely concealed. Because the massive LED array offers so much output, a little loss from the shade of the glass is almost unnoticeable.

The monitor shown in Figure 6-17 had some internal malfunction, so it will become the new home of the giant 1536 LED array illuminator, concealed behind the one-way glass mirror. This monitor was the perfect way to conceal the giant illuminator for indoor use as the appliance does not look out of place sitting on a desk and completely hides all signs of the LED array. The monitor can also be pointed towards the scene without looking suspicious.

The inside of the monitor shown in Figure 6-18 would have plenty of room for the massive LED array, as well as for a large battery pack or a power supply. To allow the monitor to be installed onsite in a hurry, we opted for a battery pack and a light-sensing switch to turn on the LEDs once the room lights were turned off. Every security job requires a unique solution, but this time a battery was better than a hard-wired DC power pack. To acquire the information you seek, creativity will be your greatest tool.

Because the LED array was approximately the same size and depth as the original LCD panel, installation was extremely easy. As shown in Figure 6-19, the LED array was simply bolted to the back casing, leaving room for the one-way glass mirror in the front and the battery pack or DC power supply in the rear. There was even plenty of room left over for a small wireless pinhole spy camera to be installed right into the lower bezel of the monitor, creating a stealthy and powerful night vision spy system that could be installed in seconds just about anywhere.

Depending on how much infrared radiation we needed, the lead acid security battery shown in Figure 6-20 would run the array for times from 30 minutes up to several hours. To limit the current and run time, we used several methods,

Figure 6-19 Fitting the LED array where the LCD screen once lived.

Figure 6-20   A rechargeable lead-acid security battery will power the LED array.

depending on the installation. Simply adding a resistive load would work, but there was some loss due to heating. A better solution was to create a pulse-width-modulated power supply that would allow the duty cycle to be varied. This worked well when precise control over brightness was necessary. The modulated power supply was based on the "pulsed LED illuminator" circuit shown earlier, but used an array of larger FET transistors as the drivers.

The completed stealth installation is shown in Figure 6-21, and turned out very well. The one-way glass mirror is a bit more reflective that the original

LCD panel, but the device still looks perfectly convincing. Only an evil genius would suspect an innocent-looking LCD monitor of being a giant high power night vision illuminator, so the device was a total success in every operation it took part in. The infrared radiation thrown from this giant 1536 LED array was so intense that we usually ended up running the LEDs at less than 50 percent of their capacity to light up a large room.

It is difficult to capture the infrared glow using a digital still camera due to the infrared filter of the image sensor, so there is only a slight glow from the LEDs shown. A security camera, on the other hand, has no such filter and will pick up the infrared radiation in the same way it would pick up 180 W of pure white light from a floodlight. Figure 6-22 shows the slight glow from the LEDs at very low power through a standard digital still camera. Besides the fact that the monitor does not work, it looks completely normal, and would not seem out of place in most installations.

The LCD monitor installation was a great success, and offered a very covert method to illuminate any room where computer equipment would not look out of place. The one-way mirror can also be used to hide LED arrays behind

Figure 6-21   The completed super stealthy giant infrared LED illuminator.

Figure 6-22   The 180 W of intense infrared radiation can only be seen by video cameras.

Figure 6-23 The massive LED array illuminates the small room like a spotlight.

photo frames, real mirror frames, and even small appliances that would have a dark reflective front panel. Other materials such as dark Plexiglas or tinted glass will also work to conceal an LED array, and many materials that seem opaque to the human eye may actually pass infrared light. Certain papers and dark plastics will seem transparent to infrared light, but you will need to use an infrared LED and video camera in order to test them.

Figure 6-23 demonstrates what the security camera would see with the stealth infrared LED array running at only 25 percent of its maximum capabilities, lighting up a room as if a large spotlight was being used. When designing your own infrared LED array, the main factors to consider are cost of parts and the size of the array, so let your imagination run wild and give your spy cameras the gift of night vision.

PROJECT 7

Pulsed LED Illuminator

ALTHOUGH THERE ARE certain limitations to how much light (or infrared radiation) can be emitted by a light-emitting diode (LED), there are tricks that can be used to push them to their ultimate maximum limits. This project will demonstrate how a simple infrared illuminator can be "pushed" a little more in order to extend the useful range of a simple night vision system using a camcorder and a low-lux monochrome camera. In order to build this project, you will need to have the datasheet handy for the LEDs you plan to use so that you can determine the amount of current the LED will withstand in pulsed-mode operation.

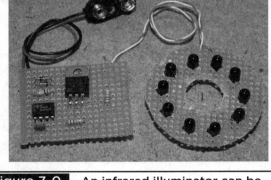

Figure 7-0 An infrared illuminator can be pushed to the max using pulsed current.

PARTS LIST

- IC1: LM555 analog timer IC
- Resistors: R1 = 100 K, R3 = 1 K, R4 = 1 K
- Capacitors: C1 = 0.01 μF
- Transistors: 2N3904 (version 1), TIP120 (version 2)
- Diodes: 1N914
- Battery: 9- to 12-V battery or pack

Pulsed-mode operation means that the LED will be turned on and off at a very fast rate using more current than it could withstand continuously. The purpose of doing this is to force the LED to output short bursts of much brighter light (or infrared radiation) than it normally would, and by keeping the pulse with duty cycle short, the LED will not overheat. Television remotes do this to create sharp intense bursts of modulated light to send out to the

receiver, and many low-voltage consumer devices do this with visible LEDs to make them appear brighter while at the same time conserving power. A pulsed-visible LED may look 10 times brighter, yet consume only half the power in pulsed-mode operation. Of course, there are limitations to pulsing LEDs, and you may find that using more LEDs or higher current LEDs to be more effective than using a pulse-mode driver.

This project will explore the strengths and weaknesses of the pulsed-mode operation of both visible LEDs and infrared LEDs, and compare both using a low-lux monochrome spy camera connected to a camcorder and small infrared illuminator.

Most newer outdoor security cameras now include an infrared ring light illuminator to enhance their ability to see in the dark. Infrared light falls just below red on the light spectrum, making up the wavelengths from about 750 to 1500 nanometers (nm). This light cannot be seen by human eyes, but it can easily be

Figure 7-1   This is a pulsed-mode illuminator taken from an outdoor security camera.

seen by many video cameras, making it useful as a covert lighting method in night vision systems. A common example of infrared light is the medium of communication between your remote control and television set. The LED on the end of your remote sends out pulses of infrared light, which is received by the infrared detector on the TV and demodulated back into data. Of course, you cannot see the pulses because they are out of our visual range, but any video camera that is not equipped with an infrared filter can see this light easily.

Figure 7-1 shows an infrared LED illuminator ring taken from a small outdoor security camera. There are 17 infrared LEDs arranged in a series parallel configuration around a small hole where the camera lens would be installed so that the light is spread evenly around the field of view. This is an older illuminator, and was also a pulse mode system, which is why there are semiconductors on the rear of the circuit board. An illuminator without a pulse-mode driver will not have any semiconductors as it is wired directly to the DC power source, giving each LED its maximum voltage and current all of the time.

You may notice that the semiconductors on the back of the illuminator shown in Figure 7-1 are completely fried, which is one of the downfalls of having more circuitry—more points of failure. This circuit was either zapped by a nearby lightning surge, or simply gave it up after

overheating, causing a massive failure in almost all of the transistors in the circuit. Luckily, the LEDs survived and found their way into our junk collection. We have a very extensive collection of infrared-enabled security cameras, and the interesting thing is that almost all of the newer ones have better night vision capabilities and do not use pulsed-mode LED drivers. Maybe the manufacturers decided that better LEDs made more sense than pushing lower quality LEDs to their ultimate maximum ratings?

There are several varieties of infrared LEDs, ranging in size, field of view, output power, and effective light color. The most commonly used infrared LEDs output 940-nm infrared light which is far beyond the human visual range, and fairly detectable by any nonfiltered video camera. There are also infrared LEDs available for the 800 to 900 nm range, and these are even better for use in night vision applications, but there will be an ever so slightly detectable red glow as the human eye can faintly detect this band of light. If you have seen an outdoor night vision security camera after dark, then you are probably familiar with this dull red glow. The infrared LEDs shown Figure 7-2 are common 940-nm infrared LEDs with a forward voltage of 1.2 volts (V).

In order to attempt any kind of pulsed-mode operation for visible or infrared LEDs, you need to look at the datasheet in order to find the maximum

Figure 7-2   Remote control infrared LEDs are commonly available.

ABSOLUTE MAXIMUM RATINGS (T_A = 25°C unless otherwise specified)			
Parameter	Symbol	Rating	Unit
Operating Temperature	T_{OPR}	-40 to +100	°C
Storage Temperature	T_{STG}	-40 to +100	°C
Soldering Temperature (Iron)[2,3,4]	T_{SOL-I}	240 for 5 sec	°C
Soldering Temperature (Flow)[2,3]	T_{SOL-F}	260 for 10 sec	°C
Continuous Forward Current	I_F	100	mA
Reverse Voltage	V_R	5	V
Power Dissipation[1]	P_D	200	mW
Peak Forward Current	I_{FP}	1.5	A

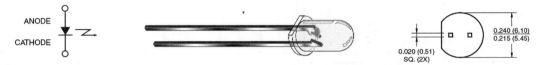

Figure 7-3 Maximum current pulse will be shown on the datasheet for your LED.

pulse current rating as well as the recommended duty cycle. Some LEDs can only handle a small increase of current during pulsed-mode operation, making them unsuitable for the task. Figure 7-3 shows the section of a typical infrared LED that details the important values needed for pulse-mode operation.

According to the datasheet segment, this LED needs 5 V and has a maximum rating of 100 milliamps (mA) for normal operation. But, in pulsed-mode operation the LED can withstand 1.5 amps (A), which is 1500 mA, or 15 times the current! This is really amazing, but don't expect the LED to output 15 times the amount of light or infrared radiation. One of the things you will realize when experimenting with pulse-driven LEDs is that visible LEDs may appear to be twice as bright to your eyes, but the difference is not that great when using them to illuminate security cameras. There is still a small amount of gain to an infrared illuminator, but you may find the added complexity of the circuit to be not worth the effort.

If your LEDs have a pulse-mode rating that is any less than 10 times the continuous current rating, then it is probably not worth your time to use them in pulsed-mode operation. Also, check your datasheet for a recommended on/off time (duty cycle), to ensure that you don't over-drive the current in pulsed-mode operation. The circuit

shown here is probably safe for most LEDs as it will deliver a very short pulse of between 8 and 10 microseconds (μs). In order to drive your LEDs up to their ultimate maximum brightness, some experimentation will certainly be necessary. Expect to send a few LEDs to the graveyard along the way!

The schematic shown in Figure 7-4 is a basic LED pulser that can send 800 mA pulses to a few LEDs, driving them with between 9 and 12 V. When pulsing LEDs, the voltages used go well beyond the specified forward voltages shown in the datasheet, and pulse on times (duty cycle) must be kept very low. This circuit plays it safe, sending a very short repeating pulse of no more than 10 μs from the LM555 timer out to the LEDs, with as much current as the NPN transistor timer circuit can handle (about 800 mA). As you add more LEDs in parallel, the current required will increase until the transistor can no longer supply all of the LEDs. Depending on the specification of your LEDs, you may be able to drive up to four LEDs, or not even push a single LED to its capacity. An LED with a maximum pulsed-mode current rating of approximately 1 A (1000 mA) is about all the transistor can drive.

Since the initial experimentation will be done using visible LEDs, it is best to start with at least four LEDs in parallel and then pull one out at a time, checking the overall brightness as well as

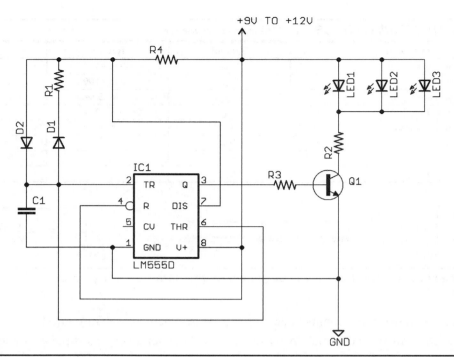

Figure 7-4 This LED pulsing circuit sends 800-mA short duration pulses.

testing the transistor for heat with your finger. The 10 or 20 ohm (Ω) resistor (R2) will limit the current enough to save your transistor from a meltdown, but the circuit should not be allowed to run hot continually or the life expectancy of the transistor will be short. This initial circuit is just a test circuit, and will demonstrate the difference between continuous low current as compared to pulsed-mode current in the visible LEDs. Later, a larger transistor will be added in order to drive a much larger infrared LED array.

If you can't find a 2N2222 or 2N3904 transistor, then you can substitute practically any small signal NPN transistor that will handle current of around 600 mA. As for the 555 timer, any of the variants of this device will work, but be aware that altering the timing capacitor or resistor values will also change the frequency and duty cycle of the pulses. As it is, the circuit is set to deliver the quickest pulse that it can (around 10 µs), and the frequency of the pulses will be approximately 1.5 Kilohertz (KHz). You can certainly alter these values, but be aware that an error will usually mean a fried LED or transistor.

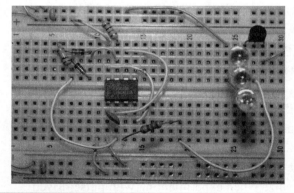

Figure 7-5 The LED pulser test circuit is initially built on a solderless breadboard.

The LED pulser circuit is shown built on a solderless breadboard in Figure 7-5, using the values from the schematic shown in Figure 7-4. The LEDs used here are visible green LEDs with values very close to the infrared LEDs we plan to use later. Our LEDs have a forward voltage of 1.2 V, a continuous current rating of 200 mA, and a pulsed-mode current rating of 1000 mA (1 A). Because the LEDs' peak current rating is much

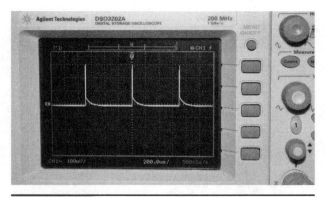

Figure 7-6 The LED pulsing circuit keeps the duty cycle to a minimum.

higher than the transistor can supply, we know that as long as the pulses are kept as short as possible, no damage will occur to the LEDs. If the transistor failed and created a dead short, then the LEDs would also be fried, so it is important to keep checking the transistor for heat by touching it after a few seconds of operation.

Once you have the timer part of the circuit built up, you can check the output on an oscilloscope, if you have one. Without any LEDs connected, take a reading from the output of the 555 timer on pin #3. The output should look like the one shown in Figure 7-6, with a frequency near 1.5 KHz and a

pulse time of around 10 μs. This extremely short duty cycle will ensure that the LEDs do not fail from overheating. The fast frequency of the pulses will make the light look continuous to both video cameras and the human eyes. For comparison, a TV remote control pulses the LEDs at 40 KHz and drives them up to a similar current level as this circuit does.

The three LEDs shown lit up in Figure 7-7 are running from the pulsing circuit, whereas the single LED is running from direct current though a current-limiting resistor. Although it is difficult to tell in the image, the three LEDs running in pulsed mode certainly seem brighter to look at, and there is an amazing shift in color from pure green to greenish blue. This effect was surprising considering all four LEDs are exactly the same model. To ensure that we did not have a mismatched LED, we replaced the single LED with one of the pulsed LEDs, and it was confirmed that pulsing this green LED caused a massive shift in color from about 500 nm (pure green) to somewhere around 470 nm (green-blue).

While doing a little research on why this bizarre color shift was happening, we found that a few other experimenters have found that sharp pulses to

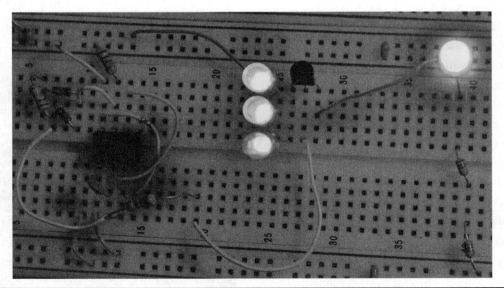

Figure 7-7 The three pulsed LEDs are brighter than the single nonpulsed LED.

LEDs can alter their wavelength. Apparently, 450-nm blue LEDs can be forced to output ultraviolet light by doing this, so this opens a few doors for future experimentation.

Overall, the results were fairly obvious to the naked eye that the pulsed LEDs were definitely brighter than the nonpulsed LEDs, even if we only drove a single LED in pulse mode. We tried various colors and styles of LEDs, and the results were mixed; sometimes the brightness change was dramatic, and sometimes it was mainly a color shift. High-brightness LEDs with a narrow field of view responded very well to the pulsed-mode operation, whereas low-power LEDs with a diffused lens did not. The good news is that infrared LEDs are similar in characteristics to high-power LEDs with a narrow field of view, so they should be the type to respond well to the pulse-mode operation. Also, since 880-nm infrared radiation is slightly better than 940-nm radiation for night vision, the slight change in the spectrum will work in our favor here (if there is any).

Because the charge-coupled device (CCD) or complementary metal-oxide semiconductor (CMOS) imager used in a video camera can see infrared light in the same way we can see visible light, night vision is made possible by illuminating an area with infrared light at wavelengths of 800 to 1000 nm. This light is undetectable by the human eye, but fully detectable by the camera, so it appears as though a standard white light source was being used to light up the scene. Not all cameras are capable of viewing infrared light though, as most color camcorders and digital still cameras include an infrared blocking filter to improve the response to the visible light spectrum. Spy cameras and security cameras on the other hand need to collect as much light as possible, so they do not include infrared filters and take advantage of their ability to see this invisible light for stealthy night vision operations.

The small spy camera shown in Figure 7-8 is a very low-lux monochrome camera made for security use, and can see any infrared light perfectly. The camera we are using here is a model

Figure 7-8   A small monochrome video camera will be used to detect the infrared light.

KPC-EX20H from the company KT&C in Japan, and is a very low-lux and high-resolution camera that has work extremely well in all of our night vision experiments. This camera is also distributed by SuperCircuits.com under the model name of PC182XS, and sells for around $100. Decent low-lux monochrome cameras can often be purchased for far less, but this one has a very low-lux rating and higher resolution CCD imager.

With the LED pulsing circuit tested and working with visible LEDs, the next step is to test it using a security camera and some infrared LEDs (Figure 7-9). Since our infrared LEDs had similar forward current specifications to the visible

Figure 7-9   Testing the pulsing circuit with the camera and infrared LEDs.

LEDs used in the last test plus a better pulsed current rating, they could be safely placed right into the circuit. Most infrared LEDs will have a decent pulsed rating of 1 A or better because it is very common to use them in pulsed-mode operation, especially when communication through modulation is necessary.

With the camera set to view the LEDs on the small liquid crystal display (LCD) monitor, the circuit was switched on to see the difference between the perceived brightness of the three pulsed LEDs as compared to the single continuous current LED. Unlike the visible green LEDs tested earlier, the difference between pulsed and nonpulsed operation as perceived by the video camera was almost unnoticeable. Actually, it appeared as though the single LED might be emitting more infrared radiation than the three pulsed LEDs. Of course, this test was done with the room lights on, and the camera may actually be compensating for contrast around the LEDs, so to be fair, the test must be done in absolute darkness and with a single LED in both modes.

With the room completely dark, the results of the test now work in favor of the pulsed-mode LED shown in the left of Figure 7-10. The pulser circuit is only driving a single LED now, and it is the same model as the one running in continuous-current mode. The pulsed LED lights up a larger area as shown in the photo, reflecting from a matte surface from a 12-inch (in) distance. It also seems that the small transistor did not have the ability to drive all three LEDs in series, so the pulsed LED also looked brighter with the lights on.

To complete this project, a small ring light infrared illuminator with 10 LEDs will be connected to a much larger driver transistor and tested using the same monochrome camera in a completely dark room. By feeding the output from the small spy cam into the video input on a standard camcorder, the result will be a simple night vision system that can capture images and record live video. By looking through the viewfinder, it is possible to navigate a completely dark room under infrared light alone.

When using a high-power transistor to drive the pulses into the LEDs, care must be taken when considering the current that each LED will have to endure. The TIP120 transistor that we are going to use to drive the larger LED array can switch up to 5 A in theory, so you will need to run at least two LEDs in series and possibly change the value of the current limiting resistor. We ran our system from 12 V and ran a dual chain of five parallel LEDs in series as shown in Figure 7-11. The current limiting resistor was changed from 10 to 75 Ω, just to play it safe.

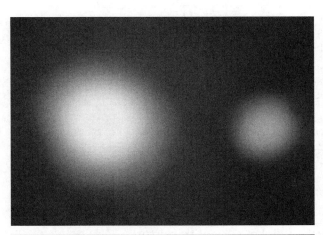

Figure 7-10   The pulsed LED on the left is definitely giving off more light.

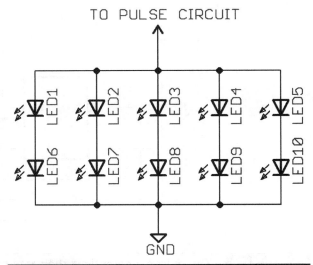

Figure 7-11   The wiring we are using for our 10 infrared LED array.

This is the point where experimentation could lead to an LED going off like a popcorn kernel, so don't be staring down over the LED when you first apply the power, or you may find yourself getting hit in the eye with an exploding LED case! The larger transistor will run with a bit of heat, but unless you are driving a massive array of LEDs, you should not require a heat sink. Again, experimentation is the goal to reaching the ultimate limits. It should be done one step at a time, adding more LEDs in series right off the start. There are some newer high-power LEDs on the market that can be pulsed well into 5 A, so you may require an even larger driver transistor.

The 10 LED ring illuminator shown in Figure 7-12 is from a previous project, but it has been rewired to match the schematic shown in Figure 7-11. Originally, we just drove this array directly with 12 V, which worked out perfectly since all 10 LEDs were in series and required 1.2 V each. The array actually worked very well with direct current, but we wanted to see how much better it would light up an interior room using 1 A pulses at 12 V.

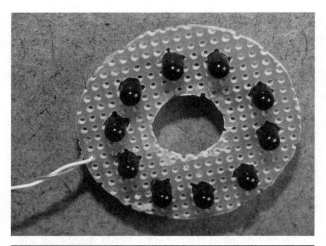

Figure 7-12   This is a 10 LED array based on the series parallel wiring.

The schematic shown in Figure 7-13 is exactly the same as the one shown in Figure 7-4 with the exception of the high-current TIP120 transistor replacing the 2N2222 transistor and the addition of more LEDs. Since this transistor can supply much greater current, the LEDs were changed from a single parallel chain to a series parallel chain to reduce the voltage in half. We were not brave

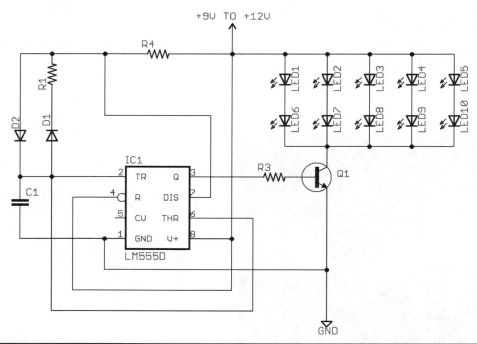

Figure 7-13   The pulser schematic is the same but uses a higher current transistor.

enough to try running all 10 LEDs in parallel as we did not want to toss 10 LEDs in the trash just in case we exceeded their pulsed current limitations. In pulsed mode, we never run as high as the stated maximum on the datasheet. Further current limiting can be done by adding a power resistor between the emitter of Q1 and ground. A value between 10 and 100 Ω will work.

The changes to the new schematic involve only swapping out the low-current 2N2222 transistor for a high-current transistor such as the TIP120. We also removed the current limiting resistor R2 and ran our LEDs in a series parallel configuration to reduce the pulsed current to each LED slightly. If you wanted to make the output variable, you could add a 100-K potentiometer in place of R3 in order to reduce the drive to the base of the transistor. Some experimentation will certainly be necessary if you plan to get the maximum output from your LEDs (Figure 7-14).

To allow the pulse circuit to be tested around the house, we rebuilt the circuit on a bit of perforated board as shown in Figure 7-15. All wiring is made by connecting the wires on the underside of the board to form traces. Perforated board is great for fast prototyping of small-to-medium circuits, and changes are easily made by simply moving wires around.

Most camcorders have a video input connector to allow recording from another video source.

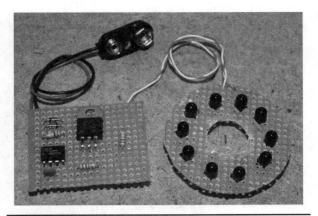

Figure 7-15 The completed infrared pulse circuit is made on a bit of perforated board.

This input may be labeled VCR Input, or External Video, and will basically replace the input from the camcorder's built in video system with your external source. This is necessary because all camcorders actually block infrared light to improve the quality of the visible color image. There is no easy way to remove the infrared filter in a camcorder, so unless you are willing to go in deep with your tools, you will need an infrared-friendly video source. In our case, it is the low-lux monochrome spy camera, which has been stuck to the side of the camera with a bit of double-sided tape.

The LED array would normally fit over the camera lens, but we just taped it to the front of the camcorder as shown in Figure 7-16 so that we

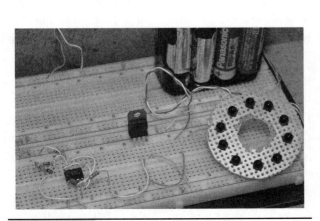

Figure 7-14 Building the high-current pulser on a solderless breadboard.

Figure 7-16 This is the test rig that will be used to compare the LED outputs.

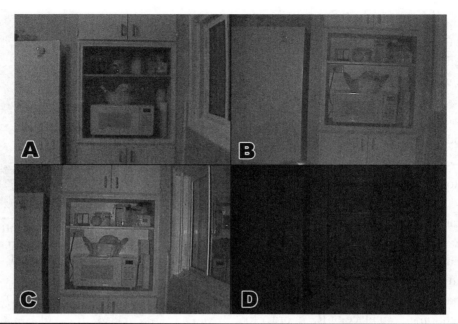

Figure 7-17 These images were captured from the camcorder recordings.

could switch back and forth from the spy camera to the original lens on the camcorder. Both the external camera and LED pulse circuit are driven from a 12-V battery pack.

The final results show that the pulsed-mode operation of the LEDs does in fact offer a stronger illumination than the same LEDs running under direct current. In Figure 7-17, (A) shows the image as seen by the camera under full room lighting. Image (B) shows the same scene using the LED array in continuous current mode, running all 10 LEDs in series from a 12-V power source. Image (C) shows the LED array running from the pulsing circuit using the high-current transistor. And finally, Image (D) shows the room as seen by the camera with only the ambient light coming into the window from the streetlight across the street.

The tests clearly show that the pulsed-mode operation of the 10 LED array offers the most illumination in the interior room. Actually, the pulsed-mode illumination is brighter to the camera than the 100 watts (W) of light from the ceiling light directly overhead! The continuous current

LED illumination is also very good, with only a slight fading of the sharp edges and highlights. With no light at all, the camera can still make out a bit of the scene, which is more than we can see with our eyes. A good low-lux camera goes a long way when it comes to night vision projects.

So, the verdict is—in pulsed-mode operation of infrared LEDs does offer some improvement, but you will have to decide if the added complexity of is worth the effort. Depending on the cost and output power of your LEDs, it may be more effective to simply add more LEDs into your array to achieve a higher output. More LEDs also has the benefit of illuminating a greater field of view, which is why most night vision camera manufacturers are now opting for the more LEDs over pulsed-mode operation. The benefit of pulsed-mode operation is that less power is consumed due to the pulse duty cycle, and that a brighter illuminator can be made in a smaller space. There also appears to be some gain due to the mysterious light spectrum shifting of running LEDs in pulsed mode as the 940-nm LEDs begin to output radiation closer to 840 nm if the pulses are very short.

PROJECT 8

Laser Night Vision

INFRARED LIGHT-EMITTING DIODES (LEDs) are the most widely used source of infrared radiation for night vision illuminators because they are inexpensive, easy to connect, and possess no safety hazards, because they are human eye safe and do not radiate much heat. The drawback to LED-based night vision illumination systems is that they are not really useable at distances of more than 100 feet (ft) no matter how many LEDs you use in the array. Filter-based night vision illuminators that change visible light into infrared light are suitable for much greater distances, but they suffer from huge energy losses due to massive heating of the filter material and because of this, they require massive amounts of current and are only suitable for outdoor use. A laser on the other hand, is capable of extremely long-distance illumination and is probably the most energy efficient source of bright light possible.

Figure 8-0 Build a long-range laser night vision illuminator.

of more than 5 milliwatts (mW). Class IIIb and Class IV lasers can output as much as 500 mW, and they are certainly not eye safe, especially when highly focused. A laser that outputs only 50 mW may seem like nothing, but be aware that instant eye damage could occur if you hit your retina with a focused beam. Using infrared lasers makes this situation so much more dangerous because you cannot see the beam, and your blink reflex will not help save your vision in the event of an accidental exposure to the laser beam.

Do not continue with any of these experiments unless you are well-aware of the dangers involved and have proper laser safety equipment and experience in using higher powered lasers. You can still create a usable short-range laser illuminator using a lower power 5-mW infrared laser diode or module, so consider starting with a Class IIIa laser if you want to experiment with laser night vision illumination.

PARTS LIST

- Laser: 5- to 500-mW infrared laser diode or module
- Camera: Low-lux monochrome composite camera
- Optics: Gun site, monoculars or binoculars
- Battery: 9 to 12 V battery pack

The main problem with using infrared lasers to create night vision illuminations systems is that there are safety issues that must be addressed, especially when using lasers with a rating higher than Class IIIa or lasers that have an output power

Figure 8-1   Laser pointers are now available that output infrared radiation.

Not that long ago, an infrared laser with enough power to burn wood was the size of a shoe box and took hundreds of watts of power. But today you can purchase a laser pointer for under $100 from many online distributors with enough power to pop a balloon at 20 ft and burn electrical tape. Make no mistake though—these lasers are not toys. Your first mistake will also be your last as a single shot into your eyes will likely permanently damage your vision. If you intend to work with high-power lasers, purchase a set of laser safety goggles and wear them in your lab when experimenting with these lasers.

The two laser pointers shown in Figure 8-1 are both extremely high-power Class IIIb lasers, capable of outputting 50 mW and 250 mW of infrared laser radiation. Calling these lasers "pointers" is really a bad term to use as they are completely invisible and not safe for pointing at objects. The laser shown in the top of Figure 8-1 will burn any black object from up to 20 ft away, and has a rated output power of over 250 mW. The laser shown at the bottom of Figure 8-1 will output about 50 mW of power and has been converted for infrared operation by changing the original visible red laser diode with one from a DVD burner. Both of these lasers work very well as long-range night vision illuminators once their beam has been spread by some type of lens.

If you are looking to purchase a laser pointer-style laser for night vision experimentation, then choose one that allows the collimating lens to be adjusted or fully removed. The goal will be to spread the beam out to an area of several feet in diameter at a distance of several hundred feet.

Laser pointers and laser modules are basically the same thing—a cylindrical enclosure containing a laser diode, front-end optics and some kind of current-limiting circuit. In the case of laser pointers, the enclosure also makes room for a battery pack, which means that the laser driver circuitry can be made simpler or that the diode can often be run directly off the battery. Laser modules are designed with much more robust driver electronics and also include power regulation so they can be run from a range of voltages. Laser modules also include much higher-quality optics and are often fully adjustable so that the beam can be focused or spread for a given distance.

This ability to defocus the beam will be the key to making a working laser illumination system, so the laser module may be the best choice for this project if your budget permits. The small laser modules shown in Figure 8-2 have various wavelengths, and output low power (5 mW) laser radiation in the red, green, and infrared bands. We also have several high-power infrared laser modules that we use in night vision systems, and they range in power from 50 mW to almost 1 watt (W) in power! A 1-W laser diode is more like a weapon than an illuminator, and special care must be taken in order to avoid serious eye damage or spontaneous fires in the lab!

If you are looking for a less-expensive source for high-power lasers, then you may consider using a bare laser diode along with some home brew driver

Figure 8-2   Laser modules are often higher quality and include adjustable optics.

Figure 8-3 DVD and CD burners are a great source of high-power laser diodes.

Figure 8-4 This laser unit contains both an infrared and visible red laser diode.

circuit or a simple battery power pack. Laser diodes are usually less expensive than laser modules, and available at much higher power levels than most laser pointers. The downside to using bare laser diodes is that some type of current-limiting circuit needs to be made, and there are no included optics. The good news is that for illumination use, optics can be found from many salvaged sources, and often laser diodes can be powered directly from a 3-volt (V) battery. The strange-looking blocks shown in Figure 8-3 are the laser heads taken from several old DVD and CD burners. This is the part that slides up and down the two linear bearings to allow the laser to read the disc surface.

You can purchase a used or brand-new DVD/CD burner combo pretty cheap these days. Inside you will find a pair of very powerful laser diodes. The DVD burner will include a high-power 150 mW or better visible red laser diode working in the 650 nanometer (nm) bandwidth, and the CD burner will include a 60 mW or better infrared laser diode running in the 780 (infrared) nm bandwidth. For night vision illumination, the infrared laser diode will be the one you want, so look for a DVD/CD writer combo, or just a CD writer. The parts shown in Figure 8-3 will be easy to remove from the unit, and contain the single or dual laser diodes.

The laser head unit shown in Figure 8-4 was pulled from an old DVD/CD writer combo unit, and contains both the high-power visible red laser

diode as well as the infrared laser diode. The infrared diode used on the CD burner side is quite a bit less powerful than the visible red DVD burner diode, but make no mistake, the infrared laser diode will still output more than 10× the light than any "eye safe" laser pointer, and both will damage your eyes if you are not extremely careful. You will have to pull the laser unit apart using a small set of screwdrivers in order to extract the actual laser diodes, and along the way you will find an array of interesting and useful optical components to use in other projects. Some of the parts are just epoxied in place, so a little prying with a small flat-head screwdriver may be necessary during the disassembly process. Be careful not to pry on the actual laser diode, though.

The bare laser diode will look like the ones shown in Figure 8-5. Often, the diode will be

Figure 8-5 Several laser diodes removed from DVD and CD burner units.

epoxied or press fit into a small metal block for better heat dissipation. It may be best to simply leave it connected, as the high-power DVD burning diode will generate a lot of heat when powered up to full current. If you want to remove the laser diode from the heat sink to make your own heat sink or to retro fit a laser pointer, it may take a bit of prying with a thin screwdriver in order to release the diode from the small metal block. The glue will be contained in a few small holes along the laser diode body. This is where you will have to pry using the screwdriver blade. If you have a very small drill bit, you can also grind out some of the epoxy to help release the diode.

The bare laser diode is shown in Figure 8-6. Most of them will look exactly the same, regardless of wavelength and output power. The small opening at the top will usually include a protective glass cover, but this is not a focusing lens, just a way to protect the actual diode from the elements. There will be as few as two connecting leads and as many as four, depending on the internal wiring of the laser diode, but all of them will function the same way.

The small can will contain both a laser diode, as well as an optical sensor that can be used to monitor the output in a closed loop driver circuit. When there are only two connecting leads, that means the steel can is tied to ground, and the leads will be the positive side of the laser diode and the optical sensor. With three leads, the wiring is often

Figure 8-6 This is the bare 60-mW infrared laser diode from a CD burner.

the same as with two leads, but will include a lead tied directly to the can. Laser diodes with four leads have separate ground connections for both the laser diode and the optical sensor. We have not yet seen a laser diode with five leads, and the three-lead package is by far the most commonly used.

If you are salvaging a laser diode from a DVD or CD burner, then you will have to do a little hacking to determine how to connect the laser diode. If you are purchasing a new laser diode or can find an Internet source that happens to list the part number on the laser diode for whatever device you salvaged it from, then you will be able to find a proper datasheet. The datasheet fragment shown in Figure 8-7 shows most of the important parameters

■ **Absolute Maximum Ratings** (Tc=25°C [1])

Parameter		Symbol	Rating	Unit
[3] Optical power output		P_o	50	mW
[2] Optical power output (pulse)		P_p	70	mW
Reverse voltage	Laser	V_{rl}	2	V
	Monitor photodiode	V_{rd}	30	V
[1] Operating temperature	[3] CW	$T_{opc}(c)$	-5 to +65	°C
	[2] Pulse	$T_{opp}(c)$	-5 to +70	°C
Storage temperature		T_{stg}	-40 to +85	°C
[4] Soldering temperature		T_{sld}	300	°C

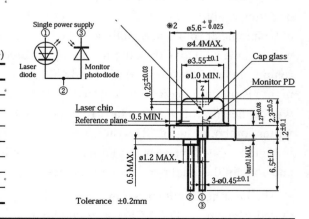

Figure 8-7 The laser diode datasheet will detail the important parameters.

for the laser diode that we pulled from an older CD burner. According to the datasheet, pin 1 is the positive supply for the laser diode, pin 2 is the output from the light-sensing photodiode, and the case is a common ground point for both. Also shown in the datasheet is the continuous output power rating of 50 mW and the peak wavelength of 780 nm, which will be in the infrared region of the light spectrum.

If you have no datasheet for your laser diode, then you will need to hack your way to figure out which leads are needed to power up the laser diode. The bad news is that laser diodes are extremely sensitive so that they can be over-powered to the point that they will fail within milliseconds of having too much current applied. The good news is that you can safely power any CD or DVD burner laser diode directly from a 3-V button battery such as a CR2032 without any damage at all and without needing any laser driver circuitry. Figure 8-8 shows the brilliant red glow coming from a random laser diode that we grabbed from our parts box. Since this output was visible red, we knew that this laser diode was from a DVD burner and had a potential output of over 200 mW. Yes indeed, you need laser goggles and a nonreflective matte surface to bounce the laser beam off of when doing this!

The reason you can safely power up most laser diodes using a 3-V lithium coin battery is

Figure 8-8   Powering up a bare laser diode using a 3-V coin battery.

because the internal resistance of the battery will limit the current to a safe level, creating a built-in laser driver without using any electronics at all. Although you will not see the full output power of a 100 mW or greater laser diode, you will still get enough output power to burn objects when placing the beam from the DVD burning laser close to any dark surface. If you are working with CD burner laser diodes, then you must do this power testing while viewing the beam on a monitor connected to a black and white security camera as your eyes cannot see the infrared radiation. Also, be extremely careful when working with infrared lasers because you can fry your retinas. You will not see or feel a thing as the invisible infrared radiation does irreversible damage. This is especially true when a laser diode is focused by a collimating lens.

To test your laser diode using the "brute force battery method," start by connecting the negative side of the battery to the can (or can-mounted lead) of the laser diode and then try the other leads using the positive side of the battery. To avoid an accidental eye flash, tape a bit of white paper over the output window of the laser diode so that the beam will be safely defused. If you do not see any visible light after trying all possible combinations with the leads and battery, you probably have an infrared laser diode, so the test needs to be redone using the video camera and monitor so you can see the invisible infrared radiation.

At 10-ft away, the intense red glow from the visible red DVD burner laser is so intense that it is almost impossible to get a good photograph (Figure 8-9). This is the output from the laser diode without any lens at all, so the beam spreads out to about as far as the distance from the laser to the object, creating a beam that covers an entire 8-ft wall at about 10-ft away. The output from the infrared laser is almost as powerful when using the coin battery, and the beam spread is similar, so we will use the visible red laser for all photography in this project.

Figure 8-9 The amazing burst of red light form a DVD burner laser diode.

Figure 8-10 This laser pointer allows the lens to be adjusted or removed.

Since beam spreading is necessary for a night vision illumination system, it may be practical to simply use the bare laser diode with no optics at all if short range and ultrabright illumination are your goals. The 50 mW infrared laser diode was actually good as an indoor illuminator, as the beam spread to cover an entire wall of most rooms we tested it in. A 50 mW infrared laser diode taken from an old CD burner and powered directly from a 3-V coin battery made a night vision illuminator that was as good as an array of 32 LEDs, yet needed only a fraction of the current.

Of course, the downside to this type of illumination system is that it may not be eye safe, so do not point it at any living creature or person, and only view the illuminated area with your viewfinder. According to the manufacturers of similar laser diode based illuminators, a beam spread like this is considered safe, but we would not take any chances. It would be best to consider the unfocused beam to be as dangerous as a highly focused beam. We do not own a laser power measuring instrument, so we always play it safe, and never allow the radiation to enter our eyes, even when completely defocused.

The laser pointer shown in Figure 8-10 is really a high-power laser module with a built-in battery holder. Being completely invisible and powerful enough to start fires at several feet from an object,

it is certainly not mush use as a pointing device! The nice thing about this laser module is the adjustable collimating lens that can be fine tuned or completely removed by unscrewing it from the main body. With a laser of this power level (greater than 250 mW), a very long-distance night vision illuminator can be made, but the unfocused beam coming directly from the laser diode will be lost after about 20 ft.

The built-in adjustable lens would allow the beam to be defocused to about 10 ft across over a 50 ft distance, but it was simply not enough beam spread for this power level. We wanted to add a telescope to a low-lux video camera and make an illumination system capable of illuminating an entire city block! Having the ability to remove the lens completely would allow other optics to be tested in order to get the resulting beam spread that we needed.

Being a hacker, we do not often have a large budget for parts, in fact we usually have no budget for parts, and optics are extremely expensive to purchase. Luckily, we are also a junk collector and have managed to salvage a large bucket full of various small and large optics taken from old appliances such as projectors, cameras, and scanners. Figure 8-11 shows some of the lenses we tested with the high-power laser pointer with the original lens removed, and many of them

Figure 8-11   A vast assortment of various optics salvaged from old electronic devices.

resulted in a good beam spread for long-distance illumination. The goal was to be able to light up an area as large as a house at a distance of about 200 ft away. It seemed as though the perfect lens for the job was a 1-inch (in) diameter focusing lens taken from an old HP scanner.

In order to get a useable photograph, we are working with our visible red laser, but will be using a similar-power infrared laser module in our night vision illuminator. Figure 8-12 shows the intense beam spread out to about 6 in by the lens we found in our scrap pile. At 10 ft, the beam was about

6-in wide, and across the room, it was about 2-ft wide, so this lens seemed to be about right for the distance we was hoping to achieve with our night vision system. Even with the beam spread out this much, we would not take any chances and consider this project to be eye safe.

Even across a 25-ft distance, the beam remains extremely bright on the wall, with the laser spot expanding to around 5 ft across (Figure 8-13). Some adjustment was possible by moving the lens closer to the laser diode, but the optimal spread seemed to happen with the lens placed right at the

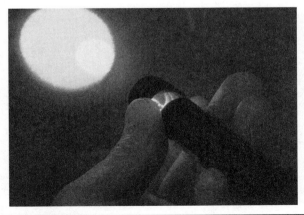

Figure 8-12   The beam spread out to 6 in at a distance of 10 ft.

Figure 8-13   Across the room, the laser spot spreads to several feet across.

edge of the laser tube, where the original focusing lens was mounted.

We wanted to make this laser illumination project in several different versions, so that we could test various laser diodes, modules, and optics using a low-lux security camera to view the scene thought a zoom lens or telescope. The final version ended up using a 4 × 32 zoom gun scope, 250 mW infrared laser module and a high-resolution camera, but we also made a self-contained dual laser system that would allow two different lasers to be aligned to the same spot. This dual laser system made it easy to interchange various diodes and modules without having to realign the camera while pointing at a distant scene.

To align two different-sized lasers to a single source, a half-inch thick piece of aluminum was drilled out to support two sizes of lasers: a small module and a typical laser pointer–sized body. This would allow many different laser modules and pointers to be installed and aligned precisely to the same spot without having to move the camera or tripod holding the aluminum block. Figure 8-14 shows the drilled aluminum block and the two laser modules that will lock into the holes.

The aluminum block was installed to the top of a small box cover that will also include a power switch to control each of the lasers (Figure 8-15). All of our laser modules and pointers work from a

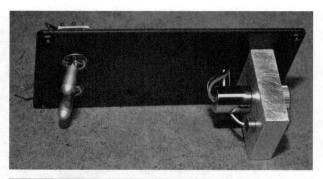

Figure 8-15   Adding power switched to control each laser module.

3-V DC powers source, so the power supply will be a pair of 1.5-V alkaline AA batteries wired in series for 3 V; 3 V seem to be a save voltage for just about any laser module or bare diode that we tested, so this worked out well.

Figure 8-16 shows the completed and self-contained dual laser beam test rig, with a high-power infrared laser module along with a visible red laser for visually targeting a distant point. This handy laser system was used in several night vision experiments along with our Laser Spy Device, a system that allows a conversation to be heard by sending the information back to a receiver that would catch the moving laser beam bounced from the target window. The two toggle switches are also specially designed so that they have to be pulled up and dropped back in place to turn

Figure 8-14   Making a support to align two lasers to the same position.

Figure 8-16   The completed dual laser and power supply system.

on the lasers. This extra safety would eliminate an accidental powering of the lasers if one of the switches was hit.

The final version of our laser based night vision illumination system will include a low-lux high-resolution camera connected to a 4 × 32 optical zoom gun sight in order to magnify the distant scene lit by the infrared radiation. The laser that will be used for illumination will be an 850 nm infrared laser with an output of 250 mW, defocused through a modified adjustable lens. The gun sight used for this version of the laser illuminator is shown in Figure 8-17. Binoculars or a low-power telescope would also work well as a camera zoom system.

Since the sight will no longer be used on a rifle, the mounting hardware was moved to the front of the magnifier tube and drilled so that a laser could be held in place using a few zip ties. Since the mounting hardware fit the gun sight perfectly, this would point the laser beam in the exact same direction that the magnifier was aimed at, making alignment of the laser illuminator fully automatic. Figure 8-18 shows the two mounting brackets installed in the front of the gun sight so that the laser would sit in the groove normally reserved for the rifle barrel.

With the laser tube fastened tightly to the rifle mounting brackets as shown in Figure 8-19, both the gun sight and the laser were now pointing in the exact same direction, so no further alignment

Figure 8-18 Modifying the rifle mounts to hold the laser in place.

Figure 8-19 The laser tube is held in place by the rifle mounting brackets.

of the night vision system would be necessary. The adjustable lends shown in Figure 8-19 is made by inserting a different glass lens into the existing adjustable collimating lens housing on order to make a beam spreader. The new glass lens was taken from a tiny spy camera lens, and fit nicely into the original lens housing so it could be glued in place. The projector lens shown spreading the beam in Figure 8-12 would have also worked, but having the ability to adjust the size of the beam was convenient.

The completed illumination and camera zoom system is shown in Figure 8-20, ready to be connected to a video monitor in order to see the scene lit by the invisible infrared radiation. Having the ability to adjust the new beam-spreading lens

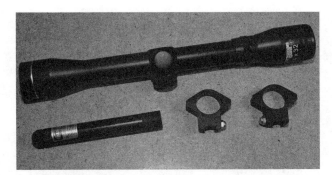

Figure 8-17 This 4 × 32 zoom gun sight will allow the camera to see long distance.

Figure 8-20 The completed long-range illuminator and camera magnifier.

Figure 8-21 Most security cameras can see infrared radiation as white visible light.

was nice, as the scene could be made either larger or brighter, depending on the position of the lens as well as the distance to the target. Having the laser in line with the gun sight also meant that alignment was automatic, so no visible laser was needed in this setup.

To see invisible infrared radiation, you need to view the scene on a video monitor as seen by a low-lux (low-light) security camera. To the security camera, infrared radiation that is completely invisible to the human eye will be displayed as normal white light, making a scene look as though a white light source was being used for illumination. The lower the lux rating on the camera, the more light it will be able to collect, so a camera rated for 1 lux or even lower will make an excellent imager for a night vision system. Unfortunately, camcorders and digital still cameras cannot see infrared light as they have built-in glass filters that block out almost all of the infrared light in order to allow the visual light spectrum to be processed much better. The good news is that all security cameras can see infrared light, and they are very inexpensive.

Another thing to note is that a black-and-white security camera will be better for use in a night vision system as they are always rated for lower lux operation, and infrared light is only shown as black and white on the monitor. Black-and-white security cameras are also less expensive than decent-quality

color cams, so you can purchase a great low-lux monochrome security camera for under $75 from many online security distributors. The camera shown in Figure 8-21 is a bit of a specialty security camera as it has very high-resolution, low-lux, and a fully-adjustable image sensor that will work in both color and monochrome. We chose this camera because it had a larger C-mount type lens that just happened to mount nicely against the viewing end of the gun sight, allowing a simple bracket to hold the camera in place. The adjustable focus of the lens allowed the image to be fine tuned for extremely high-detailed night vision imaging.

Figure 8-22 shows how easy it was to set up the C-mount lens so that an image would be

Figure 8-22 A steel clamp holds the C-mount camera lens directly to the gun sight.

projected from the gun sight viewing lens right onto the camera lens. A single steel clamp was placed between the lens and the sight, which were the same diameter, making a perfect union. Normally, we use micro spy cameras for our night vision illuminators, and having extremely small lenses means that they need some type of adapter tube in order to place them in the correct position to see through an output lens like the one on the gun sight. Binoculars and small telescopes also work well with security cameras once you find the optimal placement of the camera in relation to the viewing end of the magnifier.

Since the camera was now fastened to the end of the gun sight, it would be impossible to support the entire unit from the camera end alone, so a small metal bar was drilled out to mount to the camera tripod and hold both the video camera and gun sight in a balanced position. Having the weight balanced in the center would keep any stress away from the camera and gun sight, allowing the tripod to be adjusted without any risk of damage to the optical components (Figure 8-23).

The completed illumination and camera zoom system is shown in Figure 8-24, mounted to a camera tripod for easy adjustment and stable imaging. Now, the output from the camera can be viewed on any composite video monitor or recorded for later analysis. Because of the extremely powerful infrared laser being used, great

Figure 8-24 The completed long-range laser illuminator and camera zoom system.

care had to be taken to avoid any accidental eye exposure of the beam, even at a great distance. Window reflections can be particularly dangerous, and since human eyes cannot detect the intense infrared beam, laser safety glasses and careful placement of the laser are certainly necessary. This night vision illumination system is capable of very long distance imaging, but we only use it where there will be no danger to people or animals at the receiving end. Laser illumination systems that use even higher levels of infrared laser radiation are commercially available, and although they claim to be "eye safe," we would still not take any risks with such a high-powered laser system. At least with the beam spread out to such a large spot, this laser is no longer capable of starting small fires when the beam is highly focused!

Figure 8-25 shows the image captured using a high-resolution camera capable of seeing the infrared radiation sent from the illumination system. At 100 ft, the laser spot can be adjusted to a width of about 20 ft across, and is almost too bright for the low lux camera. Distances are really limited only to the optical range of the zoom lens and the camera, not the laser, as it can project an intense beam of invisible light even a thousand feet away. At distance beyond 200 ft, our 4 × 32 optical zoom rifle scope falls short of being able to

Figure 8-23 Making a camera tripod compatible mounting system.

Figure 8-25   From 100-ft away, the camera can see clearly using the laser illuminator.

zoom into the scene close enough to capture any detail, but the laser illumination is certainly not lacking.

For longer-range illumination and imaging, a low-lux camera fitted to the end of a 150× telescope works well, but you really have to be careful with the laser alignment and start with power levels of at least 250 mW. Such a system would be capable of a 1000-ft range. You could probably use the original collimating lens that

came with the laser module because at that distance, the beam would have spread out to at least a hundred feet. We hope you enjoyed these laser illumination experiments, and remember that any amount of laser power over 5 mW is more than enough power to cause instant and irreversible eye damage. Never use a laser illumination system on people or animals, and always wear laser safety goggles, even when viewing the image on the video monitor.

PROJECT 9

Camcorder Night Vision

CAMCORDERS ARE GENERALLY DESIGNED for well-lit scenes, using the light to create a quality color image for recording. To ensure that the image is seen by the camera in a similar way to our eyes, only the portion of the light spectrum that is visible to our eyes is processed. Infrared light falls just below red on the light spectrum, making up the wavelengths from about 750 to 1500 nanometers (nm). This light cannot be seen by human eyes, but it can easily be seen by the charge-coupled device (CCD) imaging system in the camera, allowing it to be used as a night vision viewer.

Figure 9-0 This camcorder uses a spy camera to see invisible infrared light.

PARTS LIST

- LEDs: 8 to 24 infrared LEDs in a 5-mm through hole package; wavelength of LEDs is 800- to 940-nm infrared
- Camera: Low-lux black-and-white composite video camera
- Camcorder: Any camcorder with an external video input
- Battery: 6- to 12-V pack depending on number of LEDs

Unfortunately, you cannot simply add an infrared illuminator to your camcorder and use it to capture night vision video because the CCD imager contains a glass filter that blocks out most of the infrared light. The good news is that most camcorders allow a secondary video input to be recorded, and by feeding the output from a small black-and-white spy camera into this input, you

can give your camcorder the ability to record night vision scenes that have been lit by some type of infrared illuminator. This project uses an inexpensive camcorder and a twenty-dollar black-and-white spy cam along with one of the light-emitting diode (LED) illuminators shown earlier to create a portable stealthy night vision camcorder.

Most video camcorders allow an external video source to be plugged in, essentially replacing the built-in CCD imager with some other compatible video source. This video input will often have some custom manufactured input jack with a label of "external," "line input," or "AV input." You will need the cable that came with the camcorder in order to make this project, as each manufacturer will have its own special cable for that model of camera. On the older tape-based camcorders like the one shown in Figure 9-1, there was a 1/8 jack "standard," that used a 4-ring 1/8-inch (in) male plug like the one shown in Figure 9-1, allowing any composite video source to be fed into the camera.

111

Figure 9-1   Any camcorder with an external
video input will work for this project.

You will have to identify both the external
video input and acquire the proper input cable in
order to build this project, but if you kept all of
the accessories from your camcorder, then that
odd cable will probably be still sitting in the box,
as it usually not used. Some manufacturers like to
chisel the customer out of more money by making
their own special connectors and then charging
a ridiculous amount for the cable, so check your
camcorder manual to make sure your camera
supports video input and that you can acquire the
necessary cables for a fair price.

Figure 9-2 shows the 1/8-in style connector
that was popular on many tape-based camcorders.
This connector is the same size as a typical 1/8-in
headphone jack, but has four conductor rings,
rather than the usual three as found on a headphone
connector. At the other end of this cable is a set

of standard RCA jacks to allow the camera to be
plugged into any composite video-compatible
equipment. This external video jack on most
camcorders is called a "dubbing jack" as it can be
used to record video from an external source or
send video from your camera out to another video
recorder. Check your camera's manual to ensure it
has this function.

We use many small video cameras for security
work, machine vision experiments, and robotic
projects. They range in size from half-an-inch
square to quite large depending on the features,
imaging system, and type of lens installed. For
most indoor work, a small-board camera with a
fixed medium to wide angle lens will be perfect,
but there are times when you need to see a much
larger area or over great distances, so these cameras
have several body types that allow the use of
multiple lens styles. Extremely tiny spy cameras
will not have this option, as they use a very tiny
glass or plastic lens built right into the housing.
There will be a bit of a trade-off between image
quality and size.

For most covert work, size will be the key factor
in camera selection, so you probably won't be
dealing with multiple lenses or even have a choice.
In Figure 9-3, you can see the huge difference
in size between the long-range motorized lens
telephoto camera on the left and the tiny spy
camera at the front of the photo. The tiny spy
camera has a simple fixed medium angle lens

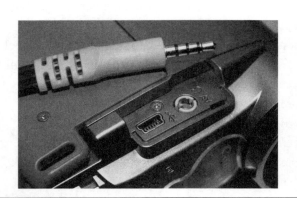

Figure 9-2   Our camcorder uses the 4-pin 1/8-in
connector and is labeled AV.

Figure 9-3   A variety of small security video
cameras.

and will run from a 6-volt (V) power source for many hours. The larger camera uses a computer to control the motorized zoom lens and requires multiple power sources and electronic control systems in order to operate. For this project, an inexpensive black and white mini camera with a medium to wide angle lens that will run from a 6- or 9-V power source will be perfect.

Security cameras and micro spy cameras will have a connection point for DC (direct current) power and a video output, and some will have an audio output or several control lines for onboard features such as gain, color, text overlay, and lens control. The most basic cameras you will probably work with most often will have only power and video connectors, or simply include three wires coming out of the tiny enclosure. The tiny board cameras with three wires will usually follow a simply color scheme of black or green = ground, red = positive power, and white, yellow, or brown = composite video output. Of course, it is always a good idea to check the manual for polarity and voltage before making any guesses.

The camera in the left of Figure 9-4 is quite advanced, having a built-in on-screen display system, many internal function parameters, lens control, and audio. This camera is somewhat large, so it has many buttons on the back side, audio and video jacks, DC-power jack, and a special connector to control a motorized lens. The basic low-lux micro

camera on the right of Figure 9-4 only has a tiny connector with a three wire cable to allow power, ground, and video output connections. Since you will be powering the camera from a battery source, the smaller camera will be the better choice.

When digging into the technical details of a security camera, one of the most important aspects will be the type of imaging device used, which will either be a CCD imager or complementary metal-oxide semiconductor (CMOS) imager. Both types of imagers convert light into voltage and process it into electronic signals. In a CCD sensor like the one shown in Figure 9-5, every pixel is transferred through a very limited number of output nodes (often just one) to be converted to voltage, buffered, and sent off-chip as an analog signal. All of the pixels can be devoted to light capture, and the image quality is high. In a CMOS sensor, each pixel has its own charge-to-voltage conversion, and the sensor often also includes amplifiers, noise-correction, and digitization circuits, so that the chip outputs digital bits. These other functions increase the design complexity and reduce the area available for light capture.

What this means is that for low-resolution and low-light imaging, CCD cameras are currently the best choice. CMOS imagers are used in very high-resolution imaging systems such as digital cameras and scanners as proper lighting is usually not a problem. If you have the choice between a

Figure 9-4 All video cameras will have connectors for power and video output.

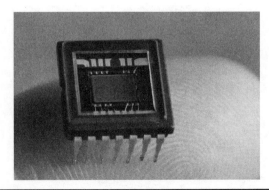

Figure 9-5 The CCD imager makes video by changing light into an analog voltage.

hundred-dollar CCD security camera with a 0.5-lux rating and a fifty-dollar CMOS camera with a 1.5-lux rating, the CCD camera will offer a far superior image and work well in a low-light or night vision application. Resolution is not much concern in a security camera, as most are already beyond the actual capabilities of the National Television System Committee (NTSC) or Phase Alternate Line (PAL) composite video standard anyhow. The lux rating will probably be the most important specification on a security camera next to the type of lens and field of view.

Lux is the measure of light or luminous power per area. It is used in photography as a measure of the intensity, as perceived by the human eye. The lower the lux rating on a camera, the better it will see in the dark. For instance, a monochrome camera with a lux rating of only 0.5 lux will see much better in the dark than you could with your naked eye, and with the help of an infrared illuminator will be able to see into the darkness and display it on a monitor as if it were midday. Color cameras require much more light to achieve a decent image, and since they usually include infrared filters, they are not the best choice for night vision applications.

The focus can be set on these micro video lenses by loosening the tiny set screw as shown in Figure 9-6 in order to turn the lens clockwise

or counterclockwise to gain the best focus for a certain installation. Most of these lenses will see perfectly from about 10 feet (ft) to infinity, but for closer focus, it will probably be necessary to make the initial adjustments. Although the optics on these tiny lenses range from good to poor, the image quality is always decent due to the low resolution of the composite video signal, along with the high quality of the CCD imaging device used. We found that a $30 color board camera will exceed the quality of a $1500 camcorder that was made five years earlier. Technology is always improving and becoming less expensive, so high-quality surveillance equipment is becoming more and more affordable for the "average" person.

This project uses the infrared radiation from an array of LEDs in order to see in complete darkness, and with the price of security gear at an all-time low, you can often purchase a complete night vision camera like the one shown in Figure 9-7 for well under $50. Inside the weatherproof aluminum shell, you will find a high-quality board camera, power supply, and another board with an array of infrared LEDs. By itself, this inexpensive security camera is a high-performing night vision camera with a range of about 50 ft, and has a very good color image for daylight operations. If you have not yet built an infrared illuminator or have a video camera for this project, consider purchasing a camera like the one shown in Figure 9-7, as it has all you need to complete this project.

Figure 9-6 Smaller video cameras have a fixed focus that can be set.

Figure 9-7 An outdoor security camera with built-in night vision LEDs.

Because the audio and video signals are prone to interference over longer distances, shielded coaxial cable is necessary in order to make the connection between the camera and the monitor or recording device. If your camera has RCA connectors at both ends, then you can use any standard patch cable to make the connection to your TV or monitor, but often you will need longer cables or have to install your own connectors. The standard coaxial cable type used for security cameras is the RG-59 type. This cable has a single insulated signal wire surrounded by a conductive shield. The "Siamese RG-59" version also includes a second shielded pair to allow camera power supplies to be located indoors at the location of the video recorder or monitor. For short connections of only a few feet, practically any coaxial wire will work just fine.

You may need to cut, strip, and solder your own connections, depending on your camera setup. Figure 9-8 shows the various stages of stripping down a coax cable to reveal the grounded outer shielding, as well as the protected inner signal wire. When connecting coaxial cable to an audio or video source, the inner wire carried the signal and the outer shield is connected to a common ground. When using coaxial cable to carry power, the positive supply is connected to the inner wire and the common ground becomes the outer shield. In almost all instances, the ground is common to all signals, including the power supply.

There are a few different types of connectors used on security cameras and composite video

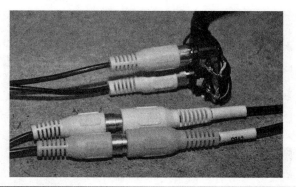

Figure 9-9 This is the most common type of composite video connector.

appliances, but the most common type is the RCA jack shown in Figure 9-9. This connector is the same one used on the back of a television set to connect older video players and camcorders. On the back of a TV or monitor, this jack will often be labeled "Video Input," "Line Input," "External Input," and will use the same connector for audio and video. Most security cameras will output composite video in either NTSC or PAL format to be connected to any TV or monitor that can accept this input. Some of the smaller spy cameras may not have any connector, just bare wires, so it makes sense to adapt them to the RCA jack as shown in the upper part of Figure 9-9 if you plan to use them with standard TVs or monitors.

If your camcorder has a mount for an external light like the one shown in Figure 9-10, then you can use the slot to create a slide-in adapter for your external camera and infrared light source. The piece of aluminum shown in Figure 9-10 is the lid

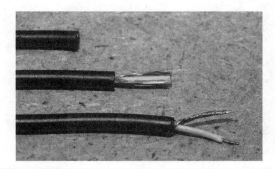

Figure 9-8 Coaxial cable is needed to carry the audio and video signals.

Figure 9-10 The external light mounting slot is going to hold the new camera.

Figure 9-11 | Marking the aluminum plate to cut the new slot.

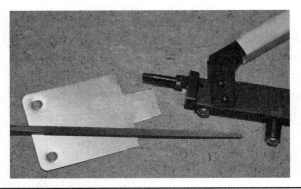

Figure 9-13 | A notching tool and a hand file is used to carve out the slot.

from a project box. It will be cut so that it fits into the light mounting slot on the camcorder, allowing the other components to be positioned over the top of the camcorder. If your camcorder does not include a light slot, then there are many other options available such as using a base that connects to the tripod screw hole or simply using Velcro on the side of the camera.

Figure 9-11 shows the aluminum lid sitting up against the light mounting slot so that two points can be marked to begin cutting a slot into the aluminum. Practically any thin aluminum, steel, or plastic can be used to fit into the slot, but these project box lids are nice because there is plenty of room to mount the components and they can be cut using a notching tool or Dremel cutting wheel.

The slot will be cut as shown in Figure 9-12 so that it fits snugly into the camcorder with very little

play. The aluminum is thinner than necessary, so the corners will be slightly bent in order to create a tighter fit so that the camera does not bounce around or fall out of the slot.

When working with thin aluminum or steel, a notching tool like the one shown in Figure 9-13 comes in very handy, as it can cut square bits out of the material. This tool can cut square holes and shapes out of material as thick as 1/16 in, and is great for cutting tabs or holes for liquid crystal display (LCD) panels, lights, or controls. A small hand file is also a must-have tool when working with thin metal or any type of project box that needs to have some type of machining done to it.

Once the slot was cut with the notching tool and then cleaned up with the hand file, it fit perfectly into the light mounting slot as shown in Figure 9-14. Because the thin aluminum plate was

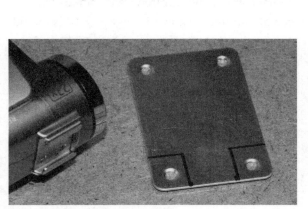

Figure 9-12 | The slot to be cut will fit into the light mount on the camcorder.

Figure 9-14 | The tab is cut and fits snugly into the light mounting slot.

Figure 9-15   The aluminum part is bent into a 90 degree angle.

Figure 9-17   The connector hole drilled through the aluminum plate.

thinner than the slot, it was bent at the front corners in order to offer a little more friction when fit into the slot.

To offer an area to mount the camera and LED illuminator, the plate was bent at a 90 degree angle as shown in Figure 9-15 so that the face would point forward along with the camera lens. The external camera does not have to be perfectly in line with the original camera lens, but it does help to have both pointing in the same forward direction, especially if you plan to navigate a completely dark room by looking through the image on the camcorder viewfinder.

Our small video camera had a connector plug in the lower bottom corner, so we had to drill a hole into the mounting plate to install the connecting wires. Figure 9-16 shows the placement of the

camera on the aluminum and the hole being marked for drilling.

Figure 9-17 shows the camera sitting on the aluminum mounting plate with the connector placed into the quarter-inch hole. Since this camera is so lightweight and small, it will only need a bit of double-sided tape to hold it onto the aluminum plate.

Rather than drilling holes for stand-offs and mounting screws, we just placed a bit of double-sided tape on the aluminum plate to hold the camera securely in place (Figure 9-18). The tape also creates an insulating barrier to keep the camera's circuit board contacts from shorting out on the conductive aluminum surface.

The camera and mounting bracket are shown sitting in the camcorder's light mounting slot in Figure 9-19, ready to have the rest of the wiring and infrared light source connected. All that is left

Figure 9-16   Marking the position of the camera connector to drill the hole.

Figure 9-18   A small piece of double-sided tape will fasten the camera.

Figure 9-19   The camcorder now has two sets of electronic eyes.

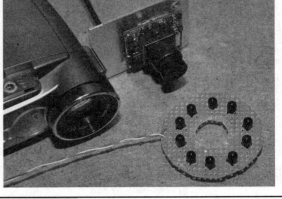

Figure 9-21   Infrared illumination will be provided by an LED ring light.

to do now is find a suitable battery pack that will power both the camera and the LED illuminator ring.

The small connecting plug used on our board camera is shown in Figure 9-20, pushed through the hole drilled in the aluminum plate. These connectors are custom-made for the cameras, so care must be taken when pulling them out of the socket. Usually, the small security cameras have three connecting wires—ground, power, and video output.

The infrared illumination ring light shown in Figure 9-21 is an earlier project made to fit over the lens of the small spy camera. The 10 LEDs are wired in series and powered by a battery pack to give each LED the proper voltage for optimal

brightness. The 940-nm infrared radiation is completely invisible to human eyes, yet shows up on the low-lux black-and-white camera as typical white light. These LEDs are the exact same type that are used in remote controls to send bursts of infrared radiation to the receiver in your TV. Any of the night vision illuminator projects presented earlier would work as an invisible light source for this project, but the ring light already fit perfectly with this small spycam.

A 12 V battery pack with a built-in switch was chosen to power both the camera and the infrared LED array (Figure 9-22). Having 12 series-connected LEDs, each requiring 1.2 V, made this battery pack the perfect choice, and the camera also

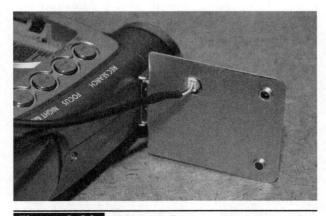

Figure 9-20   This small spy camera uses a tiny connecting plug on the back.

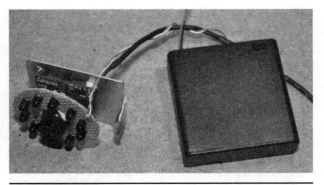

Figure 9-22   This 12-V battery pack will power the camera and the illuminator.

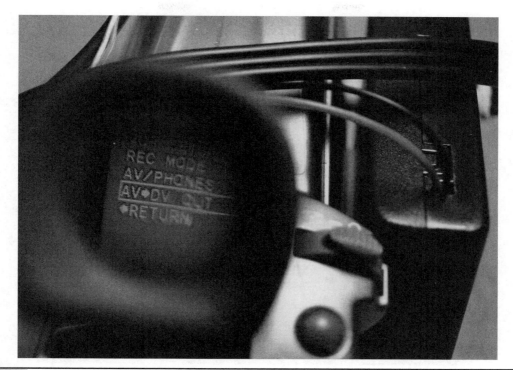

Figure 9-23   Setting up the camera to use the external camera as an input.

requires 12-V DC to operate. You could probably hack into the camcorder and pull the needed power right from the built in rechargeable battery, but this might be difficult as the lithium pack will probably have a lower voltage than necessary to operate an external camera. The camera and infrared illuminator will run for hours off of this small 1 amp (A) hour 12-V battery pack.

To get the camcorder to accept the external camera as the default video input, we had to jump into the onscreen menu system (Figure 9-23) and choose "AV Input," which turned off the camera's built-in color imaging system and replaced it with the signal coming from the black-and-white spy camera. Some cameras will automatically switch to the external input once a sync signal is detected, but you may have to play around in the menus to get your camera to switch to the external camera. Oddly, our camera only records the external source when in "playback mode," which makes it act like a mini VCR.

Figure 9-24 shows how our completed night vision camcorder conversion project looked after fastening the battery pack and wiring to the camera body. All of the components can be easily removed

Figure 9-24   The completed night vision camcorder conversion project.

Figure 9-25   **The converted night vision camcorder can see in complete darkness.**

if necessary. Switching between color daytime recording over to stealthy night vision mode only requires a few presses of the menu buttons.

This rig also offers what was once referred to as "X-ray vision," a controversial feature offered on some camcorders to remove the built-in infrared filter. We say controversial because it was found that in the daytime, certain clothing would appear opaque to the camera due to the ability of certain fabrics to pass infrared light. The company that offered this mode quickly removed it from their firmware, but many hackers figured out how to restore this functionality. Feel free to experiment with this "interesting" phenomenon, but be careful who you point your camera at!

The completed night vision camcorder is shown pointing at our parts closet (Figure 9-25 top left) as seen by the camera used to make these project photos. The top right image in Figure 9-25 is what the camcorder sees with the room lights off when using its own built-in color imaging system—not much at all. Switching over to the external

low-lux black-and-white security camera (bottom left image), the camera does a little better in the dark since the spy cam has a much better ability to see in low-light conditions. With the infrared illuminator powered on (bottom right image), the scene looks well lit, yet we cannot see anything unless we view the scene through the camcorder's viewfinder.

The nice thing about using this system as a night vision viewer is the ability to record what you are viewing, just as you could with the original imaging system on the camcorder. Navigating a room by looking through the image on the viewfinder is fairly easy to do as well, so this project is both a night vision viewer and stealthy security recoding system. For total stealth operation in complete darkness, place some black tape over the camcorders red indicator LED, but other than that, you can see in complete darkness with this unit, and your operations will be completely invisible to anyone in the room. Have fun on your covert spy missions!

Night Vision Viewer

THIS EASY-TO-BUILD NIGHT VISION VIEWER lets you see deep into the night without detection. This covert system can light up a room as if you were using a flashlight, yet only you will be able to see the light. The performance of this night vision viewer is as good as some commercially available night vision systems that cost a lot more. Using invisible infrared light, the night vision viewer can see in total darkness, indoors and outdoors, and will operate from a battery pack for several hours. This device can also be used to detect other night vision systems or as a jammer to hide your face to most security cameras. Another interesting effect of the night vision system is referred to as "X-Ray Vision," which allows the user to see through certain materials (including clothing) that may be opaque to infrared light. If covert surveillance or countermeasures is your game, then this is one piece of equipment you will definitely want in your spy gear arsenal.

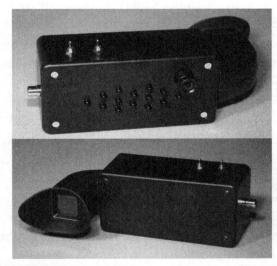

| **Figure 10-0** | Build the stealthy night vision viewer. |

The night vision viewer is built around commonly available parts, most of which can be found new at any electronics store, or taken from dead video appliances. This project is well within the reach of anyone with a desire to do a little hardware hacking, and even includes a basic guide to getting started in electronics. There is a lot of room to add your own modifications as well, so you can create your own unique spy gear in order to further your cause. The truth is out there, and now you will be able to see it even in total darkness!

The night vision system consists of three main parts: a low-light (lux) camera, an invisible light (infrared) illumination system, and a video viewfinder. The three components work together in order to extend your range of vision into the infrared spectrum, which is beyond what the

PARTS LIST

- LEDs: 8 to 24 infrared LEDs in a 5-mm through hole package Wavelength of LEDs is 800 to 940 nm infrared

- Camera: Low-lux black-and-white composite video camera

- Camcorder: Any camcorder with a CRT-based viewfinder

- Battery: 6- to 12-V pack depending on number of LEDs

human eye can see, allowing you to see in complete darkness. Because the camera sees this infrared light as if it was visible light, the image looks no different on the screen than it would if you lit the dark area with a typical flashlight.

The first part of this project will be the acquisition of the viewfinder, which is nothing more than a tiny composite video monitor that can be powered from a battery. These small video screens can be removed from older camcorders or purchased new from many security supply outlets. Figure 10-1 shows an older tape-based camcorder, which includes a cathode ray tube (CRT) based viewfinder that can be removed from the rest of the camera. Also shown in Figure 10-1 is a small ready-to-use viewfinder purchased from an online security store. These ready-to-use viewfinders are sometimes referred to as "camera testers," or "micro monitors." These units will have a jack (RCA jack) marked "video input" on the back of the case, so you can just feed your camera video in as if you were connecting to a standard monitor.

If you do plan to hack up an old camcorder for the viewfinder, it is important that it be a "tube-based type" that includes all of the electronics in the actual viewfinder housing. A clear indication of this is the 3- or 4-inch (in) long housing and the white screen, which is actually made of glass. Newer camcorders use an LCD screen, which is

rarely hackable, and they can be easily identified by their small size and color screens. CRTs are strictly black and white and glow blue when they are first powered up.

These older tape-based camcorders can often be purchased inexpensively from second-hand stores, yard sales, or online (check eBay, Kijiji, Craigslist, etc.). Even if the tape mechanics are broken, the viewfinder will probably be fine, and a simple power up will tell you right away. If you see the tube begin to glow, then it probably works fine. We will now cover the viewfinder extraction operation, which will vary depending on your camera model and age.

The viewfinder will be powered by the camera's main circuit board, and normally only requires a single DC voltage between 9 and 12 volts (V) to operate. The viewfinder will also have a video input and ground line, as well as several other functions such as a power light-emitting diodes (LED) or function switch that can be ignored. In the end, you will only need three of the wires coming from the viewfinder (power, ground, and video input). Start by removing the multitude of tiny screws from the camcorder shell until you reveal the main board where the viewfinder connects as shown in Figure 10-2.

Normally, the viewfinder will have a removable plug that can be pulled right out of the main board, making it easier to probe for voltage if necessary. Having removed many of these viewfinders, we can assure you that there is no standard whatsoever

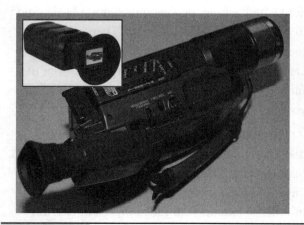

Figure 10-1 An older camcorder with a CRT viewfinder.

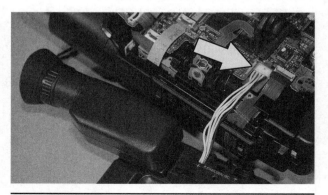

Figure 10-2 Identifying the connection point.

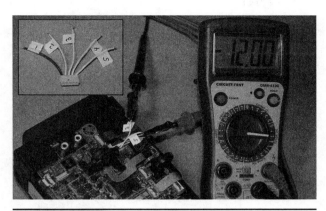

Figure 10-3 Live testing to decode the viewfinder wiring.

when it comes to number of wires, or wiring colors. Black and red almost always do not indicate positive and negative, so a little hardware hacking will likely be needed in order to get the viewfinder to power up and run without the rest of the camera.

If your viewfinder is soldered directly to the board, cut the wires so there is a bit of wire left on the main board in case you need to power up the camera to aid you in finding the power and ground lines later. If all the wires are the same color (often the case), then find some small labels to stick on the wires, or make a diagram of each wire so you can match it back up to the viewfinder.

Figure 10-3 shows one of the many alternative methods that you can use to help identify the positive (power) and negative (ground) wires of your viewfinder—by checking every wire on the main board while the camera is powered on. This can be done with or without the viewfinder connected, as most cameras will still power up with it disconnected from the main board. Figure 10-3 (inset) shows how our connector looked, having four white wires and one blue wire. Again, there is almost always no logic to the colors or even the number of wires, many of which are totally redundant.

Viewfinder voltages range from 6 to 12 V, and will often run fine from any voltage between 9 and 12 V due to the built-in regulation, which make them easy to connect to an external battery source. Figure 10-3 shows our multimeter registering 12 V, but negative,

meaning that we have the correct two power wires, but they are reversed according to the colors of the jacks on our meter. Sometimes you have to test every possible combination of wires to get a voltage reading, but if your camera is powered up and fine without the viewfinder connected, then you will eventually find the magic pair. If you have a dead camera, or one that does not like operating with the viewfinder out of the circuit, then there are several other ways to decode the cryptic viewfinder wiring.

Another way to decode the wiring of your viewfinder if you cannot probe the camera board is to use a "brute force" hack attack. This operation is a bit more dangerous, as you may apply power in reverse to the viewfinder circuit, but by using a current-limiting resistor or power supply, chances of damaging the board are greatly reduced. Figure 10-4 shows another CRT viewfinder connected to a breadboard, which contains a 9-V battery and a 100-ohm (Ω) resistor.

By running the battery in series with the resistor, you reduce the current draw in case of a short circuit or supply reversal, so damage is not likely to occur. Often the internal circuitry in the viewfinder is very robust and may include a power regulator or clamping diode, so this operation is normally safe. If using a battery, choose dry cell over an alkaline as it will deliver much less current if shorted or reversed. Also, a power supply with an amperage control is great, as you can limit the current to a

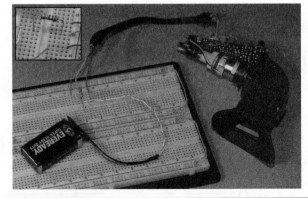

Figure 10-4 Brute force testing to power up the viewfinder.

hundred milliamps and then look for spikes as you randomly test the wiring combinations. A power reversal will pin your current reading or meter instantly, alerting you of the situation.

To test your viewfinder using the brute force method, bare all of the wires and plug them into a breadboard as shown in Figure 10-4. Using a current-limiting power supply, or a dry cell (nonalkaline) battery and a 100-Ω (or close) resistor, start trying combinations until you see your viewfinder begin to glow. Sometimes you can listen for the high-voltage transformer buzzing, but seeing the screen light up is your best bet. Also, don't touch the circuit board while the viewfinder is running, as there are a few kilovolts coming from the tiny high-voltage section. Due to the extremely low current, you won't hurt yourself, but we guarantee you won't enjoy a shock from 10,000 V either! If you can identify the high-voltage wire, then drop a small neon bulb from the output to the metal cage on the HV section, and it will glow the instant you have power connected properly.

After a bit of brute force hacking, we decoded our viewfinder's wiring map as: yellow = power, red = ground, and orange = video input. This is practically the opposite of what you might expect for wiring colors, but hackers like us are not usually a concern during the manufacturing process!

The dull blue glow shown in Figure 10-5 is what you will see when your CRT viewfinder finally

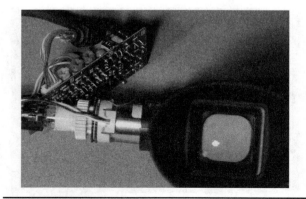

Figure 10-5   **The CRT will glow blue once powered up.**

powers up for the first time on battery power. There won't be any image yet, but the glowing blue tube is a great sign that you are very close to having the viewfinder wiring fully decoded. After a few seconds of being powered on, remove the battery, and then feel around the tiny circuit board for any hot spots. If your board seems hot, try a lower voltage until the screen no longer glows or until it shrinks below the screen borders. Usually 9 to 12 V will run these viewfinders without any hassle, but the odd one may be designed for only 5 V.

If you have failed to decode your viewfinder using the last two methods, then you will need to explore the actual circuit board for clues. Looking for large capacitors is a good start, as the negative side of the can will always be going to ground. Often, there is just one common ground, and it may be connected to several wires on your connector. On mine, there were four or five redundant ground connections. Another easy way to decipher the wiring is to identify the small integrated circuits (IC) on the board and then dig up the datasheet to figure out which pins are VCC (positive) and VSS (ground). Almost all viewfinders will have a single large IC, which will be called a "television on a chip" or "complete NTSC video processor IC." We have not found a viewfinder we could not decode, but sometimes a little hacking and experimentation will be needed, especially with multiple wires of the same color. Feel free to scan or post an image of your viewfinder board in our forum, and you may find someone else with the same unit already decoded.

Once you have your viewfinder powered up and glowing, the next step is to determine which of the wires the video input is. The video input will accept any composite video signal like the output of a VCR, camera, or video game. Cut an old RCA jack and bare the wires at one end as shown in Figure 10-6 (inset) so you can take the output from your video player as a test signal. Don't worry, you won't hurt your video machine during these test since it has a high-impedance input, which is protected from DC voltages. Place the jack into the

Figure 10-6 Testing the wires to find the input line.

"video out" plug and then set your player to send video.

The coaxial cable will have a center wire and a stranded wire around it, making the stranded wire negative (ground) and the center wire the signal wire. Connect the stranded wire to your viewfinder's ground wire (battery negative), and then try the signal wire on all the other viewfinder wires until you see an image on the tiny screen. Figure 10-6 shows our small DVD player displaying a movie on its screen, as well as onto the viewfinder after finding the video input line on the viewfinder.

If you fail to see any video, then either your power lines are not correct, or your video source is not the proper kind of output. The video machine will have an RCA jack labeled "Line Out," "Video Out," or "Composite Out." This is the signal you need to use.

The second part of your night vision system will be the ultra low light (lux) security camera. These inexpensive camera modules run from 9 to 12 V, and output a standard composite color or black-and-white video signal. There are hundreds of suppliers on the Internet, and you can expect to pay between $10 and $100, depending on the quality of the camera.

Since this is a night-only project, a pinhole camera is not necessary, and since the camera

sees infrared as monochrome, a color camera is also not necessary, reducing the cost further. What does matter is the lux rating of the camera, and this should be as low as possible. A lux rating of less than 1 will work fine for this project, and you should acquire a camera with a nonpinhole lens and a lux rating as low as 0.001 lux, which makes the camera almost a night vision system already.

One type of camera that will not work in this application is a typical camcorder, as it has a filter to block out infrared light for improved image quality. You could probably dig into the guts and remove the tiny glass infrared filter, but the bulky, power hungry camcorder includes so many options you simply do not need, that the best bet is certainly a tiny CCD video camera module like the ones shown in Figure 10-7.

A web search for "board cameras," "CCD camera" or "camera module" will yield you many sources to purchase online, and if you stick to the black and white cameras with a standard lens, expect to pay under $100 for a very low-light camera. Try to find a camera that will match the power pack you plan to use for your viewfinder as well; then you can avoid adding a regulator. Most board cameras will work on 9 or 12 V and only require a few milliamps, so this is an easy requirement.

To ensure that your camera is working properly, using the manufacturer-supplied cables, plug it into

Figure 10-7 Several small CCD camera modules.

the required power source and a composite monitor so you can test the image and focus the tiny lens if necessary. Usually, the lens is adjusted for all but really close focus, but it doesn't hurt to loosen the tiny set screw (if there is one) and fine tune the lens for the sharpest image at a distance of about 10 feet (ft) or so. The lens will typically have threads and you can adjust it by screwing it closer or further away from the CCD image sensor.

Figure 10-8 shows our tiny black-and-white low-lux camera looking at the battery, which is displayed on a small video screen. Our camera is rated for 12 V, but runs just fine on a 9-V battery. Below 8 V, our camera starts to show a degraded image, so it is probably regulated internally for 5 V.

To test your camera's response to invisible infrared light, take any TV or DVD remote control and point it at the camera while you view the output on a video screen. As shown in Figure 10-9, the remote will send bright bursts of infrared light into your camera so you can see it on the video screen. Your eyes cannot see this light at all, but the camera sees it as if you were holding a flashlight. This is the general idea of the night vision system, although it will use many more infrared LEDs to illuminate a much larger area.

The single LED on your TV remote seems bright to your camera, but does not cast enough light to

Figure 10-9 ·Testing the CCD cameras response to infrared light.

be used in a night vision system. To light up a large enough area to navigate, a night vision system will need at least 10 or more infrared LEDs. The good news is that the infrared LEDs are easy to source and can be purchased for a few pennies per unit. Infrared LEDs are typically found in 5-mm plastic cases like the ones shown in Figure 10-10, and will have various plastic colors as well as output strength and angle of view. Some infrared LEDs are completely black, but to the camera, they will seem transparent. The pile of infrared LEDs shown in Figure 10-10 are the same type that are found in TV remotes, although they usually only have one or two. Read entire project text to understand the effects of adding more LEDs to your system before you choose how many you will want to use.

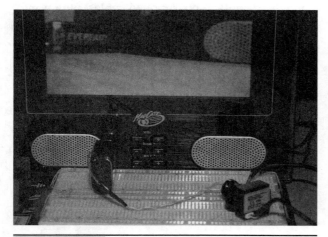

Figure 10-8   Testing a camera with a video monitor.

Figure 10-10   An assortment of infrared LEDs.

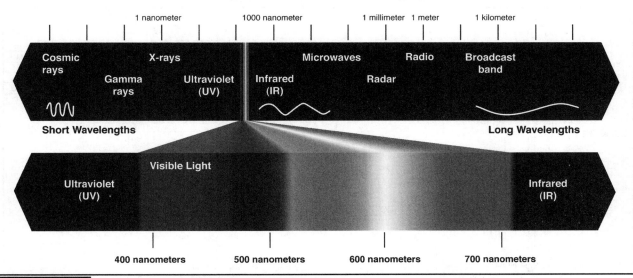

Figure 10-11 The visible and infrared light spectrum.

The light spectrum diagram shown in Figure 10-11 also includes the two far ends of the spectrum where our eyes cannot function; ultraviolet and infrared. Light as a wave is measured in nanometers, with infrared light being the longer wave and ultraviolet the shorter. Our eyes are most sensitive to green light, and we fail to see any light longer than about 740 nm. For this reason, stealth night vision devices should always use infrared light beyond 750 nm in order to achieve complete invisibility to the human eyes. Some infrared LEDs are closer to visible red, falling in the 800 to 880 nm range, and although the camera is also more sensitive to this light, we can see it as a dull red glow. Because the low-lux cameras are so good at seeing most infrared light, it is best to stay with the commonly available 940-nm infrared LEDs that are always used in remote control applications.

Some security cameras use the 880-nm infrared LEDs for illumination because the slight red glow acts as a warning to those who see them, and it gives a little bit more illumination over a long distance. Security cameras often employ color cameras, which are not nearly as good in the dark as their black-and-white counterparts, so this advantage is important. If you are using a low-lux black-and-white camera, the 940-nm LEDs will be just as bright as the slightly visible 880-nm series, so they are the best choice for complete stealth.

When choosing your infrared LEDs, it is important to check the datasheet for the required information. The things you want to know are reverse voltage, continuous forward current, wavelength, and beam pattern. The reverse voltage is the actual voltage you will apply to the LED to make it illuminate. Maximum continuous current is basically the maximum power rating of the LED. This is not to be confused with pulse current. Beam pattern is the angle that the light will come from the plastic lens, although this is not that important since most infrared LEDs have basically the same radiation pattern.

Of all of the technical specs, reverse voltage is the most important one, since you will be placing many LEDs in series in order to run them from a single higher voltage source. You need to know this voltage because exceeding it will likely fry your LEDs, in the same way that being too low will greatly reduce your output power. It's also important that all of your infrared LEDs be from the same source and have identical ratings.

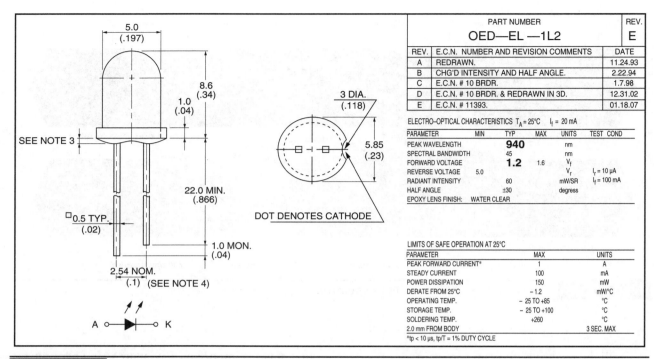

PART NUMBER		REV.
OED—EL —1L2		E

REV.	E.C.N. NUMBER AND REVISION COMMENTS	DATE
A	REDRAWN.	11.24.93
B	CHG'D INTENSITY AND HALF ANGLE.	2.22.94
C	E.C.N. # 10 BRDR.	1.7.98
D	E.C.N. # 10 BRDR. & REDRAWN IN 3D.	12.31.02
E	E.C.N. # 11393.	01.18.07

ELECTRO–OPTICAL CHARACTERISTICS T_A = 25°C I_f = 20 mA

PARAMETER	MIN	TYP	MAX	UNITS	TEST COND
PEAK WAVELENGTH		**940**		nm	
SPECTRAL BANDWIDTH		45		nm	
FORWARD VOLTAGE		**1.2**	1.6	V_f	
REVERSE VOLTAGE	5.0			V_r	I_r = 10 μA
RADIANT INTENSITY		60		mW/SR	I_f = 100 mA
HALF ANGLE		±30		degress	
EPOXY LENS FINISH:	WATER CLEAR				

LIMITS OF SAFE OPERATION AT 25°C

PARAMETER	MAX	UNITS
PEAK FORWARD CURRENT*	1	A
STEADY CURRENT	100	mA
POWER DISSIPATION	150	mW
DERATE FROM 25°C	– 1.2	mW/°C
OPERATING TEMP.	– 25 TO +85	°C
STORAGE TEMP.	– 25 TO +100	°C
SOLDERING TEMP.	+260	°C
2.0 mm FROM BODY		3 SEC. MAX

*tp < 10 μs, tp/T = 1% DUTY CYCLE

Figure 10-12   **A typical infrared LED datasheet.**

Figure 10-12 shows a typical infrared LED datasheet, stating a reverse voltage of 5 V and a continuous current rating of 100 milliamps (mA). Pulse current is shown at 1 amp (A), but you will need some fancy electronic trickery in order to get that much current into the LED without making a mini pyrotechnic show.

There is another source for infrared LEDs that you might want to consider if complete covert operation is not important to you. As shown in Figure 10-13, a typical outdoor night vision security camera will have a decent quality color camera as well as 20 or more infrared LEDs that you can remove from the case. The color camera will not work nearly as well in total darkness as a black-and-white camera, which is why the included LEDs are normally 880 nm rather than 940 nm. They will certainly work, but there will be a dull red glow that is visible to anyone looking at your night vision system from less than 20-ft away. In a completely dark environment, you will compromise your position with 880-nm LEDs.

Figure 10-13   **An alternative to individual LEDs.**

One advantage to using a pre-made night vision camera like the one shown in Figure 10-13 is that you only need to supply power, which is often 12-V DC, so you only need to attach the viewfinder to make a working system. Improvements can be made by replacing the color camera with a better black and white camera, and you could go as far as replacing the 880-nm LEDs with 940-nanometer (nm) LEDs, but it is probably easier to build your system from scratch since LED power requirements will vary.

Before we put the night vision system together, we want to jump off the track a bit and show a few alternatives that can help build up your covert night viewing arsenal. You can generate a lot of infrared light by passing a white light bulb or full spectrum light source through what is called an "infrared pass filter." This system works the same way as a colored light filter, except that only the camera will see the light that passes through the filter. In the 1990s, Sony released a camcorder that had a function they called "night shot," and it was later discovered that this function could actually see through some clothing, depending on the material and lighting conditions. Although the function was removed from all newer camera models, this effect became an Internet "cult classic" known as "X-Ray video" and several companies began supplying adapter lenses they dubbed "X-Ray lenses." These lenses are basically infrared filters like the one shown in Figure 10-14.

Although the X-Ray effect is interesting to experiment with, the more useful purpose of such a lens is to light a large area with infrared light for your night vision camera to see. Depending on the type of lens (glass or plastic), and the light source, you will be able to illuminate areas as small as a room or as large as an entire backyard. It really depends on how much heat your filter can take. You will have to search various Internet sources and scientific supply stores for the appropriate filter, but start by looking for the phrase "infrared pass filters."

The glass filter we sourced was fine with a 50-watt (W) incandescent light source, so a suitable tin can was fitted with a socket as shown in Figure 10-15. A lot of heat will be generated by this system if it is sealed, so make sure that your filter can withstand the heat or you will have to lower the bulb wattage, move it further from the filter, or even make vent holes and install a cooling fan. If light escapes the can, then it will be visible to your subjects, so keep that in mind. If you want to use this type of infrared light source in a portable night vision system, then replace the bulb with one that runs on batteries. Perhaps one of those million candle watt flashlights could be used. (We have not tried that yet.)

Figure 10-14 A 940-nm infrared pass filter glass.

Figure 10-15 Making a housing for a bulb and the filter.

Figure 10-16   The completed high-power illuminator.

Figure 10-17   A 5-mw infrared laser module.

The lens was fitted to the end of the can as shown in Figure 10-16, completing the low-cost long-range infrared illuminator. This system was great for illuminating an outdoor scene for a security camera. To extend the life of the filter, you could also wire the light to a motion-sensing security light so that it only runs when there is motion detected. Since this type of illuminator is power hungry, it is best used in a fixed installation, as a large power source would be needed in order to make it portable.

Lasers can also make effective night vision illuminators, but they are best for very long-range systems and will require your camera to have a magnifying system or zoom lens. Much like the way a visible laser pointer can reach hundreds of feet, so can an infrared laser, as it is basically the same thing. A tiny point is not much use in a night vision system, so the laser collimating lens has to be removed so that the beam can be spread out a lot more. This has the effect of lowering the output power, but also makes it much safer to work with when dealing with dangerous laser light. If you do not have the appropriate laser safety equipment or understand how to work with lasers, then do not attempt these experiments as eye damage can occur, especially if working with lasers that have an output power of more than 5 mW (yes, milliwatts!).

The tiny laser module shown in Figure 10-17 will make a 2-mm wide spot on the wall at 20-ft away

as viewed from the video camera (the only way to see that it is working). By unthreading the tiny lens, the beam can be made much larger, illuminating an area several feet in diameter from about 10-ft away in total darkness. This illumination is not nearly as bright as the LED system, but is good for experimenting with directional or longer range systems that have zoom lenses or have been retrofitted to a telescope or binocular.

An infrared laser with an output power of 500 mW can illuminate a very large area from a 100 ft away or more, but your camera will definitely need to have a long-range zoom or be attached to a device such as a telescope or binocular to be useful. High-power laser modules like the one shown in Figure 10-18 can be

Figure 10-18   A very high-power infrared laser module.

purchased at many electronics supply outlets, or you can attempt to hack the infrared laser module from a green laser pointer as we did. To make a green beam, an 808-nm infrared laser is sent through a series of lenses, filters, and crystals in order to alter the wavelength, so a much higher power laser diode is needed. We broke into a 5-mW green laser pointer and ended up with the 500-mW 808-nm laser module shown in Figure 10-18. It took many hours to carefully file the brass casing away to extract the infrared driver module.

Be very cautious when working with any laser at power levels beyond 5 mW, as instant and inseparable eye damage will occur if you are careless. This danger become even worse due to the fact that you cannot see the beam, and your eye will not blink if you hit yourself. A 500-mW laser will give you an instant blind spot, so keep that in mind. Read up on laser safety first!

It is fairly easy to make a high-power laser diode cast a large beam, and although there are expensive precision lenses available to do this job, we have found great success using random lenses taken from scrap cameras and video systems. Figure 10-19 shows a few of the assorted lenses

that were found to make a decent beam when placed in front of the high-power laser diode and viewed on the video screen. Our goal was to connect a video camera to a telescope or binocular for some long-range covert surveillance at 100 ft or more, so it took a bit of experimenting with the position of the lens in front of the diode.

The lens was fastened to the high-power laser module so that it would light up an area of about 20 ft² at a distance of about 50 to 100 ft (Figure 10-20). Of course, only a camera connected to a telescope would work, but that was the original goal. You may also notice that the light cast from the laser diode is an odd rectangular shape when cast for such a long distance. This effect is normal due to the way a laser diode is constructed and will actually work in your favor.

The high-power illuminator is hungry for power, so a large battery pack was added, and the unit was placed into a box with a switch (Figure 10-21). It's fine to use larger batteries of the same amperage, but never more voltage. Although the beam is many hundreds of times less focused, the illuminator is still to be treated with caution, only viewing the beam on the monitor and never using it for "live" covert operations where living eyes may be in the beam. There are actually laser illuminators just like

Figure 10-19   Various lenses taken from old cameras.

Figure 10-20   Beam spreading lens fit to the laser module.

Figure 10-21   Completed high-power laser illuminator.

this one for sale on the market the claim to be "eye safe," but we still wouldn't take any chances.

To work with the long-range laser-based illuminators, you will need to extend the range of your video camera. Binoculars or a telescope work perfectly for this once you determine the optimal position and focus for the camera. To mount the camera to a pair of binoculars, cut a hole in the lens cap and simply glue the camera to it. Figure 10-22 shows the small video camera mounted to the binoculars through the lens cap. This is nice because you can still use the other side to aim the camera before switching on the laser illuminator. You may need to adjust the camera focus or carefully choose the spot behind the lens to get optimal focus.

If your illumination system can reach a hundred feet or more, a telescope will be needed in order to zoom into the area that is illuminated by the laser. The rig shown in Figure 10-23 works well with the high-power laser illuminator and the camera is held in place by gluing it to one of the Barlow lens adapters. Some focusing of the camera may be necessary, and you may only get a small area of view, but this system does work well for going the long distance, if you can light up the scene. Even without infrared illumination, these low-lux cameras can outperform the human eye for low light viewing any day.

Now, back to the night vision system. Once you have collected all of the components that make up the night vision system, you can source a cabinet that will hold everything, including a battery pack. Black is always best when attempting to be stealthy as this will avoid reflections from any ambient light sources if you are moving around. Flat black is even better as it ensures a zero reflective surface.

Figure 10-22   The lens cap makes a good camera mount.

Figure 10-23   A very long-range night vision camera.

With proper clothing, you could slip through the night completely undetected by your targets, even at close range.

One of the typical Radio Shack black boxes was found to hold all of the parts, and it is $5 \times 2.5 \times 2$ in in size. Remember to account for your battery (or battery pack), a few switches as well as a bunch of wiring when looking for a project cabinet. It may be best to build the entire circuit before looking for a cabinet, and then you will know how much room you need inside the cabinet. Don't try to cram things in too tight, or it will be highly annoying to change the battery or make adjustments later (Figure 10-24).

The actual wiring and number of LEDs in your system is completely dependent on your battery pack voltage and the maximum voltage ratings of a single LED. The idea is to have all of your LEDs run from the same voltage that is needed for the rest of the system, which will most likely be either 9 or 12 V. Since most LEDs are rated between 1 and 2.5 V, you will need to wire them in series and parallel in order to achieve the best voltage. Also, keep in mind that more LEDs will require more power from your battery pack, and the range of your system will not increase greatly as you add more LEDs, only the width of the area you are illuminating. Feel free to experiment with LED placement and quantity, but any number of LEDs from 10 to 20 will be just fine for the night vision viewer.

Look at your LED datasheet and find the absolute maximum voltage rating. For our infrared LEDs, it was 1.4 V. Now, divide your battery (or battery pack) voltage by the LEDs' maximum voltage rating, and that number will be your series chain quantity. By placing LEDs in series, you divide the voltage to a safe level. For our single 9 V battery: $9/1.4 = 6.428$. Of course, we cannot have 6.428 LEDs in our system, so the number is rounded up to 7 to go slightly under the maximum voltage rating rather than exceeding it. It is always best to be slightly under the maximum rating because slightly over will kill your LEDs in a hurry. So with 7 series LEDs, each one will see 1.23 V ($9/7 = 1.285$), and that is a safe voltage.

Since only 7 LEDs might not cast enough light, we decided to add another series chain of 7 LEDs to the system. Once you have a single series chain, you can simply multiply it and add a second chain in parallel with the other one. By doing this, both chains divide the voltage correctly to the safe level, and your system uses twice the current. Figure 10-25 shows the two series LED chains connected in parallel so that every single LED will only see 1.23 V. We could add more series chains in parallel with the first two, but 14 LEDs was a good trade-off between battery life and illumination. Again, feel free to experiment, but always be mindful of your LEDs' maximum voltage rating.

You can install your LEDs in any pattern you like, but try to spread them out as much as possible on your project case for maximum coverage. Also remember that you need to leave room for the camera and any switches that you plan to install. To make the night vision system look more professional, a paper template was made using a ruler or a computer paint program taped to the lid or box shell to help when drilling out the LED

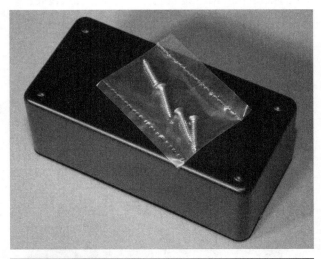

Figure 10-24 Choosing a cabinet for all of the parts.

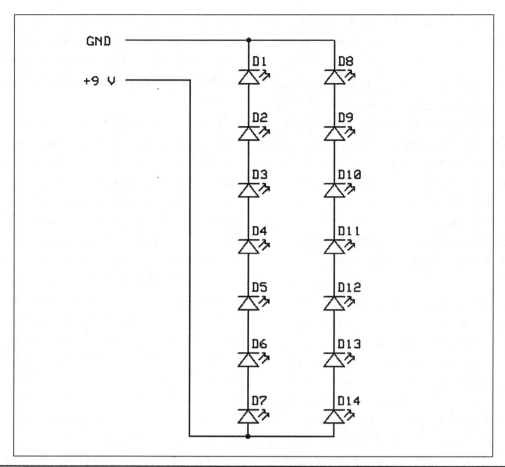

Figure 10-25 Series and parallel LED wiring.

holes. Figure 10-26 shows the template taped to the lid and ready to be drilled.

If you are not sure which drill bit is needed to make a hole that the LEDs will fit snugly into, start with an obviously smaller bit and then work your way up to determine the best size. It's also best to drill all holes initially with the smaller bit in your set to keep the drilling aligned to the hole center. Figure 10-27 shows the 14 LED holes drilled from the lid of our project box.

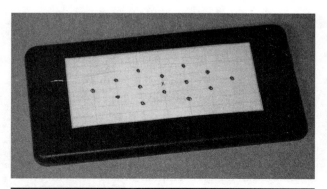

Figure 10-26 Using a template to drill the LED holes.

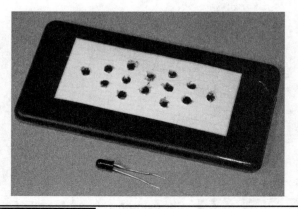

Figure 10-27 Drilling the LED holes.

Figure 10-28   LED holes drilled.

Figure 10-30   Install the LEDs in the same orientation.

The edges of the drilled holes can be cleaned up after drilling by using a much larger bit turned by hand to take away the edges a bit. Figure 10-28 shows the box lid ready to have the 14 LEDs press fit into the holes.

It's a good idea to trim the leads of each LED to a minimal length as shown in Figure 10-29 so you have more room to work with inside your cabinet. You might have known that the longer lead of an LED is the positive lead, but there is another way to check the polarity of the led. The flat side of the bottom edge is the negative side. This is shown in Figure 10-30. If your LEDs are all in line with each other, you could even bend the leads to each other to create some of the series wiring as well.

Install all of the LEDs so that the positive and negative connections all face the same orientation. As shown in Figure 10-30, the flat side of the case

on an LED is the negative side, so that makes it easier to wire all of the LEDs without having one or more in reverse. If you get a single LED in a series chain reversed, the entire chain will fail to light. No damage will occur, but it simply won't work.

The LEDs do not have to be evenly spaced around the camera as long as the camera points in the same direction of the LEDs. We installed the camera in the top left corner to allow more room for the other components as shown in Figure 10-31. One thing to keep in mind when assembling your system is which side of your camera as well as your viewfinder is up. It would be really annoying to put the entire project together and then get an upside down image! Also, some viewfinders have a "mirror image" setting, so if you plan to navigate a room in complete darkness, make sure that the image is not reversed horizontally.

Figure 10-29   LED leads trimmed to a minimal length.

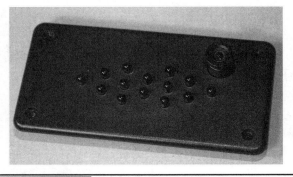

Figure 10-31   Installing the camera.

Figure 10-32   Installing the viewfinder.

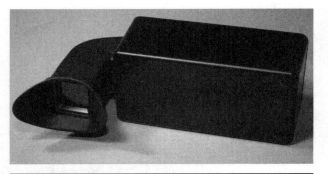

Figure 10-34   Rear view of the night vision system.

Although you can use the viewfinder from either eye, it is best to position it so that the box crosses your face, placing the camera lens in the center of view, rather than off to one side. It will take some time to easily navigate an obstacle-filled room using only the viewfinder, but this task is made a lot easier with the camera looking out from between your eyes. Also, keep in mind which way the picture will be when you install the viewfinder so that it is not upside down. Figure 10-32 shows how our viewfinder was installed onto the box—by cutting out a slot so that it would slide in and stay secure. Hot glue also works well for holding any loose parts together or even the circuit board to the base of the box.

Our night vision system is a left eye version since that is our good eye! The camera is positioned so that it will end up between our eyes when we have the night vision system up to our face. Figure 10-33 shows the basic system with the three main components installed.

Figure 10-34 shows the rear of the unit with the viewfinder installed. When positioning your viewfinder, you also have to consider your nose, as it may end up pressing against the back of the box if the viewfinder is not back far enough. If you can't make room for your face, a side mount may be better.

When you're at the stage to wire up the viewfinder, camera, LEDs, and battery pack, take your time and consider adding an extra switch to turn off the LEDs and the main system separately. By doing this, your night vision system becomes a multi-purpose device, able to see in the dark, as well as hunt down other night vision security cameras. As will be shown in following photos, the night vision viewer can see other security cameras when its own infrared LEDs are turned off, so it functions as a kind of night vision camera detector. The LEDs have been wired according to the schematic shown in Figure 10-35.

Figure 10-33   The basic hardware installed.

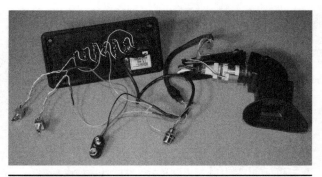

Figure 10-35   Wiring up all of the components.

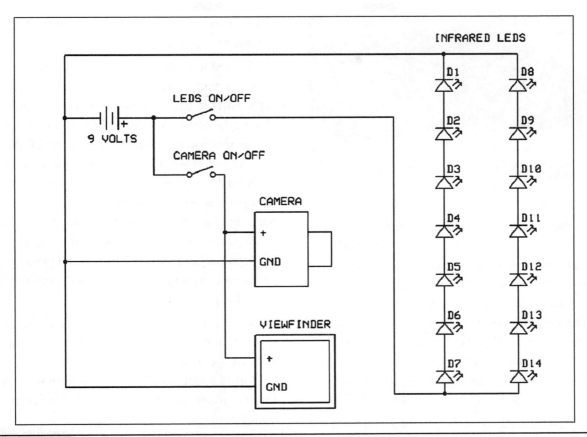

Figure 10-36   Wiring the components to the power source.

We also added an RCA output jack so that we could record the video output into a camcorder to demo this unit on our Web site as well as record certain covert activities. Figure 10-35 shows the system completely wired, including the two power switches, battery clip, and video output jack.

Figure 10-36 shows a typical power wiring diagram for the night vision viewer if you decide to add switches to both the video section and the LED section. Your wiring diagram may differ depending on the number of LEDs in your system, as well as the required voltages for the camera and viewfinder. It is certainly okay to use more than one battery or even a voltage regulator if you need to. A single 9-V alkaline battery will power our system for a few hours, but a pack of AA batteries would run it for many hours at the expense of using a larger cabinet. Rechargeable batteries are great for this project, but keep in mind

that they often have lower voltages than the type of battery they are intended to replace. A rechargeable 9-V battery outputs around 8.6 V typically.

Once all of the wires and switches were added, the box that once looked a bit large was almost completely filled. Figure 10-37 shows all of the

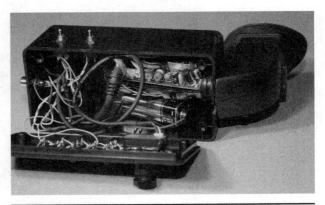

Figure 10-37   Fitting all the parts into the box.

Figure 10-38   The completed night vision system.

parts being jammed into the box after checking that the unit actually functioned. Loose parts such as the viewfinder circuit board and battery should be secured with a bit of double-sided tape or some other means so that they don't bounce around. Another thing to do is insulate any part of the circuit board from any other metal parts such as the battery case. A bit of tape or cardboard placed on the back of the board works well for this. You can also fill voids with some bunched-up paper towel or cotton balls.

After precariously cramming all of the components into the box, the unit is tested to ensure that nothing was shorted or broken during the installation. Figure 10-38 shows the completed

night vision system, ready to go hunting deep into the pitch darkness.

The first test you should do with the night vision system is to stand in front of a mirror to see if the infrared LEDs are working. Figure 10-39 (Frame 1) shows the brilliant beam of light that the camera will see when looking at the LEDs at such a close range. Frame 2 shows our lab with all of the lights turned off. Frame 3 is a shot of a dark parking lot at night and Frame 4 is a vicious, yet oblivious watchdog who can only hear us but not see us. Okay, that's just our buddy "The Prince of Dogness," and we don't recommend you try to sneak past any animal because it will probably end badly for you!

When traversing an indoor space using only night vision, take your time and get used to the field of view. It took us a day or two to move smoothly around a closed space without banging into furniture and door frames, but it can be done.

If you added an optional switch to turn the LED array on or off, then you can also use your night vision viewer to hunt down other sources of

Figure 10-39   A few tests with the night vision viewer.

Figure 10-40   Using the device to hunt down security cameras.

Figure 10-41   Ready to cut through the darkness!

infrared light such as hidden spy cams or security cameras. Figure 10-40 shows how a night vision camera looks to a still camera (using the flash at night) and how it looks using the night vision viewer. The still camera sees a dull purple glow from the security camera, but the night vision viewer sees it as a brilliant beam of white light. Covert cameras like this can be spotted from a great distance due to the amazing sensitivity of the low-lux camera to infrared light.

The night vision viewer actually has a third function as well—to swamp out security cameras so they cannot see you. By switching off the viewfinder and just leaving the infrared LEDs on, you can simply point the unit at a security camera as you walk past and it will completely wash out the picture, making it impossible to see who you are. This is great when you value your own privacy (Figure 10-41).

We hope you enjoyed building the night vision viewer, and may all of your covert operations succeed! Don't get into too much trouble with your new toy, and if you do, always remember, "You never heard of us or this Web site!"

PART THREE
Telephone Projects

Phone Number Decoder

WHEN YOU PRESS THE NUMBERS on your phone to make a call, the tones you hear are called Dual Tone Multi Frequency tones (DTMF), and they represent 1 of 16 possible numbers on the keypad. The DTMF tones are made up of frequency pairs ranging from 697 to 1633 Hertz (Hz). There are actually eight frequencies that form a 4 × 4 matrix of 16 possible tones by mixing the pairs together. DTMF signaling has proven to be a very robust method of sending data over the telephone system, and it has become a universally excepted standard.

Figure 11-0 This device will decode the DTMF tones from the phone line or audio source.

PARTS LIST

- IC1: CM8870 DTMF signal decoder
- IC2 (optional): 74LS154 4 to 16 line decoder
- IC3 (optional): 74LS240 inverting buffer/driver
- X1: 3.58 MHz crystal for CM8860
- LEDs (optional): 16 low-current LEDs with 1-K resistors
- Resistors: R1 = 100 K, R2 = 100 K, R3 = 300 K
- Capacitors: C1 = 0.1 µF ceramic

In order to decode the tones back into a 4-bit (16 possible combinations) signal, some type of high-speed and accurate signal processor is needed. A fast microprocessor can be programmed to decode DTMF signals, but it will take a lot of processing power and careful analog signal control in order to make a reliable DTMF decoder in software. There are, however, inexpensive single integrated circuit (IC) solutions for touch tone

decoding. This project is based on one called the CM8870 DTMF decoder chip. Several methods of decoding DTMF tones back into digital pulses will be shown here, allowing devices to be controlled over the phone line or allowing you to spy on numbers being dialed live over the phone or from prerecorded audio clips.

The CM8870 DTMF decoder IC shown in Figure 11-1 only needs an external 3.579-megahertz

Figure 11-1 The CM8870 DTMF decoder chip is a standalone DTMF decoding solution.

(MHz) crystal oscillator and a few passive components in order to decide DTMF signals on the fly. There are actually several makers of this chip, so when you are scouring for this part, the name "8870 DTMF Decoder" will be a close enough description. The reason for the odd 3.579-MHz crystal is that this value is the same as the NTSC color burst frequency used in analog television and is very common and easy to source. The 8870 DTMF decoder does all of the difficult work for you, taking an analog audio signal at its input and then converting the decoded tones back into a 4-bit binary value that can represent 1 of 16 possible values. Most phones only have nine buttons on the keypad, but there are 16 DTMF combinations.

The 8870 DTMF decoder sheet shown in Figure 11-2 includes an example circuit as well as a table showing how the binary output represents 1 of 16 possible DTMF tones. There are actually five output pins used on the 8870 decoder—four of them represent the binary data and a fifth pin toggles from low-to-high and then back to low every time a valid DTMF signal has been decoded.

This allows your receiving circuit or device to know when a number has been repeated. Without this data-receive pin, you would have no way to know if the same key has been pressed multiple times as the last data on the 4-bit output will simply remain the same. So every time a new DTMF tone has been decoded and sent to the 4-bit binary port, the data ready pin will toggle to high for a short time. On our 8870 version, the binary data output pins are marked Q1, Q2, Q3, and Q4, and the data-ready pin is marked as STD on the DTMF decoder chip.

To get DTMF data into the 8870 decoder chip, pins 1, 2, 3, and 4 are set up as an analog input as shown in the datasheet example. The audio signal is processed by a robust signal processor inside the 8870, and any DTMF tones heard will be registered on the data output port. The input is very versatile and can be directly connected to the phone line or to any audio playback device such as computer sound card or digital recording unit. Basically, any device capable of sending an audio signal can be fed into the 8870 DTMF decoder, and as long as the signal quality is decent, the DTMF tones can be decoded. Cordless phones and analog cell phones

CALIFORNIA MICRO DEVICES ▶ ▶ ▶ ▶ ▶   **CM8870/70C**

CMOS Integrated DTMF Receiver

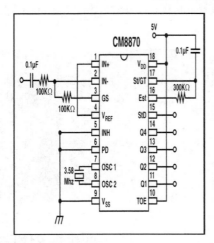

F$_{LOW}$	F$_{HIGH}$	KEY	TOW	Q$_4$	Q$_3$	Q$_2$	Q$_1$
697	1209	1	H	0	0	0	1
697	1336	2	H	0	0	1	0
697	1477	3	H	0	0	1	1
770	1209	4	H	0	1	0	0
770	1336	5	H	0	1	0	1
770	1477	6	H	0	1	1	0
852	1209	7	H	0	1	1	1
852	1336	8	H	1	0	0	0
852	1477	9	H	1	0	0	1
941	1209	0	H	1	0	1	0
941	1336	.	H	1	0	1	1
941	1477	#	H	1	1	0	0
697	1633	A	H	1	1	0	1
770	1633	B	H	1	1	1	0
852	1633	C	H	1	1	1	1
941	1633	D	H	0	0	0	0
-	-	ANY	L	Z	Z	Z	Z
L = logic Low, H = Logic High, Z = High Impedance							

Figure 11-2   The CM8870 DTMF decoder datasheet showing a recommended circuit.

that are not scrambled are particularly vulnerable as a cheap radio-frequency (RF) scanner connected to the DTMF decoder will show the listener every button pressed in real time.

There are several ways you can build your DTMF decoder. You can feed a line from the telephone jack to the 8870 to decode tones pressed on any phone on the same line, or you can feed the output from some audio source such as an RF scanner, digital recorder, or computer sound card directly into the DTMF decoder. You could build a DTMF decoder that uses both methods, including an RJ-11 telephone jack and an audio input jack. For testing purposes, it is much easier to feed the signal from your telephone system into the DTMF decoder, so that you can use the phone keypad to verify the operation of your circuit.

Using the 8870 DTMF decoder IC and a breadboard, build up the circuit as recommended in the data sheet, which will include a 3.579-MHz crystal resonator and a few passive components (Figure 11-3). For the 300-K resistor, a more common value of 270 or 220 K will work just fine. The 0.1-microfarad (μF) ceramic capacitor and 100-K resistor in the input removes all DC offset, leaving only the AC audio signal for the DTMF detector. This setup allows a wide range of audio devices to be fed into the input without having to worry about impedance matching or the level of the

output. You could even run an electret microphone into the input and still get a valid DTMF decode by simply placing the speaker from an audio recorder next to the microphone. The 8870 DTMF decoder really works well on a wide range of devices.

To test the 8870 DTMF decoder using your phone line, you will need an RJ-11 female socket like the one shown in Figure 11-4 running into the breadboard. You can also cut the end off a standard phone cord and then insert the wires directly into the breadboard. A standard residential phone cord will have four wires, but only the two innermost wires are used. There will be a red wire and a green wire, and these two wires will carry the signal to your telephone terminal block. Be aware that there is a low-amperage 40-volt (V) potential on the phone line that jumps up to over 100 V when the phone is ringing. There is very little current in the phone line, but the high voltage is enough to give you a painful tingle if you happen to be touching both wires as the phone begins to ring. Keep this in mind as you work with live phone cables!

Also shown in the breadboard is a set of five light-emitting diodes (LEDs) that connect through 1-K current limiting resistors and then to the five output pins on the 8870 DTMF decoder chip. When a valid DTMF tone is heard, the data-ready LED will flick for a few milliseconds, and then the

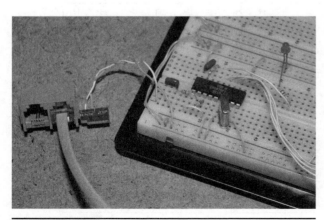

Figure 11-3 Breadboarding the DTMF decoder circuit using the CM8870 chip.

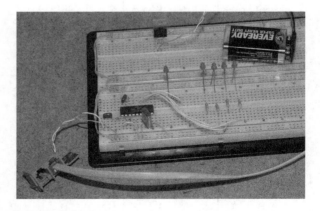

Figure 11-4 An RJ-11 telephone jack feeds the audio from the phone into the decoder.

four data LEDs will change to represent a binary number from 1 to 16. Figure 11-2 shows the results of the binary outputs (Q1 to Q4) and how they correspond to one of the 16 DTMF tones.

The output from the 8870 DTMF decoder is a 4-bit binary value, which is the perfect method of transferring the data to a microcontroller or computer. Of course, this binary value is not much use for controlling external devices with a relay or driver transistor unless you decode it back from 4-bit binary data to a single digital input-output (IO) line. An easy way to get from 4-bit binary data back to a single digital IO line is by using a 74154

4-line to 16-line decoder chip. The 74154 decoder takes 4-bit binary data and converts it to a 16-pin output port. Unfortunately, the data is inverted so that only 1 out of 16 outputs is off at any one time, so if you need to use the IO line to switch on another device, you will have to invert each of the 16 IO lines. Looking at the datasheet for the 74154 decoder shown in Figure 11-5, you can see that only one of the 16 outputs will be off at any one time. The 74154 is a common logic IC, and will have several other part numbers such as 74LS154, 74HC154, DM54154. For this project, any of the 4-16 decoder variants will work just fine.

54154/DM54154/DM74154
4-Line to 16-Line Decoders/Demultiplexers

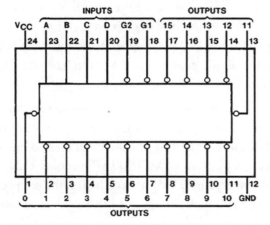

Features

- Decodes 4 binary-coded inputs into one of 16 mutually exclusive outputs
- Performs the demultiplexing function by distributing data from one input line to any one of 16 outputs
- Input clamping diodes simplify system design
- High fan-out, low-impedance, totem-pole outputs
- Typical propagation delay
  3 levels of logic 19 ns
  Strobe 18 ns
- Typical power dissipation 170 mW
- Alternate Military/Aerospace device (54154) is available. Contact a National Semiconductor Sales Office/ Distributor for specifications.

Inputs						Outputs															
G1	G2	D	C	B	A	0	1	2	3	4	5	6	7	8	9	10	11	12	13	14	15
L	L	L	L	L	L	L	H	H	H	H	H	H	H	H	H	H	H	H	H	H	H
L	L	L	L	L	H	H	L	H	H	H	H	H	H	H	H	H	H	H	H	H	H
L	L	L	L	H	L	H	H	L	H	H	H	H	H	H	H	H	H	H	H	H	H
L	L	L	L	H	H	H	H	H	L	H	H	H	H	H	H	H	H	H	H	H	H
L	L	L	H	L	L	H	H	H	H	L	H	H	H	H	H	H	H	H	H	H	H
L	L	L	H	L	H	H	H	H	H	H	L	H	H	H	H	H	H	H	H	H	H
L	L	L	H	H	L	H	H	H	H	H	H	L	H	H	H	H	H	H	H	H	H
L	L	L	H	H	H	H	H	H	H	H	H	H	L	H	H	H	H	H	H	H	H
L	L	H	L	L	L	H	H	H	H	H	H	H	H	L	H	H	H	H	H	H	H
L	L	H	L	L	H	H	H	H	H	H	H	H	H	H	L	H	H	H	H	H	H
L	L	H	L	H	L	H	H	H	H	H	H	H	H	H	H	L	H	H	H	H	H
L	L	H	L	H	H	H	H	H	H	H	H	H	H	H	H	H	L	H	H	H	H
L	L	H	H	L	L	H	H	H	H	H	H	H	H	H	H	H	H	L	H	H	H
L	L	H	H	L	H	H	H	H	H	H	H	H	H	H	H	H	H	H	L	H	H
L	L	H	H	H	L	H	H	H	H	H	H	H	H	H	H	H	H	H	H	L	H
L	L	H	H	H	H	H	H	H	H	H	H	H	H	H	H	H	H	H	H	H	L
L	H	X	X	X	X	H	H	H	H	H	H	H	H	H	H	H	H	H	H	H	H
H	L	X	X	X	X	H	H	H	H	H	H	H	H	H	H	H	H	H	H	H	H
H	H	X	X	X	X	H	H	H	H	H	H	H	H	H	H	H	H	H	H	H	H

H = High Level, L = Low Level, X = Don't Care

Figure 11-5   The 74154 will decode a 4-bit number back to 1 of 16 outputs.

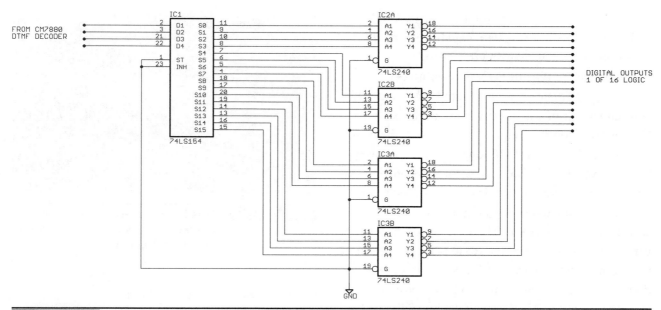

Figure 11-6 Decoding the 4-bit binary data back to positive 1 of 16 logic.

To invert all of the outputs on the 74154 decoder so that they can be used to trigger external devices such as relays or logic pins, you will need 16 inverters. You could use any type of logic inverter such as a 7404, but the easiest method is to add a pair of octal inverting buffers onto the outputs to invert them and offer some driving current for a small relay or driver transistor. The schematic shown in Figure 11-6 uses a pair of octal inverting buffers (74LS240) to invert the 74154 output signal so that only 1 of the 16 possible outputs is high at any given time. Since the 74LS240s have 8 inverters per chip, you will need two of them to invert all 16 IO lines from the 74154 decoder. Now, the output from the 8870 can be fed through the 4-16 decoder and inverters so that only a single IO line will be high for each of the 16 possible DTMF tones decoded.

Once the 4-16 decoder and inverters are connected to the DTMF decoder, you now have a system that will turn on 1 of 16 possible IO lines corresponding to the numbers on the telephone dial pad. Of course, the telephone only has 12 digits, so four of the IO lines will be unused unless you have made your own DTMF tone generating system as well. The circuit shown breadboarded in Figure 11-7 will switch on one of the 16 LEDs each time a digit is pressed on the phone. This system forms the basis of a remote control system that can turn on devices via telephone command. This basic system can also decode incoming touch tones and display them in real time once you have each LED labeled to show which of the dial pad buttons it represents.

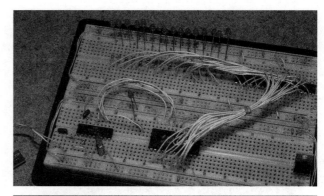

Figure 11-7 Each of the 16 LEDs corresponds to a single digit on the telephone dial pad.

DTMF signaling is very robust, so you could make a remote control system capable of switching 16 different devices by using this decoder circuit and some type of DTMF generator. A microcontroller can easily generate DTMF tones if you send a pair of square waves out of the digital IO pins based on the frequencies shown in the data sheet (Figure 11-2). DTMF tones can be sent over low-bandwidth radio links, over long-distance wires, or even through the air using infrared pulses. The 8870 is able to decode DTMF tones very accurately, even when the signal-to-noise ratio is very high. This is one of the reasons it has been in use on the telephone system for such a long time. Figure 11-8 shows one of the 16 LEDs turned on after pressing a number on the telephone dial pad.

If you intend to use a microcontroller to respond to the DTMF data or want to display it on some type of LCD screen, then you do not need to use the 4-16 line decoder or the inverters. The 4-bit binary data can be sent directly to the input pins on any microcontroller as it is, which will reduce the number of components needed to make a microcontroller-based DTMF control system. The next part of this project will take the binary data right from the 8870 DTMF decoder and display it in real-time on an LCD screen.

LCD screens are very easy to connect to just about any microcontroller. The small 16 character by two line LCD screen shown in Figure 11-9 needs only a single serial input in order to display characters. LCD screens that use serial, parallel,

Figure 11-9   This is an easy-to-interface 16 character by two line LCD screen using a serial input.

or I2C are easy to work with because many microcontrollers already support these languages, or have routines built into their compilers to add LCD support. We had this serial LCD in our parts box, and it was a perfect choice to display the data decoded in real-time by the 8870 DTMF decoder chip. Now, all we had to do was feed the five IO lines (4-bit binary and data ready) into some kind of microcontroller and then write a simple program to send out the serial data to the LCD screen as tones are decoded.

We decided to leave the 4-16 line decoder, inverters, and LEDs on the breadboard and then add the LCD screen and microcontroller into the mix so that both systems were still functioning at the same time. The little breadboard was running out of room so we grabbed an Arduino and just popped the wires into the breadboard from the 8870 and then out to the Arduino header. Normally, we would use a bare microcontroller on the breadboard, but for convenience it's hard to beat these little prototype boards as they are ready to rock as soon as you plug them in.

Arduino is an open source hardware platform that allows an Atmel microcontroller to be programmed through the USB port using a very easy-to-use IDE and C-type language (Figure 11-10). We purchased the Arduino at SparkFun.com, and had it up and running in a few minutes by simply plugging it into the USB port on the computer. The pin headers on the Arduino board allow access to the AVR digital IO

Figure 11-8   Testing the 16 LED DTMF decoder system using the telephone dial pad.

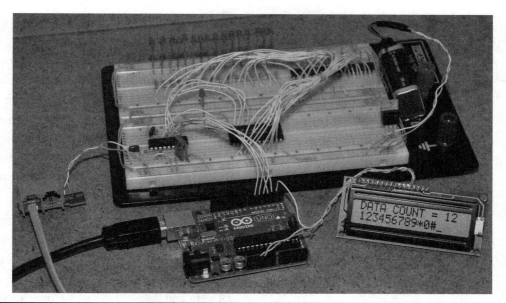

Figure 11-10   An Arduino UNO will receive DTMF data and send serial data to the LCD.

ports so connecting to a breadboard is just a matter of plugging in the necessary wires. The program is also very simple. It just waits for the data ready line on the 8870 to toggle and then reads the 4-bit binary data and then writes the corresponding dial pad digit out to the LCD screen using the serial write command. The program also keeps count of the number of DTMF tones decoded since the last reset.

Here is the Phone Number Decoder Arduino source code:

```
/////////////////////////////////////
// CM8870 TOUCH TONE DECODER BY RADBRAD
FOR LUCIDSCIENCE.COM
/////////////////////////////////////

// PROGRAM VARIABLES
int dtmf;
int cntr;
String dial(16);

void setup() {
dial = "";

// CM8870 BINARY INPUT PORT
pinMode(2, INPUT);
pinMode(3, INPUT);
```

```
pinMode(4, INPUT);
pinMode(5, INPUT);

// CM8870 BINARY DATA VALID PIN
pinMode(6, INPUT);

// INITIALIZE SERIAL LCD SCREEN
Serial.begin(9600);
delay(100);

// CLEAR THE LCD SCREEN
Serial.print(12, BYTE);
delay(100);
Serial.print(12, BYTE);
delay(100);

// TURN ON BACKLIGHT
Serial.print(17, BYTE);
delay(100);

// SEND THE READY MESSAGE
Serial.print("* SYSTEM READY *");

}

// DISPLAY DATA FROM DTMF DECODER
void loop() {
```

```
// WAIT FOR DATA VALID SIGNAL
if (digitalRead(6) == HIGH) {

    // DECODE CM8870 DATA
    dtmf = 0;
    if (digitalRead(2) == HIGH) dtmf =
    dtmf + 1;
    if (digitalRead(3) == HIGH) dtmf =
    dtmf + 2;
    if (digitalRead(4) == HIGH) dtmf =
    dtmf + 4;
    if (digitalRead(5) == HIGH) dtmf =
    dtmf + 8;

    // CLEAR THE LCD SCREEN
    Serial.print(12, BYTE);
    delay(100);

    // DISPLAY THE COUNTER
    Serial.print("DATA COUNT = ");
    Serial.print(cntr);

     // CARRIAGE RETURN
    Serial.print(13, BYTE);

    // DISPLAY DTMF DATA
    if (dtmf == 1) dial = dial + "8";
    if (dtmf == 2) dial = dial + "4";
    if (dtmf == 3) dial = dial + "#";
    if (dtmf == 4) dial = dial + "2";
    if (dtmf == 5) dial = dial + "0";
    if (dtmf == 6) dial = dial + "6";
    if (dtmf == 8) dial = dial + "1";
    if (dtmf == 9) dial = dial + "9";
    if (dtmf == 10) dial = dial + "5";
    if (dtmf == 12) dial = dial + "3";
    if (dtmf == 13) dial = dial + "*";
    if (dtmf == 14) dial = dial + "7";
    Serial.print(dial);

    // INCREMENT THE COUNTER
    cntr++;

}
}
```

The DTMF display program is extremely simple, but does demonstrate how robust and easy it is to interface the 8870 DTMF decoder to any microcontroller. With a few more lines of code and an external relay, you could easily look for the telephone ring signal, have the microcontroller pick up the line, and then wait for DTMF commands to be sent, creating an entire telephone-operated automation system. A lot can be done with the 12 digits on a phone dial pad, especially if the microcontroller is also sending back some type of audio feedback as you dial in your commands.

The completed DTMF display system is shown in Figure 11-11, counting and displaying the digits pressed on the LCD as we type them on the telephone dial pad. The five IO lines running from the breadboard into the Arduino board are the 4 bits of data and data-ready lines from the 8870 DTMF decoder. We still have the 4-16 line decoder, inverters, and LEDs connected on the breadboard, but the microcontroller does not need any of that hardware in order to operate, just the 8870 and its passive components. Even with a very noisy phone line, We can press numbers on the dial pad as fast as we want, and the LCD never misses a single digit. The 8870 DTMF decoder IC is very robust and will accept practically any type of audio signal.

If your intent is to "capture" a phone call and then later display the DTMF data on the LCD screen, then you only need to feed the output from your recording device into the input section of the

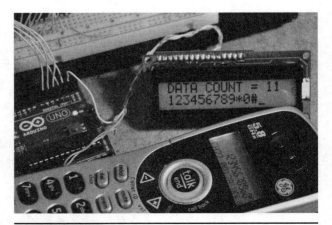

Figure 11-11   Decoding DTMF tones from the phone and displaying them on the LCD screen.

8870 decoder, and it will operate as if connected directly to the phone line. You can even use an electret microphone and just let it listen to a nearby audio signal that contains a DTMF tone and if the signal is loud enough, the 8870 will have no problem registering and decoding the data. It's actually quite difficult to make the 8870 DTMF decoder fail; you need a really bad signal for it to give up on the conversion.

To make a versatile DTMF decoding unit that will display digits in real time as they are captured, a circuit that includes both an RJ-11 female telephone jack and some kind of audio input jack as shown in Figure 11-12 will allow decoding from a variety of formats. For live telephone use, the RJ-11 cord is plugged into any wall jack and will show digits pressed from any phone in the house or from any remotely located phone on the current call. The 1/16 audio jack allows a small digital audio recorder or computer sound card to be plugged in for decoding of DTMF signals already captured and stored for analysis.

The 8870 DTMF decoder is so robust that we were able to feed the audio from our computer into the decoder unit and pull telephone numbers out of

Figure 11-12   The completed DTMF decoder accepts both a phone cord and an audio input.

YouTube clips that contained the sound of telephone dialing! It was also possible to feed the audio from the headphone jack on an RF scanner into the device and see people dial telephone numbers on older cordless phones that did not include a scrambling system to make them secure. So, the next time you enter your password or banking information over a phone system, ask yourself if your line is secure from eavesdropping. You just never know who may be listening to your call. With a simple DTMF decoder, your passwords and credit card numbers are extremely easy to decode.

PROJECT 12

Spammer Jammer

WHY IS IT STILL LEGAL for telemarketers to invade a person's privacy baffles our mind! Here you are either trying to relax during dinner or busy with a soldering iron on a 144-pin FPGA and then right at the worst moment possible…rrrrrring! So, you drop whatever you were doing to go get the call and to your absolute disgust, it's another spammer trying to sell you some useless product, or worse—an automated message telling you to "Hold on for an important message." Can you imagine the nerve? They so blatantly destroy *your* peace and quiet to put *you* on hold as if *your* time is not nearly as important as the cheesy redirect they are about to spew into your already angry ears! Oh, did we mention that of all things in this life that we find annoying, phone spammers top our list?

Figure 12-0 | This little box will give those pesky telemarketers an earful of fun!

PARTS LIST

- IC1, IC2: LM555 analog timer
- Resistors: R1 = 10 K, R2 = 10 K, R3 = 4.7 K, R4 = 220 K, R5 = 10 K
- VR1: 50-K variable resistor
- Capacitors: C1 = 1 µF, C2 = 0.1 µF, C3 = 0.01 µF, C4 = 0.1 µF, C5 = 0.1 µF
- Battery: 9- to 12-V battery or pack

This simple project will give those tele-spammers exactly what they are trying to give you—an earful of useless and highly annoying noise. "You're mean, they are just doing their jobs." Well, let us tell ya buddy, they can do some other job that doesn't involve ticking us off, otherwise they will become victims to whatever we decide to feed

into our own phone line back at them! If you are like us and have no mercy for those who choose to invade your privacy, then this little box will be right up your alley as it sends a very loud warbling alarm sound back into your phone lines, giving the spammers an earful they won't forget. You can even adjust the tone quality from a steady police-like siren to a belching screech that sounds like a robotic cat fight. Even though the spammers will probably continue to call you back regardless of being on those useless "Do Not Call" lists, you will at least have some enjoyment at their expense with this device.

This device can be made in two versions—one that jacks right into your home phone line for maximum volume level, and a portable unit that just feeds sound into the mouthpiece of any portable phone. The wired version is certainly the most effective version as it can deliver the sound to

153

the spammer at a level you could not achieve by screaming into your phone. Because the spammer jammer feeds the audio signal directly into the phone line, it bypasses all audio conditioning circuitry in your phone handset and spews out the sound at the maximum volume possible. Having a direct phone line connection also means that it works on every phone in the house connected to that line.

If you are not afraid of the "Phone Police," then you can hack into your phone line by simply cutting the end of any standard phone cable that includes an RJ-11 connector at one end. This four conductor connector will be used to connect the spammer jammer into the phone line, so you need the RJ-11 male jack at one end and bare wires at the other end (Figure 12-1).

When you cut the end off your phone cable and expose the wires, you will have either two or four wires. Only the two center wires are used in most residential phone systems. There is a good chance these two wires will be red and green in color. You can cut the two outer wires short and then melt away a bit of the insulation from the two inner wires in order to solder them to your device. You may find it difficult to strip the ends of the wires in the phone cord as they are made of some bizarre braded wire strands, but it can be done. Melting away the insulation is a lot easier than trying to strip them with a knife. Figure 12-2 shows the phone cable adapted for a breadboard by soldering the two innermost wires onto a dual pin header.

Figure 12-1   This standard phone cord has an RJ-11 connector at one end.

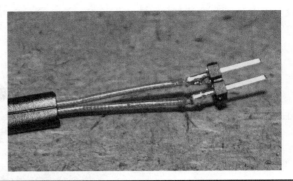

Figure 12-2   Only the two center wires are used in most residential phone systems.

Oh, we almost forgot to warn you! The phone line carries 40 volts (V) when idle, and over 100 V when ringing, so don't try to strip those wires with your teeth with the cord plugged in, or you will get a painful lesson in electrical theory! The current is very low in a phone system, but the 100-V ringing line can still give you a good tingle. Plug the cord in after your device is completed.

If you plan to mess around with telephone hardware hacks on a regular basis, then the schematic shown in Figure 12-3 will be very useful as it allows just about any device to send or receive audio or data to and from the phone line. The pair of 0.1-microfarad (μF) ceramic capacitors block any direct current (DC) from passing and allow your device to send audio or listen to audio on the phone line. This project includes this interface in the schematic already, but you may find it useful to add the capacitors right at the end of your hacked phone cord so that you can simply jack into your breadboard or any other project that may need to connect to your phone system.

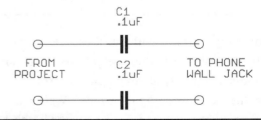

Figure 12-3   A very simple and effective way to interface "hack" into your phone line.

Figure 12-4 This interface will connect your phone line to your breadboard.

Of course, all of this telephone hacking is strictly against the rules and terms of service of your communications provider, but we are certainly willing to break a few rules in order to fight back against telephone spammers! It is a good idea to leave your evil gadgets disconnected from the phone line when not in use, though.

The multipurpose telephone line interface is built using a pair of 0.1-μF capacitors to remove the DC from the phone line. Figure 12-4 shows the interface prepared on a small breadboard for use

in other projects. The interface is so simple that you could just solder the capacitors right to the end of the telephone cord and then place some heat shrink tubing over them, leaving the free leads of the capacitors to act as plugs for your breadboard. Without this interface, 40 or even 100 V would be fed directly into your electronic projects, likely causing a total failure of any transistors or ICs.

The spammer jammer schematic shown in Figure 12-5 is based around a pair of 555 timers that work together to create a dual tone siren. The first 555 oscillator circuit (IC1) acts like a low-frequency pulse generator, changing the tone of the second 555 oscillator circuit (IC2) back and forth like an ambulance siren. The rate that the tones are changed can be controlled by the variable resistor VR1 so that the sound can be slow like a police siren or very fast like an angry screech. All of the sounds are highly annoying and very loud when fed directly into the phone line!

You can mess around with the actual tones of the siren as well by changing the values for C3, R4 and R5 in the second oscillator circuit. The output from the second oscillator is then fed through the telephone interface capacitors (C4, and C5) to

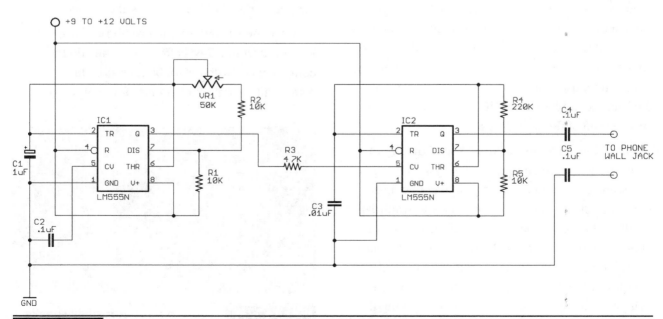

Figure 12-5 The spammer jammer schematic is based on a pair of 555 timers.

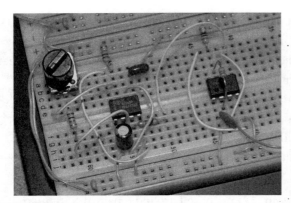

Figure 12-6 The spammer jammer circuit built on a solderless breadboard.

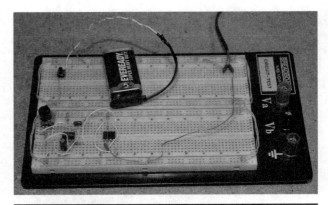

Figure 12-7 The phone line is plugged into the breadboard for live testing.

introduce the sound into the found line as an AC audio signal. If you don't want to interface to your phone system and need portable operation, a piezo buzzer can be connected from pin 3 on the second 555 timer (IC2) to ground to produce directly audible sounds.

Before you commit to any wiring, build the circuit on a solderless breadboard (Figure 12-6) so that you can tailor the sound to your liking. As is, the sounds can be adjusted from a siren to a digital screech, but you could certainly make a lot of other sounds by altering some of the components. Actually, only one 555 oscillator is needed to produce the sound, but it would only be a single frequency rather than a warbling siren. Add another 555 oscillator into the mix and you could produce an amazing variety of sounds. When you are testing your sound, use a piezo buzzer connected to pin 3 of the second 555 timer (IC2) and ground, or a speaker and 100-ohm (Ω) resistor.

Once you are happy with the sound quality of your noise-making circuit, test the system using the adapted telephone cord if you plan to connect to your phone system. You don't have to wait for a call, just pick up your phone and then press the button to engage your oscillator. The sound from the spammer jammer will be so loud that it will overpower the dial tone and become the only thing that you hear. What you hear is the same thing that the person at the other end of the phone line will hear so when you are blasting spammers, keep

the phone away from your own ear! Figure 12-7 shows the completed prototype with the phone line interface installed on the top right of the breadboard.

If you plan on making this project portable for use with a cell phone, then take the output from the second 555 timer oscillator (IC2) from pin 3 and ground out to either a piezo buzzer or small speaker (Figure 12-8). For an 8-Ω speaker, add a 100-Ω resistor in parallel. This will allow you to hold the device over the mouthpiece of your phone and blast spammers on your cell phone. Of course, the audio will not be nearly as loud as a direct connection to the phone line, but it will annoy them nonetheless!

Since the spammer jammer will run for a long time from a 9-V battery, you can add the components, switch, variable, resistor, and battery into a small project box for completion

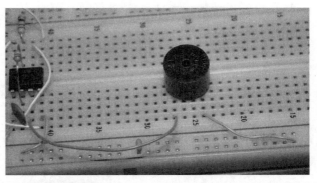

Figure 12-8 A piezo buzzer or small speaker can be used for direct audio transfer.

Figure 12-9   Adding the components and battery into a project box.

Figure 12-11   This versions of the spammer jammer has direct audio and a phone interface.

(Figure 12-9). We made our unit with both the piezo buzzer and a jack to plug into the phone line so it could work on the home phone system as well as cell phone. You will also want to add a pushbutton switch between the battery and circuit so that it only sends out a tone when you press the button. It only takes a few seconds for the telemarketers to get the point, so you only need a momentary push button switch.

Perforated board is great for small low-frequency projects like this. This board can be purchased in 6-inch square sheets for only a few bucks and will allow you to mount all through hole semiconductors into the holes for soldering wires to the underside. Figure 12-10 shows most of

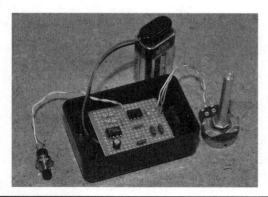

Figure 12-10   The perforated board circuit is installed in the project box.

the parts ready to be installed in the small plastic box after building up the circuit on a small bit of perforated board.

To make this project more versatile, we added the small piezo buzzer into the project box as well as a 1/8 mono jack that would connect to the adapted RJ-11 telephone cord. With the telephone jack plugged in, the switch built into the 1/8 jack cuts off the piezo buzzer and send the audio directly into the phone line for maximum delivery. For portable phone use, the small hole on the side of the box is an output from the piezo buzzer, allowing the audio to be directed into the mouthpiece of the phone (Figure 12-11).

Using the circuit shown in Figure 12-5, the spammer jammer can be made to emit a sound similar to a siren when the variable resistor is turned one way, and will continue to speed up when turned the other way until the sound that comes out is like an annoying digital mess that is truly obnoxious to listen to. Having this device plugged directly into the phone line allows the sound to be fed at maximum volume level to the telemarketers for an earful they will not easily forget. Don't get mad, get even!

PROJECT 13

Phone Voice Changer

THIS PROJECT CONVERTS an old desktop telephone into a versatile audio mixing station that lets you route your telephone calls through an effects processor in order to create a state of the art voice changer. By using a real-time computer voice filter or a professional quality effect box, you can change your voice in ways that will make you sound like a completely different person. You can make a man sound like a woman, a girl sound like a man, a man sound like an elderly lady, or any possible combination imaginable with results that will fool anyone. Unlike those "spy toy" voice changers that make you sound like a funny cartoon, a real vocal effects unit or computer vocal filter will alter a voice in a perfectly convincing manner, allowing fine control over both the formant (gender) and the pitch of your voice. Sure, you can have a lot of fun with evil and chipmunk voices as well, but if you really want to mask your voice identity in a convincing manner, then this useful device will allow you to connect any microphone compatible audio processing unit into your phone so you can alter your voice in real time.

Figure 13-0   This black box allows routing effects through your phone system.

There are many extremely powerful audio processing programs available for a computer that allow a person to alter his or her voice by talking into a microphone. Many of them are inexpensive or even free. Music stores also offer digital effect boxes that are designed for vocal processing, and these have the same functionality as the computer programs, but do their processing in a dedicated digital signal processor (DSP). We will be using both the computer software voice changer as well as the "effect box" version of the vocal processor to show how each one can be connected to the phone system using this project.

To alter your voice in real time, you must have a microphone connected to the input of a digital signal processor that can break your voice down into a digital signal, apply a complex algorithm to it, and then feed it back to an audio output without any noticeable delay. These audio processors are widely available in the form of computer programs or musical effect boxes, and all of them will allow you to speak into a microphone and hear an altered

PARTS LIST

- Phone: Any working corded desktop phone for parts

- Voice changer: Boss-VT1 or similar format changer

- Headphones: Mono or stereo headphones with jack

- Misc: Plastic project box, toggle switch

Figure 13-1   This basic desktop phone will be converted into an effects mixer.

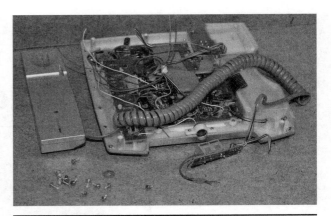

Figure 13-2   Test your phone to make sure it works and then tear it all apart.

voice as an output. Okay, that's great, but how do we get this altered voice into the phone system so that you can have a conversation with your newly disguised voice? Well, we will make a high-quality telephone audio input/output mixer using a cheap desktop phone, that's how!

Any hard-wired desktop phone will work for this project, and the fewer features the phone has, the easier it will be to hack. The cheap desktop phone shown in Figure 13-1 is a prime candidate for hacking as it has no built-in answering machine, and will be easy to take apart. A cordless phone will not work for this project—you need a phone that plugs directly into the wall. When you are sourcing a phone to hack, look for one that has as few extra features as possible, and preferably one that does not require an external power supply. Just a plain old desktop phone that still functions is all you need.

Once you have found a desktop phone that works and can be sacrificed in order to become reborn as something better, remove all of the screws and spread the innards all over your workbench! Take the phone apart as much as you can, taking note of any wires you absolutely have to cut in order to get the main circuit board removed from the plastic case. The phone we chose shown in Figure 13-2 came apart fairly easy, but we did have to cut a few wires to remove the plastic casing. The good news is that most of the cut wires will not be needed in the final version, but do take

note of where they were in case they are needed to make the phone function later.

The goal is to tear your phone completely apart and reduce it into the smallest functional size possible. Remove all of the screws and reduce your phone to the bare circuit board as shown in Figure 13-3, taking note of any wires you have to cut that lead to other smaller printed circuit boards. Most phones will only have a single circuit board inside, but this one had several as it seemed to be an earlier model that was "upgraded" to have a hands-free option and built-in calculator. If you have to cut the wires leading to the wall jack or handset, take special note of the color and placement of these wires.

The telephone will be converted into an audio mixer device capable of taking the output from

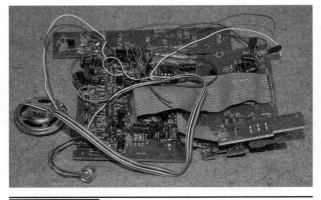

Figure 13-3   Breaking down the phone until only the circuit board remains.

your voice processing hardware (or computer) and inserting it into the phone line. The device will also let you listen to the audio from the phone line using a set of standard headphones. So basically, the phone will be converted to use a line input in place of the receiver's mouthpiece and a set of headphones in place of the receiver's earpiece. You will use the device to talk on the phone by speaking into the microphone plugged into your voice changer and listen to the call with a set of headphones. You will hear the person on the line as well as your own altered voice on the headphones so that you can hear yourself the same way they hear you. To make all of this audio mixing magic work as it should, this hack relies on the electronics included in the phone.

There are three sets of wires that you need to be concerned with: the telephone jack pair, the headset earpiece pair, and the headset mouthpiece pair. Starting with the telephone jack pair, you can find them by following the two wires leading from the RJ-11 telephone socket as shown in Figure 13-4. There are only two wires needed in a residential phone system, and these correspond to the two center pins on the 4-pin RJ-11 female jack at the back of the phone. Some phones don't have a jack, only the cord with a male connector at the end, but the wiring is the same, using only the two center wires. Usually (but not always), these wires will be colored green and red for (tip and ring),

so remember the polarity and position of the wires on the jack socket if you have to cut it to remove the circuit board from the phone. If you reverse the "tip" and "ring" wires, you may have a noisy hum on your phone when you use it again.

Once you have removed the circuit board entirely from the phone, solder any wires back where they belong and plug the mess into the phone hack to make sure it still functions properly. The receiver hook switch will probably be "up" already since the receiver no longer has a cradle, so the instant you plug the phone into the wall you should hear a dial tone. The ugly mess of circuit boards and soldered wiring shown in Figure 13-5 is actually fully functional, just as it was before we took everything apart. Now, you can begin reducing the phone to the minimal working size.

There are actually several ways you can go about converting your phone into a universal telephone audio input mixing device, so read through the entire project before you continue. We wanted to reduce the hacked phone down to a size small enough to jam into a small black box, and did not plan to make use of the dialer anymore as we would use the main phone to make new calls. Our final goal is a small black box that basically adds an audio input and output into the phone system and has a single switch to "jack" it into a call in progress. This way, we can just use our normal phone as we always do, but switch on the voice changer if needed. To make

Figure 13-4 The RJ-11 telephone jack will have a pair of wires leading to the main board.

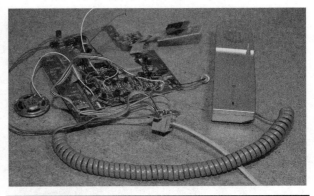

Figure 13-5 This ugly mess is actually a fully functioning desktop telephone!

a call, we would dial using the normal telephone. You can opt to leave the dialing capabilities on your hacked phone, which would make it a complete working telephone capable of making new calls, as well as mixing the voice changer audio.

The receiver "hook switch" will cut the telephone from the line when the receiver is pressing down on the switch. So, the hook switch is a "normally closed" switch that places the phone on the line when it is in the on "off hook" position. These hook switches are designed with two or more poles, and require some object to sit on them to engage them into the "off hook" position. Since the completed project will no longer make use of the original hand held receiver, this hook switch is no longer of any use. To allow the new device to be placed on or off hook (on or off the line), a single pole toggle switch will just be added between one of the telephone line wires to cut the circuit from the telephone line.

To remove the function of the original receiver hook switch, you have two options: simply leave the switch in the up position "on hook," or decode the wiring and solder the necessary wires together that will place the phone on the hook. Since a new toggle switch will control the phone being on or off the hook, you no longer need the receiver hook switch, and will need to make it permanently "off the hook" so the phone is always engaged as long as it is connected to the line. The hook switch was attached to the small carrier board shown in Figure 13-6, and although

we could have left it just the way it was, we wanted to remove it completely in order to reduce the phone to its smallest possible configuration.

In order to decode the pins on the hook switch, you will need to unsolder it from your phone's circuit board so you can test the points using an ohmmeter. The hook switch had its own small carrier board as shown in Figure 13-7, so we just removed the wires connecting it to the main board. Remember to take note of the color and position of any wires when removing the switch so you know which points or wires need to be soldered back in order to make your phone stay on the hook when complete. The goal will be to decode the on hook operation of the switch and short the appropriate wires or traces on the main board.

There may be as few as two and as many as eight wires coming from the hook switch, so you will need an ohmmeter to help decide the pins on the more complex hook switch. Since the goal is to emulate the on hook position of the switch, it will remain in the upright position (receiver off the cradle) when decoding the pins. Take an ohmmeter and test all of the combinations until you have figured out which wire pairs are tied together. As shown in Figure 13-8, a short circuit (zero ohms) means that those two pins are tied together when the switch is in the on hook position. Basically, you are going to make a schematic of your switch and then copy the on hook wiring onto your telephone's circuit board.

Figure 13-6   Identifying the receiver hook switch to decode the wiring.

Figure 13-7   The receiver hook switch is a two- or three-pole normally closed switch.

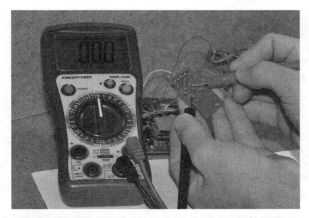

Figure 13-8   Using an ohmmeter to decode the pinouts on the hook switch.

Figure 13-10   Soldering the hook switch points together to fix the phone on the hook.

After a bit of prodding, we figured out the wiring diagram of the receiver hook switch. Figure 13-9 shows that pins A and D as well as E and C needed to be shorted together in order to fix the phone into the permanently on hook configuration. This pinout was actually fairly common on several of the phones we hacked for this use over the years.

Once you have decoded the function of your receiver hook switch, you can simply bridge together the appropriate points or traces on the

phone's circuit board in order to make the phone stay on the hook. The phone had a ribbon cable leading to a smaller circuit board that carried the hook switch, so we just soldered the necessary two pairs of wires together at the end of the cable as shown in Figure 13-10. Now, when we plug the RJ-11 jack into the phone socket, the phone is instantly on the hook and we hear a dial tone. Again, you could just leave the hook switch in the upright position, as that would also leave the phone permanently on the hook as well.

Figure 13-9   Making a wiring diagram for the receiver hook switch.

With the receiver hook switch removed from the board as well as the other unnecessary fixtures such as indicator lights, and even the dialer pad, the mess of wires shown earlier in Figure 13-3 has been transformed into the much neater bundle shown in Figure 13-11. If you do not plan on using this system to dial the phone, then you can probably just remove the dialer pad from the main circuit board without affecting the operation of the phone. If the dialer pad is on a connector, then pull it out and test your phone to ensure you get a dial tone and can hear your own voice in the earpiece as you talk into the mouthpiece. Some hacking may be necessary when removing the other components such as the dialer from the main board, but it can certainly be done with a little work. On this phone, we were able to just cut the bundle of wires that wind from the main board to the dialer pad, and this had no effect on the operation of the phone.

With the telephone reduced to the smallest possible configuration necessary to talk on the line, you can now move onto the next step, which involves the conversion of the handset into an audio input/output device. Since the telephone already has a working audio amplifier and mixer on the board, all you have to do is remove the mouthpiece microphone and the earpiece speaker and replace them with input jacks. You will be feeding the output from your voice changer into the circuit that normally amplified the handset microphone, and you will be feeding the output from the handset speaker out to a set of headphones. So, the result is a telephone that works the same as it was originally intended, but in a headset configuration rather than with the handset. By exploiting the audio mixing circuit in the telephone, you do not need to worry about designing your own complicated telephone-compatible audio amplification and filtering system.

Open the handset as shown in Figure 13-12 so you can isolate the two pairs of wires that connect the microphone and speaker to the main circuit board. Inside a typical desktop telephone handset, you will usually find a small electret microphone in the mouthpiece section, a small speaker in the headset section, and a swell block of steel or lead in the center. This metal block is installed there to make your handset feel heavier, which instills a false sense of quality into the minds of the gullible (us consumers)! Take note of the color and placement each pair of wires so you can trace them back to the main circuit board. The goal is to replace the original microphone and speaker with your own input/output (IO) jacks.

Unless your phone is from the dark ages, it will probably have a typical electret microphone for the mouthpiece and a tiny speaker for the earpiece. The electret microphone is the small metal can shown

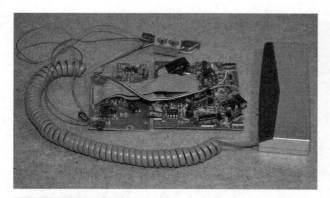

Figure 13-11 The ugly mess is growing smaller as we hack away at the circuit board.

Figure 13-12 Opening the receiver unit to convert it into an input/output device.

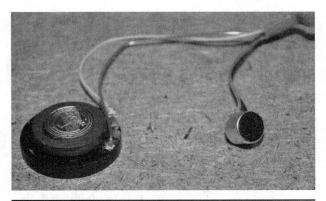

Figure 13-13 The electret microphone and speaker from the original handset.

in Figure 13-13. It includes its own high-gain amplifier right inside the small metal container. Take note of the polarity and color of the wires coming from both the speaker and the microphone as this will be important when interfacing to your effects unit or computer sound card output.

Sometimes there will be an obvious red and black wire on a polarized device, but other times there will be no logical color scheme or wires of all the same color. Rather than guessing that the red wire on the microphone is positive, look at the underside of the metal can closely and identify the pin that also connects to the side of the metal can. As shown in Figure 13-14, the pin that is connected to the metal can is also the negative or ground

lead. Once you remove the electret microphone to install the audio input connector, the signal wire from your device will be the positive lead, and the shielded ground wire will be the negative lead. The small speaker may also have a positive and negative lead, but it will actually work fine either way around.

The wiring from the headset can easily be traced back to the main circuit board, especially if the color coding is easily visible. Our headset had two silver wires, so we used the ohmmeter again (Figure 13-15) to trace the wires back to their points on the main circuit board. Once you have identified the points for both wire pairs (microphone and speaker), cut away the headset wires and leave enough wiring to install your new connectors. If the headset wires are not long enough, just solder in a new pair of wires for each device.

The phone base hacking is almost complete, but before you add the input connectors, it's a good idea to plug the original electret microphone back into the system to make sure you installed the new wiring properly. With the microphone plugged into the end of your input wire pair as shown in Figure 13-16, you should be able to hear your voice on another phone in the house once the hacked phone is plugged into the wall. If you can't hear your voice on the other phone or on this phone with the headset speaker reconnected then you probably

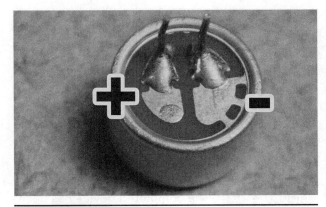

Figure 13-14 The electret microphone has a positive and negative side.

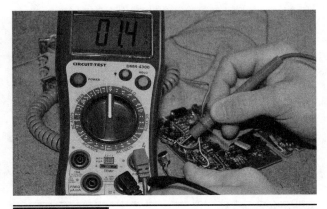

Figure 13-15 Tracing back the headset wiring to the main circuit board.

Figure 13-16 Testing the phone using the original electret microphone.

have the microphone polarity reversed. Once you have verified that the hacked phone is still working normally, IO connectors can then be installed.

Another test you can do with your system is to feed a low-level audio signal into the microphone input as shown in Figure 13-17. This basically creates a music on hold system, playing the music from the MP3 player into the phone system. You should be able to hear the music on both the headphones jacked into the audio output on your hacked phone as well as any phone connected to the same telephone jack in your house. If you have a cell phone, call your home line and then plug in the hacked phone; it will pick up the line, and you will hear the music on your cell phone. Because the audio mixer in the phone has its own filter and

preamplification circuit, the music that you input only needs to be at a very low level. For MP3 players, the lowest volume setting above zero is perfect. Now that your IO functionality has been tested, the IO jacks can be installed to finish the telephone audio mixing device.

Compare to the mess of boards and wires shown in Figure 13-3, the completed telephone audio mixing device shown in Figure 13-18 sure went on a successful diet! The device is now small enough to mount in a plastic project box, and performs the function of mixing an audio input into the phone line and at the same time echoing it back to the listener along with the audio from the caller on the other end of the conversation. This hacked phone is now officially doing the job of a device called a data access arrangement (DAA), which is a special mixer that relays audio or data to and from the telephone network. You can actually purchase DAA modules, but they are somewhat complex to connect and will certainly cost you more than a cheap old desktop telephone.

Figure 13-18 shows the completed telephone audio mixing device with the two boards taped together to save space. The device has a total of six wires (three pairs) coming out of it for the audio input (was the mouthpiece), audio output (was the earpiece), and wall jack plug.

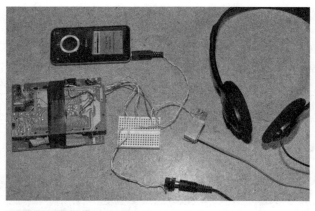

Figure 13-17 Using an audio source to test the functionality of the device.

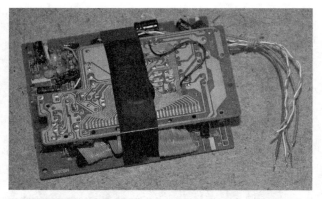

Figure 13-18 The completed device has been reduced to its smallest possible size.

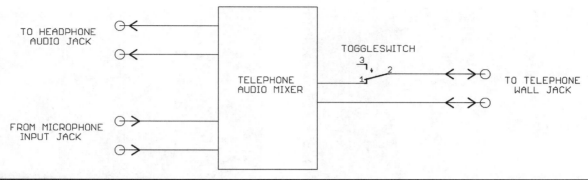

Figure 13-19 A wiring diagram showing the input output connectors.

Although your system may differ due to the design of the telephone used for parts, the wiring diagram will look something like the one shown in Figure 13-19, with three sets of wiring pairs connecting your completed telephone audio mixing device to the audio input, headphone output, and telephone wall jack. Notice the toggle switch installed in series with one of the RJ-11 connector lines—this essentially replaces the built-in receiver hook switch, allowing the telephone audio mixing device to be switched on or off the hook. You can answer calls by flipping the switch to the on position.

After a few hours of hacking the old desktop phone, the resulting telephone audio mixing device is finally ready for a new home inside the small black box shown in Figure 13-20. When you are

choosing a project case for this project, do not use a conductive metal box—we learned this the hard way after drilling all of the mounting holes and installing the board. Because the connectors will share a common ground, this could (did in our case) cause the system to fail as the phone circuitry may not be okay with a common ground for the input and output. We had to make a second installation, but everything worked fine once in the plastic box.

Now that you have your completed telephone audio mixing device ready to use, you only need to supply a microphone, headphones, and some kind of vocal effect that will alter your voice. The easiest and least expensive option is to find a real-time voice changer software for your computer and run the output from your computer soundcard into the input on the telephone audio mixing device. The only drawback to using a computer is the fact that you have to unplug your computer speakers to use the telephone voice changer and portability is certainly not so great. Another option is to use a dedicated vocal effects unit such as the Boss VT-1 voice transformer effect box shown in Figure 13-21.

Make no mistake, the Boss VT-1 is no toy, and can alter a human voice so realistically that we have been able to fool people that have known us for years! Imagine their surprise after a 15 minute prank call that that elderly lady voice was actually us all along! The Boss VT-1 can change a speaker's gender, alter the pitch, add amazingly realistic

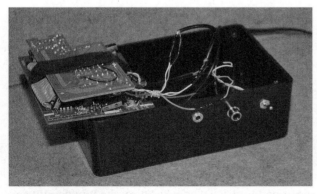

Figure 13-20 The completed telephone audio mixing device getting a new home.

Figure 13-21 The is the amazing Boss VT-1 voice transformer effect box.

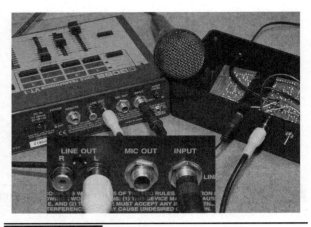

Figure 13-22 Connecting the telephone audio mixing device to an effect box.

room echo, and even make robotic voices. Since this thing was designed just for vocal use, it does an amazing job as a voice masking device. We bought the Boss VT-1 from eBay for about $100, but there were similar effects available for about the same cost that had very good vocal effects. When you are looking for a voice changer, the word "formant changer" is the key. "Formant" is the shift between the male and female vocal tract and this is not just a simple pitch change. Changing only the pitch means you can sound like a cartoon chipmunk or Satan himself, and although that can be great fun as well, it will not convince anyone on the phone. Effect units that are designed for vocalists will usually have this format control, but you should always test the effect at the store or at least look for an online review or sound demo before spending any money. Effect units called "Vocoders" may also have a formant control, and they will allow a diverse variety of voice alterations to be made. We have several vocal effect boxes as well as the best of the best for computer filters, but can honestly say nothing at all touches the Boss VT-1 as far as being the best gender-changing and most realistic voice-masking effect of all time. If you can find one for less than $200, it is definitely worth the money.

To connect an external audio effects processor to the telephone audio mixing device, you will need to identify the input and output connectors on the back of the effect unit. Most effect devices for use with guitars or microphones will have 1/8-inch (in) mono audio jacks, labeled "microphone input" and "audio output" or something similar. The Boss VT-1 also included a stereo audio "line level" RCA connector pair as shown in Figure 13-22, so we used the left channel to connect to the telephone audio mixer. Line level or microphone level audio inputs and outputs will work fine with the telephone mixer, but do not connect the telephone audio mixer to any output designed to directly drive a speaker as this will severely overdrive or even damage the preamplifier in the telephone mixer circuit. An output designed for a headphone will be okay, but if it has a volume level control, set it all the way down at first and adjust it later if necessary.

To use a real-time voice-changing software with your telephone audio mixing hardware, you will need to connect to the "line output" and "microphone input" on the computer. Desktop computers will have these jacks on the rear as shown in Figure 13-23, and laptops will usually have them on the side. If the connectors are not clearly marked, then usually the pink-colored connector is the microphone input, and the green connecter is the line output. Since some computer soundcards are designed to drive a speaker directly, you should set the volume on your computer all the way down at first and then adjust it while you are speaking into the microphone in order to set it to the

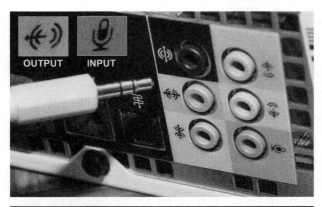

Figure 13-23   Connecting the telephone audio mixing device to a computer.

perfect level for the telephone audio mixer. When you have the microphone and volume levels set correctly, your voice will sound natural and distortion free on the telephone to both you and the caller on the line.

One of the more popular voice changers available at the time of writing this was one called "AV VCS Voice Changer," and it is available as a trial download or as a full upgrade for a small fee. The quality of this software is actually very good. It can change a person's voice just enough to sound different and convincing, or it can warp your voice into all sorts of hilarious cartoon characters. This software works in real time so you can hear the results as you speak into the microphone. Once your microphone and volume levels are set up to work with the telephone audio mixer, you will see the record indicators at the top of the software window (Figure 13-24) move as you speak into your microphone. You will hear your altered voice over the headphones as well as the caller on the other end of the line. The voice-changing software also lets you design your own voice presets and

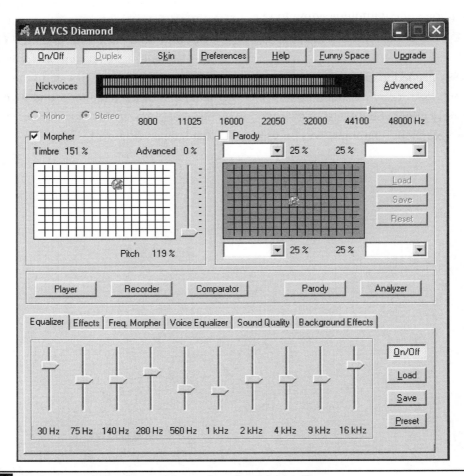

Figure 13-24   Using a real-time computer voice changing software.

save them so you can become a multitude of characters by simply pressing a button.

The completed telephone voice changer allows a person to remain anonymous over the phone by routing the audio signal through a real-time vocal effect unit or computer voice changer. Because the audio mixing and filtering are done by the original audio processing electronics that came with the phone, the conversation sounds completely natural, as if using a standard telephone. By using a vocal effect box or software program, you can alter your voice in very convincing ways that will fool anyone, even those who have known your normal voice for years. You can also have some fun with "over the top" vocal effects such as echoes, chipmunk voices, evil voices, and robot voices. The audio mixing portion of the telephone voice changer is also a very versatile input/output device that will allow you to send or receive audio data to and from the phone line without needing to use a data access arrangement (DAA) or filtering system.

If you have ever needed to disguise your voice on the telephone for any reason, then this device will certainly allow you to do it in a very convincing way. Telephone spammers will be sorry

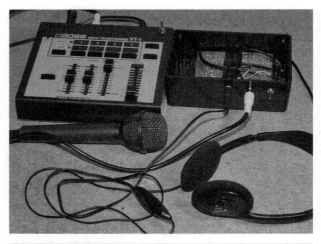

Figure 13-25 The completed telephone voice changer using the Boss Voice Transformer.

they ever dialed your number when you answer them with Satan's angry voice, and your friends will never know when you are pranking them until you can't hold back your laughter! Of course, there are also legitimate reasons why you might want to keep your identity private to the person on the other end of the telephone, and with a good voice changer connected to this device, you can make that happen (Figure 13-25).

PART FOUR
GPS Location Tracking

GPS Data Receiver

GPS is short for "Global Positioning System." As the name implies, it is a globally available positioning and time system that uses a radio fix on multiple orbiting satellites. A GPS receiver will operate wherever there is an unobstructed view (line of sight) to four or more GPS satellites so that it can receive the radio signal from each satellite. The receiver uses the information it receives in the radio signal to determine the distance to each satellite. The position of the receiver is calculated by an algorithm that includes both the information and strength of the radio signal received from each satellite. With this information, exact time, and position data such as latitude, longitude, height from seal level, moving speed, and direction can be computed and displayed to the user.

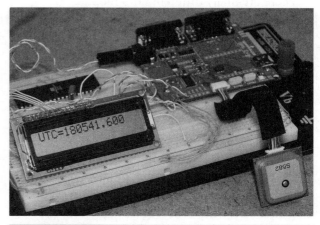

| Figure 14-0 | This project will connect a microcontroller to a GPS module. |

PARTS LIST

- GPS1: Any GPS module with a digital serial output
- Eval1: Optional sparkfun GPS evaluation board
- IC1: ATMEL ATMega32p or comparable microcontroller
- XT1: 14.755-MHz crystal for microcontroller USART clock
- LCD 1: 16 character serial input text LCD
- LED 1: Red LED to display GPS status
- R1: 1-K current limiting for LED 1

Having all of this location information available instantly from anywhere on the planet means that your next project can know exactly what time it is, where it is, how fast it is moving, and in what direction it is moving at any time. Robots can self-locate for autonomous operation and surveillance equipment can keep track of a person or vehicle location in a very precise manner. The good news is that you can add GPS capabilities to your project for under $100 and there are no licensing fees or user fees. The not so good news is that a consumer grade GPS is really only accurate to a radius of about 25 feet (ft), although for the most part the accuracy will be much better. If you own a handheld GPS unit, then you have already seen the limitations of these devices. Of course, even at 25 ft, the accuracy is more than enough to track a person or object along a city street and then display that data on a program such as Google Earth or Google Maps.

A GPS module is a self-contained GPS receiver that does all of the difficult signal processing and computation for you. These inexpensive and amazing 1-inch (in) square boxes will lock onto all of the satellites in range and then start sending out the location and time data in a simple-to-read string that can be received by a microcontroller using a few input-output (IO) lines. This project will explore the basics of connecting one of these GPS modules to a microcontroller in order to receive the data and decode it into a usable format.

There are many different GPS modules available, and like all things, some are good and some are not so good. Being a fairly new technology, improvements are constantly being made, so you must do a little research to ensure that you find a GPS module that uses the most current technology. Some of the more important aspects to consider when choosing a GPS module will be: the output format, time to fix, antenna type, and available support. Support and documentation cannot be underestimated when dealing with these GPS modules. They will require some very precise communication to and from your microcontroller in order to operate properly. With good documentation and a support community, you should have no problem getting your GPS module working in a few hours, but don't expect a happy ending if you find a budget receiver with little documentation, or you plan to be the first one to get that latest receiver working. We know—this is our second attempt at a GPS-based project, and our fist attempt was a frustrating failure because to having a module with almost no documentation and support!

The best source for GPS receivers and evaluation boards that we have found is the Internet-based electronics hobby supplier Sparkfun.com. They offer a wide range of reasonably priced GPS units along with full documentation and a forum where users can post questions and offer working source code for various models and microcontrollers. SparkFun also offers an evaluation board that takes care of the power supply, connector issues, and level translation between the serial port on the GPS and your microcontroller or computer. We found the evaluation board to be a great help

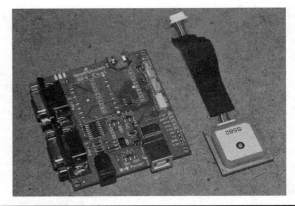

Figure 14-1   A GPS module and evaluation board supplied by SparkFun.

when working with the GPS module as we knew any errors in communication would be in our own source code rather than in the actual hardware. The GPS module (SanJose FV-M8) and evaluation board (GPS-08334) we ordered from SparkFun are shown in Figure 14-1. The total cost of both items was under $140.

The FV-M8 GPS module made by SanJose Navigation is a powerful 32-channel GPS receiver with a 5-Hertz (Hz) output rate and a built-in antenna. This GPS module would be considered high end, but for under $100, it is a bargain for those who want to add fast response GPS navigation to a project. Having a built-in antenna also makes things much easier. This GPS was able to find a fix in less than a minute, even in a basement lab. The evaluation board from SparkFun allows several GPS modules to plug right in and take power from a USB port or external DC source. The evaluation board also includes a USB-to-serial converter and a level translator for serial communication with a PC. Using free software such as "Mini-GPS" made it extremely painless to get the module up and running to verify its operation.

GPS receiver modules have come a long way over the last few years. The tiny 1-in square FV-M8 receiver module from San Jose Navigation is ready to use as soon as power is supplied (see Figure 14-2). Once powered up, the GPS module

Figure 14-2 The SanJose FV-M8 from SparkFun is a fully functional GPS in a tiny box.

will search for satellites and then send a 1-Hz "heartbeat" pulse down one of the input/output (IO) lines to show that it has a valid fix. The module will also send out serial data containing all of the relevant GPS information so that a microcontroller or terminal can decode the data. So, to add GPS capabilities to your project, you really only need three wires: one for power (3.3 volts [V]), one for ground, and one to receive the serial data! A few years ago, we tried to add GPS capabilities to a robot project using a much earlier GPS module, and failed after wasting a week trying to decode the data sheet. Nowadays, things are much easier.

Most GPS modules can also be configured via serial communications, but unless you need to change the default settings, the modules can be used without any prior setup or configuration. Serial communication speed, data transmit rate, and data format are some of the more common configurations that can be changed in most GPS receivers, but for this project we will simply work with the default settings as indicated in the datasheet for the GPS module.

The first step in connecting a GPS module is to supply the needed voltage and then connect the receiver and possibly the transmit line to your microcontroller. The data sheet will show you the pinout of the GPS connector, as well as all of the timing and communications parameters that are

needed. If you are planning to use a development board that will accept the connector for your GPS module, then the IO pin configuration is not important at this time, but eventually you will probably want to reduce your project down to just the GPS module and a microcontroller. Figure 14-3 shows the section of the datasheet that indicates the purpose of all eight wires coming out of the GPS module. Out of the eight possible wires, we only needed three of them to get the data into the microcontroller: power (VIN), ground (GND), and transmit (TX1).

A GPS module is fairly easy to connect to a microcontroller as long as you supply it with the proper DC power source and know exactly what type of data to expect from the transmit line. If the GPS module includes a "heartbeat" IO pin, then you can simply add a current-limiting resistor and an LED to create a visual indicator that will tell you when the GPS module has found a fix and is sending serial data. On our GPS module, the heartbeat IO line is labeled on the datasheet as "1PPS," which means one pulse per second. On power up, this IO line will send out a constant voltage and then blink once per second as soon as the GPS has locked on to as many satellites as it needs in order to create a valid location fix.

To verify the operation of this GPS module, we ordered the evaluation board (Figure 14-4) from SparkFun, as it included a socket for the GPS and several other popular models. This socket will supply the required DC power to the GPS module and carry all of the other IO lines out to both a solder pad header as well as an RS232 level translator that will allow the data to be received by any computer with a serial port. For PCs without a serial port, a USB-to-RS232 converter is also included on the board so that data can be received on a virtual com port on the PC. Basically, with the evaluation board you cannot go wrong when it comes to connecting the GPS module up to a computer to read the serial data or send commands to the module. It's all plug and play.

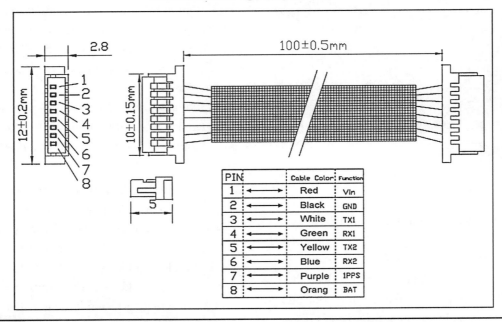

MTK-3301 GPS Receiver Series

Model: FV–M8 GPS Receiver

USER'S GUIDE SANAV™

PIN		Cable Color	Function
1	←→	Red	VIn
2	←→	Black	GND
3	←→	White	TX1
4	←→	Green	RX1
5	←→	Yellow	TX2
6	←→	Blue	RX2
7	←→	Purple	1PPS
8	←→	Orang	BAT

Figure 14-3 The datasheet will detail the IO lines as well as default settings.

Connecting the GPS module to the evaluation board and to a computer was just a matter of plugging in the cables. Since we had a 9-pin serial cable on our test bench for use with the Atmel STK500 programmer, we just moved it from the programmer's socket to the SparkFun evaluation board. We then plugged the FV-M8 into the mating socket on the evaluation board and then added a 12-V DC jack that had the correct polarity as indicated in the evaluation board datasheet (Figure 14-5). For USB operation, the board will take power from the

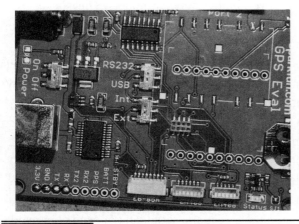

Figure 14-4 Initial testing will be done using the evaluation board.

Figure 14-5 Connecting the FV-M8 GPS module to the evaluation board.

USB bus, so no external DC power source would be necessary.

As soon as the power switch was placed in the ON position, the status LED on the evaluation board lit up, showing that the GPS module had the proper DC power applied. Since we were using the 9-pin serial connector rather than the USB connector, we set the communication switch on the evaluation board from "USB" to "RS232." The 9-pin serial port labeled "Port1" was then connected to the serial port on our PC using a standard "straight through" serial cable, not a "null modem" cable.

Once we had the evaluation board connected and powered up, the status LED went from solid to a 1 second pulse within about 60 seconds, indicating that the GPS now had a valid fix. Surprisingly, this was faster than our handheld Garmin GPS when both were powered up at the same time. The GPS module was able to gain a fix in under a minute even in a basement lab, so this was a good sign that it would be quite usable in a robot project or even a tracking device. The next step was to display the GPS data on the computer. This step was not as easy as we thought it would be.

We downloaded several free GPS data display programs from the SparkFun website as well as from the GPS module manufacturer's website so we could view the data in real time. According to the FV-M8 data sheet and sales documentation, it would power up and begin to send serial data strings at 4800 BPS at a rate of 5 Hz (5 strings per second). Well, that seemed simple. The indicator LED was already pulsing at 1 Hz, indicating that the GPS already had a valid fix. For the next four hours, we attempted to receive this 4800 BPS data on all of the free GPS programs as well as with the built-in HyperTerminal program, but could not get any of the programs to receive the serial data. We tried several cables, and even went back to the USB port, but could not get any terminal program to receive the 4800 BPS serial stream. Since the datasheet clearly stated that the module would send

4800 BPS data on power up, and would not retain any serial configuration speed changes without a backup battery, we began to suspect that the evaluation board was nonfunctional.

The next day, we began to simply hack away on the project, trying random switch settings and other baud rates, and wouldn't you know it—the data sheet was wrong! The FV-M8 GPS module defaults to 38400 BPS, not 4800 BPS as stated in the datasheet, so we were now up and running. Sadly, we cut the cables in frustration and was planning to wire up a new RS232 IC when we suspected the evaluation board was not working. It's not often that datasheets have bogus information, but if you are using this same GPS module, be aware that it defaults to 38400,8,N,1 and not 4800,8,N,1 as stated in the datasheet.

Once the PC was set to 38400 BPS, the GPS data began to display on the screen as shown in Figure 14-6. The "Mini-GPS" program presents the satellite fix data in a configuration similar to many handheld GPS units, showing position and signal strength for all satellites it can receive from. You can also set the default parameters of the GPS module, changing the frequency and speed that serial data is sent. We tried changing the serial speed back to 4800 BPS, and this worked fine, but without a backup battery, the default speed went right back to 38400 BPS. We decided to use only the default settings that the GPS module had at power up, so that if power was lost, it would not be a huge task to get the module communicating with the microcontroller again. Sure, we could set the GPS for 4800 BPS, and program the microcontroller for the same speed, but then what would happen if the GPS lost power or if the backup battery failed? The microcontroller would be unable to reprogram or even communicate with the GPS module, and it would need to be removed and reprogrammed from a terminal set for 38400 BPS. Consider this issue when adding a GPS module permanently to some other project that has to communicate with it.

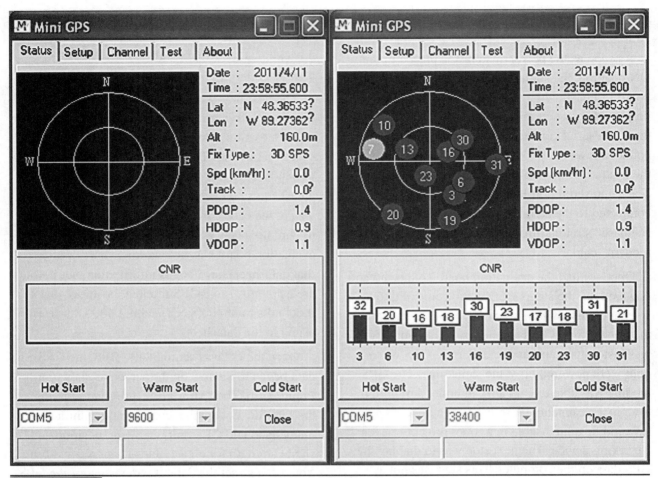

Figure 14-6 Receiving the GPS data on the "Mini-GPS" PC program.

LCD screens are very easy to connect to just about any microcontroller. The small 16 character by two line LCD screen shown in Figure 14-7 needs only a single serial input in order to display

Figure 14-7 This is an easy to interface two-line serial LCD screen.

characters. LCD screens that use serial, parallel, or I2C are easy to work with because many microcontrollers already support these languages, or have routines built into their compilers to add LCD support. We had this serial LCD in the parts box, and it was a perfect choice to display the data sent from the GPS module. Of course, the LCD screen cannot accept data directly from the GPS module, so a microcontroller must be used to first read the serial data from the GPS module and then send it out to the serial LCD module after formatting it into a compatible string.

Once you have verified that the GPS module is working properly and sending out serial data at a speed and format that you know, connecting it directly to a microcontroller is easy. Any microcontroller that can accept the transmitted serial data at the required baud rate can read the data from the GPS and then use it internally. Your project could be as simple as a Basic Stamp or Arduino reading only the TX line from the GPS module to acquire location data or more complex, controlling the functionality of the GPS module through its serial receive IO pin. In this experiment, we will simply read the serial stream as sent by the GPS module default configuration and then echo the time and location data back to the LCD screen.

Figure 14-8 shows the simple schematic that will accept GPS data into an Atmel 324P microcontroller and then send it back out to a serial LCD screen. We chose the ATMega324 because it had a dual hardware USART, and plenty of

program memory for expanding the project into something more interesting like a robot navigation system. In the schematic, "CON-1" shows the solder pads on the SparkFun evaluation board and how they relate to the connector for the FV-M8 GPS module. Eventually, the GPS module will be connected directly to the microcontroller, but for prototyping, the evaluation board ensures that there will be no wiring errors while working on the microcontroller source code.

Both the microcontroller and the GPS module will run from 3.3 V, so no level translation is necessary between the RX and TX serial lines. If the microcontroller were to be powered from 5 V, then some GPS modules may require level translation in order for serial communications

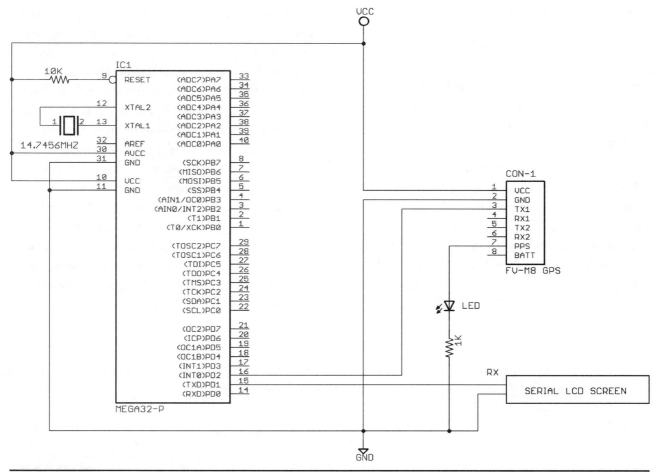

Figure 14-8 Using a microcontroller to receive GPS data and send it to an LCD screen.

to work between 5 and 3.3 V. Check your GPS module datasheet to ensure that it can be connected to your external project at whatever voltage your microcontroller is running from; 3.3 to 5 V Level translation is probably not necessary for receiving date from the GPS as the microcontroller, but going to the GPS, you have to be careful, as the higher voltage could damage the GPS module.

Since the serial LCD screen needed to communicate at 9600 BPS and the GPS module was communicating at 38400 BPS (its default setting), we needed to choose a microcontroller with a dual serial USART and then find a clock oscillator that would work for both baud rates. When working with lower-speed serial communications such as 2400 BPS or 4800 BPS, you can often get away with any clock speed, but for higher data rates such as 38400 BPS, there are "magic" frequencies that will ensure that you end up with an error-free communication system: 14.7456 was one of these well-known "magic" serial communication clock speeds, so it was easy to source this crystal. The canned crystal shown in the breadboard in Figure 14-9 was one of several that we had in the salvaged parts bin, and probably came from an older PC modem card.

If you are planning on communicating with your GPS at a lower baud rate, then you will not have to be so selective of your microcontrollers clock speed, but be aware that there will be a certain error level that may or may not cause a slowdown in your serial link from the GPS module. If you are constantly missing serial strings from the GPS module, then you may have to either adjust the serial transmit

Figure 14-9 An ATMega324p is chosen because it has a dual hardware USART.

speed or choose a clock frequency that gets closer to the desired board rate with fewer timing errors.

I wrote the basic GPS data receiver program in C for the ATMega324p microcontroller using AVR CodeVision, but it is very easy to follow and convert to just about any language. This program represents the minimal program needed to decode the GPGGA NMEA serial data string into its individual parts and then send them back out to the serial LCD screen. Most GPS modules output data in the NMEA (National Marine Electronics Association) format, which is a collection of standardized data formats that contain comma separated values following an identifier code. You will definitely have to get familiar with the data format sent by your GPS module and chose one of the many possible strings to decode. A Google search for "NMEA strings" or "NMEA sentence" will yield all of the information you will ever need for every possible NMEA string format.

```
//
//************************************************************************************
// **** PROGRAM = SIMPLE GPS DATA RECEIVER (C) 2010 LUCIDSCIENCE.COM
// **** TARGET = ATMEGA88P
// **** CLOCK SPEED = 14.7456 MHZ CRYSTAL
//
//************************************************************************************
#include <mega324.h>
#include <delay.h>
```

```
//
*****************************************************************************
// **** GLOBAL VARIABLES TO HOLD NMEA DATA
//
*****************************************************************************
char fix[16];
char sats[16];
char height[16];
char utctime[16];
char altitude[16];
char latitude[16];
char longitude[16];
char horizontal[16];



//
*****************************************************************************
// **** SEND USART0 DATA TO SERIAL LCD @ 9600 BPS
//
*****************************************************************************
void LCDWRITE(char lcd) {
while(!(UCSR0A & (1<<5)));
UDR0 = lcd;
}



//
*****************************************************************************
// **** READ USART1 NMEA DATA TO STRINGS @ 38400 BPS
//
*****************************************void GPSREAD() {
char cntr;
char data;

// $GPGGA, 223611.000, 4821.9234,N, 08916.4091,W, 1,   8,    1.17,   190.1,M, -35.0,M, ,*6F
// $GPGGA, UTCTIME,    LATITUDE,    LONGITUDE,     FIX, SATS, HORIZ, ALTITUDE, HEIGHT    CHECKSUM

// CLEAR LAST STRING DATA
for (cntr = 0; cntr < 16; cntr++) {
fix[cntr] = 32;
sats[cntr] = 32;
height[cntr] = 32;
utctime[cntr] = 32;
altitude[cntr] = 32;
latitude[cntr] = 32;
longitude[cntr] = 32;
horizontal[cntr] = 32;
}
```

```
// WAIT FOR STRING MARKER ($GPGGA)
while (cntr != 6) {
cntr = 0;
while (!(UCSR1A & (1<<7)));
data = UDR1;
if (data == '$') cntr++;
while (!(UCSR1A & (1<<7)));
data = UDR1;
if (data == 'G') cntr++;
while (!(UCSR1A & (1<<7)));
data = UDR1;
if (data == 'P') cntr++;
while (!(UCSR1A & (1<<7)));
data = UDR1;
if (data == 'G') cntr++;
while (!(UCSR1A & (1<<7)));
data = UDR1;
if (data == 'G') cntr++;
while (!(UCSR1A & (1<<7)));
data = UDR1;
if (data == 'A') cntr++;
}
while (data != ',') {
while (!(UCSR1A & (1<<7)));
data = UDR1;
}

// READ UTC TIME STRING
data = 0;
cntr = 0;
while (data != ',') {
while (!(UCSR1A & (1<<7)));
data = UDR1;
utctime[cntr] = data;
cntr++;
}
utctime[cntr-1] = 32;

// READ LATITUDE STRING
data = 0;
cntr = 0;
while (data != ',') {
while (!(UCSR1A & (1<<7)));
data = UDR1;
latitude[cntr] = data;
cntr++;
}
```

```
data = 0;
while (data != ',') {
while (!(UCSR1A & (1<<7)));
data = UDR1;
latitude[cntr] = data;
cntr++;
}
latitude[cntr-1] = 32;

// READ LONGITUDE STRING
data = 0;
cntr = 0;
while (data != ',') {
while (!(UCSR1A & (1<<7)));
data = UDR1;
longitude[cntr] = data;
cntr++;
}
data = 0;
while (data != ',') {
while (!(UCSR1A & (1<<7)));
data = UDR1;
longitude[cntr] = data;
cntr++;
}
longitude[cntr-1] = 32;

// READ SATALITE FIX FLAG
data = 0;
cntr = 0;
while (data != ',') {
while (!(UCSR1A & (1<<7)));
data = UDR1;
fix[cntr] = data;
cntr++;
}
fix[cntr-1] = 32;

// READ SATALITE COUNT
data = 0;
cntr = 0;
while (data != ',') {
while (!(UCSR1A & (1<<7)));
data = UDR1;
sats[cntr] = data;
cntr++;
}
sats[cntr-1] = 32;
```

```
// READ HORIZONTAL STRING
data = 0;
cntr = 0;
while (data != ',') {
while (!(UCSR1A & (1<<7)));
data = UDR1;
horizontal[cntr] = data;
cntr++;
}
horizontal[cntr-1] = 32;

// READ ALTITUDE STRING
data = 0;
cntr = 0;
while (data != ',') {
while (!(UCSR1A & (1<<7)));
data = UDR1;
altitude[cntr] = data;
cntr++;
}
data = 0;
while (data != ',') {
while (!(UCSR1A & (1<<7)));
data = UDR1;
altitude[cntr] = data;
cntr++;
}
altitude[cntr-1] = 32;

// READ HEIGHT STRING
data = 0;
cntr = 0;
while (data != ',') {
while (!(UCSR1A & (1<<7)));
data = UDR1;
height[cntr] = data;
cntr++;
}
data = 0;
while (data != ',') {
while (!(UCSR1A & (1<<7)));
data = UDR1;
height[cntr] = data;
cntr++;
}
height[cntr-1] = 32;
}
```

```
// ********************************************************************************
// **** PROGRAM VARIABLES AND PORT SETUP
// ********************************************************************************
void main(void) {

// MAIN PORGRAM VARIABLES
char temp;
char ct;

// SET USART0 TO TX @ 9600,8,N,1
UCSR0A=0x00;
UCSR0B=0x08;
UCSR0C=0x06;
UBRR0H=0x00;
UBRR0L=0x5F;

// SET USART1 TO RX @ 38400,8,N,1
UCSR1A=0x00;
UCSR1B=0x10;
UCSR1C=0x06;
UBRR1H=0x00;
UBRR1L=0x17;


// ********************************************************************************
// **** PROGRAM INITIALIZATION
// ********************************************************************************

// CLEAR LCD SCREEN
LCDWRITE(12);
delay_ms(500);

// TURN OFF CURSOR
LCDWRITE(22);
delay_ms(500);

// TURN ON BACKLIGHT
LCDWRITE(17);
delay_ms(500);

// STARTUP MESSAGE
LCDWRITE('G');
LCDWRITE('P');
LCDWRITE('S');
LCDWRITE(32);
LCDWRITE('S');
LCDWRITE('T');
LCDWRITE('A');
LCDWRITE('R');
```

```
LCDWRITE('T');
LCDWRITE('U');
LCDWRITE('P');
LCDWRITE('.');
LCDWRITE('.');
LCDWRITE('.');
ct = 0;


// *******************************************************************************
// **** MAIN PROGRAM LOOP
// *******************************************************************************
while (1) {

// READ GPGGA DATA FROM GPS
GPSREAD();

// CLEAR LCD DATA
LCDWRITE(128);
ct++;

// DISPLAY DATA : FIX STATUS
if (ct == 1) {
LCDWRITE('F');
LCDWRITE('I');
LCDWRITE('X');
LCDWRITE('=');
for (temp = 0; temp < 14; temp++) {
LCDWRITE(fix[temp]);
}
}

// DISPLAY DATA : SATALITE COUNT
if (ct == 2) {
LCDWRITE('S');
LCDWRITE('A');
LCDWRITE('T');
LCDWRITE('=');
for (temp = 0; temp < 14; temp++) {
LCDWRITE(sats[temp]);
}
}

// DISPLAY DATA : UTCTIME
if (ct == 3) {
LCDWRITE('U');
LCDWRITE('T');
LCDWRITE('C');
LCDWRITE('=');
```

```
for (temp = 0; temp < 14; temp++) {
LCDWRITE(utctime[temp]);
}
}

// DISPLAY DATA : LATITUDE
if (ct == 4) {
LCDWRITE('L');
LCDWRITE('A');
LCDWRITE('T');
LCDWRITE('=');
for (temp = 0; temp < 14; temp++) {
LCDWRITE(latitude[temp]);
}
}

// DISPLAY DATA : LONGITUDE
if (ct == 5) {
LCDWRITE('L');
LCDWRITE('O');
LCDWRITE('N');
LCDWRITE('=');
for (temp = 0; temp < 14; temp++) {
LCDWRITE(longitude[temp]);
}
}

// DISPLAY DATA : HORIZONTAL
if (ct == 6) {
LCDWRITE('H');
LCDWRITE('O');
LCDWRITE('R');
LCDWRITE('=');
for (temp = 0; temp < 14; temp++) {
LCDWRITE(horizontal[temp]);
}
}

// DISPLAY DATA : ALTITUDE
if (ct == 7) {
LCDWRITE('A');
LCDWRITE('L');
LCDWRITE('T');
LCDWRITE('=');
for (temp = 0; temp < 14; temp++) {
LCDWRITE(altitude[temp]);
}
}
```

```
// DISPLAY DATA : HEIGHT
if (ct == 8) {
LCDWRITE('H');
LCDWRITE('E');
LCDWRITE('I');
LCDWRITE('=');
for (temp = 0; temp < 14; temp++) {
LCDWRITE(height[temp]);
}
ct = 0;
}

// DELAY ONE SECOND
delay_ms(1000);

// END OF MAIN LOOP
}
}
```

By default, the FV-M8 GPS module sends several strings at a rate of 5 per second and one of these strings is the GPGGA string, which we found to be easy to understand. The serial data string for the GPGGA sentence will look like this:

```
$GPGGA,223611.000,4821.9234,N,08916.4091
,W,1,8,1.17,190.1,M,-35.0,M,,*6F
```

Each of the numbers or characters following the "$GPGGA" identifiers represent the received and calculated GPS location data as:

```
$GPGGA,UTCTIME,LATITUDE,LONGITUDE,FIX,SA
TS,HORIZ,ALTITUDE,HEIGHT,CHECKSUM
```

So, all your program needs to do is read the serial data stream and look for the "$GPGGA" identifier text and then parse the comma-separated data values until the end of the line. We wrote this GPS receiver test program in the most basic way possible so that it would be easy to follow the string decoding routine for translation into another language. There are no time outs or error checks being done by the string decoder routine;

it just waits for the next "$GPGAA" identification and then reads the following comma-separated values into usable string variables. These values are then echoed to the serial LCD screen at one field per second so you can see the data sent from the GPS module. As shown in Figure 14-10, the program will start by sending the message "GPS LOADING…" and then wait for the next GPGGA data string to arrive for parsing.

Figure 14-10 The microcontroller is now receiving serial data from the GPS module.

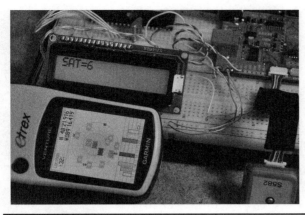

Figure 14-11 Comparing the GPS module data with the information on the handheld GPS.

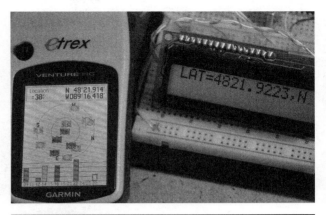

Figure 14-12 The GPS module and handheld GPS comparing latitude values.

To verify the operation of the FV-M8 GPS module and compare its accuracy against a good quality handheld GPS unit, we let both GPS systems gain a fix and then displayed the data side by side as the program ran through its loop. Figure 14-11 shows the info screen on the Garmin handheld GPS, which is being used to verify the data strings received from the FV-M8 GPS module and echoed to the LCD screen. Both GPS systems managed to lock onto six satellites within 60 seconds, and the GPS module later found eight satellites from our indoor lab while the handheld stayed at only six. Both GPS systems show almost identical coordinates, but the handheld seemed to respond to movement a little quicker than the GPS module as we moved them both across the room. So far it was a close tie between the handheld commercially available GPS unit and the FV-M8 GPS module.

After moving both across the room and back, the latitude values were very close, being 48-21.914-N on the Garmin handheld and 48-21.9223-N on the FV-M8 GPS module. When we plugged this data into Google Earth, it showed the center of our house, so it was certainly correct data. The difference between the 2 GPS modules was so small that we could not determine which one was more accurate using Google Earth or Google Maps (Figure 14-12). Actually, the GPS module had one

decimal digit more accuracy than the handheld, but this is probably overkill considering the margin of error in GPS positioning. So far the GPS module was getting a great score, able to perform at least as good as or better than a good quality handheld GPS unit. The next test to be done will be the elimination of the evaluation board to get the GPS module talking directly to the microcontroller.

We often do the prototyping on a breadboard. The connector on the GPS module was certainly too small for any breadboarding work. The female connectors shown in Figure 14-13 mate with the connector that came with the GPS module, and would be perfect for installation onto a circuit board. We decided to make a breadboard

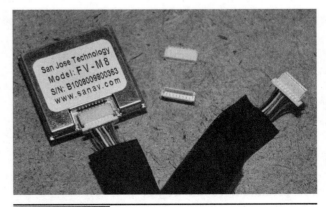

Figure 14-13 The GPS module came with a small 8-pin connector.

Figure 14-14   Converting the GPS cable to work on a breadboard.

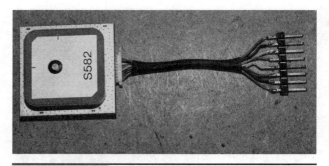

Figure 14-15   The completed breadboard compatible GPS cable.

compatible connector so that we could further explore the functionality of the GPS module. Warning—ugly cable hacking to come!

The cable that came with the GPS had a male 8-pin connector at each end, so we decided that if we simply cut the cable in half, we could make two new cables—one with breadboard-compatible pins and another that could later be soldered to whatever circuit board we planned to make our GPS project on. The cable shown in Figure 14-14 was cut in half and then one side was separated so that each wire could be soldered to a breadboard compatible header pin. Since SparkFun also sold these small cables and sockets separately, we could always acquire another one ifwe decided to use the original socket on a printed circuit board later on.

Once the eight colored wires were soldered to the header pins, the GPS cable was now ready to be plugged into any solderless breadboard for rapid prototyping. The colors of each wire were fairly obvious, with red being VCC and black being GND, so it would be difficult to connect the power in reverse. Besides power and ground, the only other wire we planned to use, from the possible eight, was the serial transmit line (white wire). We connected all eight wires to the header pins just in case we wanted to experiment with changing the default parameters on the GPS module using the serial receive line. Figure 14-15 shows the completed breadboard compatible GPS module and cable.

With the evaluation board no longer included in the project, the size was greatly reduced as shown in Figure 14-16. We removed the LCD routines from the original source code and programmed a smaller ATMega88 microcontroller to receive the NMEA serial data strings from the GPS module. The heart beat LED was added from the "1PPS" line on the connector so that we could see that the GPS module was properly powered up and how long it took to get a valid location fix. This functionality is built into the GPS module, so only a current-limiting resistor was needed between the "1PPS" IO line, ground, and the LED. The GPS module serial transmit line "TX" was connected directly to the serial receive pin on the microcontroller so that the hardware USART could be used to receive the serial data at 38400

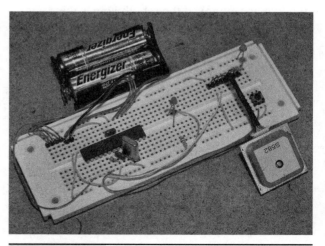

Figure 14-16   Interfacing the GPS module directly to a microcontroller.

baud. Once again, the "magic" crystal frequency of 14.7456 was chosen so that serial communications would be error free.

The program was modified so that when we pressed the pushbutton, the latitude value from the GPGGA string was memorized. Each time the memorized latitude value was found to match the current received value coming from the GPS module, an LED would be lit. We only took the first 7 digits of the Latitude data, so that the accuracy would be about 25 ft in any direction. A GPS is capable of higher accuracy, but only under ideal conditions such as being outdoors and stationary for some time. The goals of this test was to determine if the system could remember which room in the house we pressed the memorize button. After several tests around the house, it was determined that yes, this simple prototype could indeed remember which room we were in as long as we gave it a few seconds to gain a more accurate fix. While moving, the accuracy was limited to one end of the house or the other, but once stationary in a room, the accuracy seemed to be within 20 ft or so. Not bad at all!

This project proves that an inexpensive GPS module can certainly be used with a microcontroller to create a fairly accurate navigational system capable of determining time, latitude, longitude, direction, height, and location within about 20 ft from just about any place on the planet. Along with basic machine vision, and an obstacle-avoidance system a robot could be made to navigate the outdoor environment using the data received from the GPS module. A small and accurate stealth tracking system could also be made that would allow the user to later visually inspect the route on Google Maps or Google Earth. The possibilities are endless when your projects are given the ability to know exactly where they are on the planet at any given time, and as GPS technologies become more and more accurate, indoor robotic navigation will eventually be possible. Thanks for stopping by, we must now go back to our secret lab located at 296407.42 mE, 5350996.54 mN, 1080 ft.

PROJECT 15

GPS Tracking Device

THIS PROJECT COMBINES an 8-bit microcontroller along with a GPS module to create a simple tracking device that can record and play back location data to a mapping program like Google Maps or Google Earth. This project demonstrates the basic method of serial communication between the GPS module and microcontroller, as well as how to decode the data stream being sent from the GPS module. To make this project easy to follow, the source code and hardware are both made to be as minimal as possible, leaving a lot of room to build a much more powerful tracking system.

GPS is short for "Global Positioning System." As the name implies, it is a globally available positioning and time system that uses a radio fix on multiple orbiting satellites. A GPS receiver will operate wherever there is an unobstructed view (line of sight) to four or more GPS satellites so that

it can receive the radio signal from each satellite. The receiver uses the information it receives in the radio signal to determine the distance to each satellite. The position of the receiver is calculated by an algorithm that includes both the information and strength of the radio signal received from each satellite. With this information, exact time, and position data such as latitude, longitude, height from seal level, moving speed, and direction can be computed and displayed to the user.

A GPS module is a self-contained GPS receiver that does all of the difficult signal processing and computation for you. These inexpensive and amazing 1-inch (in) square boxes will lock onto all satellites in range and then start sending out the location and time data in a simple to read string that can be received by a microcontroller using a few input/output (IO) lines. This project will explore the basics of connecting one of these GPS modules to a microcontroller in order to receive the data and record it for later use in a computer mapping program such as Google Earth.

GPS receiver modules have come a long way over the last few years. The tiny 1-in square FV-M8 receiver module shown in Figure 15-1 is from San Jose Navigation, and is ready to use as soon as power is supplied. Once powered up, the GPS module will search for satellites and then send a 1-Hz "heartbeat" pulse down one of the input/output lines to show that it has a valid fix. The module will also send out serial data containing all of the relevant GPS information so that a microcontroller or terminal can decode the data. So, to add GPS capabilities to your project,

Figure 15-0 This project demonstrates a simple AVR based GPS tracker.

193

Figure 15-1   The San Jose FV-M8 GPS module from SparkFun.

you really only need three wires: one for power (3.3 volts [V]), one for ground, and one to receive the serial data. All of the difficult work of receiving the RF signal for triangulation is done inside the module for you.

The GPS module came with a small cable and connector set that would be perfect for installation onto a printed circuit board, but not much use for breadboarding work. We decided to convert the cable for breadboard use by cutting it in half so that we could solder a row of header pins to the wires (Figure 15-2). This would essentially give us two cables when done, as the GPS compatible connectors would remain at one end of each half of the cable after cutting.

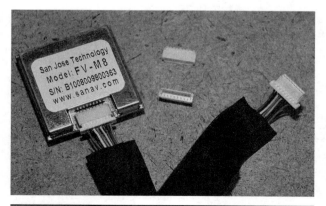

Figure 15-2   Converting the GPS connector for breadboard use.

According to the FV-M8 datasheet, all of the signal lines were either DC or digital, so there would be no issues with modifying the connector or cable length. The data sheet will show you the pinout of the GPS connector, as well as all of the timing and communications parameters that are needed. If you are planning to use a development board that will accept the connector for your GPS module, then the IO pin configuration is not important at this time, but eventually you will probably want to reduce your project down to just the GPS module and a microcontroller. Figure 15-3 shows the section of the datasheet that indicates the purpose of all eight wires coming out of the GPS module. Out of the eight possible wires, we only needed three of them to get the data into the microcontroller: power (VIN), ground (GND), and transmit (TX1).

The cable that came with the GPS had a male 8-pin connector at each end, so we decided that if we simply cut the cable in half, we could make two new cables: one with breadboard-compatible pins and another that could later be soldered to whatever circuit board we planned to make our GPS project on. The cable shown in Figure 15-4 was cut in half, and then one side was separated so that each wire could be soldered to a breadboard compatible header pin. Since SparkFun also sold these small cables and sockets separately, we could always acquire another one if we decided to use the original socket on a printed circuit board later on.

Once the eight colored wires were soldered to the header pins, the GPS cable was now ready to be plugged into any solderless breadboard for rapid prototyping. The colors of each wire were fairly obvious, with red being VCC and black being GND, so it would be difficult to connect the power in reverse. Besides power and ground, the only other wire we planned to use from the possible eight, was the serial transmit line (white wire). We connected all eight wires to the header pins just in case we wanted to experiment with changing the

MTK-3301 GPS Receiver Series

Model: FV–M8 GPS Receiver

USER'S GUIDE SANAV™

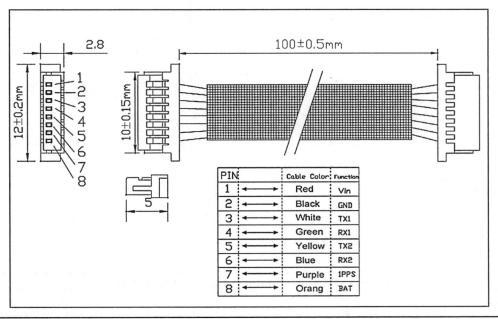

PIN		Cable Color	Function
1	←→	Red	Vin
2	←→	Black	GND
3	←→	White	TX1
4	←→	Green	RX1
5	←→	Yellow	TX2
6	←→	Blue	RX2
7	←→	Purple	1PPS
8	←→	Orang	BAT

Figure 15-3 The datasheet will detail the IO lines as well as default settings.

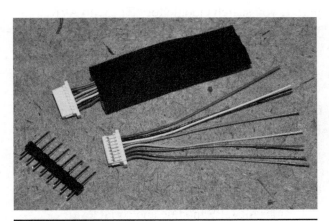

Figure 15-4 Converting the GPS cable to work on a breadboard.

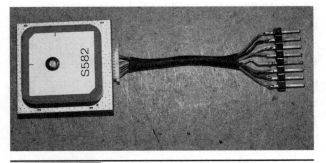

Figure 15-5 The completed breadboard compatible GPS cable.

default parameters on the GPS module using the serial receive line. Figure 15-5 shows the completed breadboard compatible GPS module and cable.

The goal of this project is the bare minimal simplified system that will record GPS coordinates at a set rate into the microcontroller's built-in memory so that they can then be played back and viewed on a mapping program like Google Earth. To get the GPS module to communicate with the microcontroller, the hardware serial port (USART) on the AVR ATMega324p is set as a transmitter/ receiver for 38,400 baud, which is the default serial transmission speed of the GP module on power up.

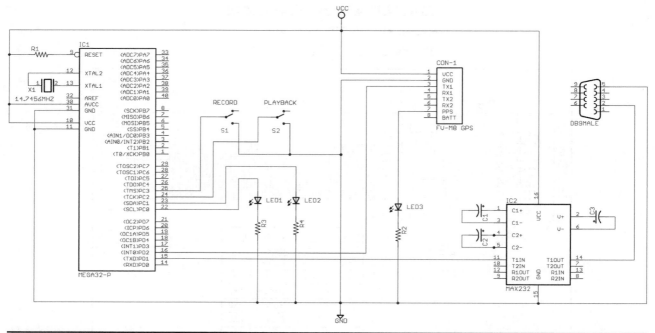

Figure 15-6 Basic GPS tracking system schematic using an ATMega324.

On the 324p, the serial receive pin is on PORTD.2 as shown in the schematic (Figure 15-6).

To communicate with the computer, the serial transmit line on the 324p (PORTD.1) is fed to an RS232 level converter so that the voltage expected by the PC is correct. The MAX232 from Maxim-IC is an easy to use serial level converter that will take the 3- or 5-V serial output from the microcontroller and boost it up to the correct 12-V level expected by the computer's serial port. Without the level converter, it is unlikely that the PC would be able to receive the serial data, and even if it did, there would be a lot of errors and bogus data received. As you can see in the Schematic shown in Figure 15-6, the MAX232 level converter only needs three external capacitors in order to function, so it is a very easy to use and inexpensive solution to allow any microcontroller to talk both ways to a PC serial port.

Since the old 9-pin PC serial port is now considered retro, you may want to use the USB port instead. A similar serial level converter and translator integrated circuit (IC) called the

FT232 from FTDIChip will do the same work as the MAX232, but for translation between the microcontroller's serial port and the PC's USB port. If you search Google for "FT232 to AVR," you will see that like the MAX232, only a few external capacitors and resistors are needed to allow a microcontroller to talk to the PC using a virtual USB serial port. Choose whichever serial communications method is available on your PC, but remember—without the level translator (MAX232 or FT232), you cannot expect to receive serial data from the microcontroller reliably or even at all.

Besides the GPS module and serial level converter, the schematic only has a few other parts: two push button switches and three optional indicator light-emitting diodes (LEDs). The two buttons trigger record and playback, so that GPS location points can be stored into the microcontroller's internal SRAM at set intervals. When in playback mode, the GPS information is streamed out to the PC at high speed in the standard NMEA $GPGGA format so that programs like

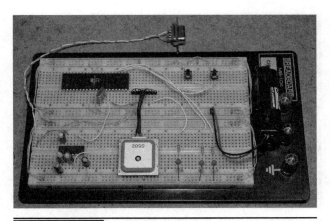

Figure 15-7 The GPS tracking system built on a solderless breadboard.

Google Maps or Google Earth can read and display the data. Basically, this project is an ultrasimplified yet functional GPS tracking system that will record location points until it fills whatever memory you have available. The status LEDs show both the record and playback mode status as well as the GPS lock status.

The basic GPS tracking system is shown in Figure 15-7, first built on a solderless breadboard for easy prototyping. If you have not yet read the "GPS Data Receiver" project, that would be a good start as it offers a good intro on working with GPS modules and microcontrollers. As you can see, the AVR324p is only using a few of the 32 available input/output pins, but we needed a microcontroller with a decent amount of internal SRAM in order to store the GPS data, so this meant using a large IO package. Realistically, you will need a much larger storage memory for GPS points than what is available on any microcontroller, but this project is just a simplified demonstration that leaves a huge amount of room for improvement.

Since both the GPS module and microcontroller will run fine on a 3-V power supply, we chose to use a pair of AA batteries in order to later create a small self contained tracking system that could be stuck to a vehicle using a magnet mounting system. The MAX232 level converter is being underpowered on 3 V, but we have never had any

PC communication issues with running them this way. If you are a stickler for specifications, then there is actually a low-power (3 V) version of the MAX232 level converter available, but only in a surface mount package.

Since the FV-M8 module will be sending out serial data at 38400 BPS (its default setting), we needed to choose a microcontroller with a hardware serial USART and then find a clock oscillator that would allow the serial speed to be set with minimal error. When working with lower speed serial communications such as 2400 BPS or 4800 BPS, you can often get away with any clock speed, but for higher data rates such as 38400 BPS, there are "magic" frequencies that will ensure that you end up with an error free communication system. 14.7456 was one of these well known "magic" serial communication clock speeds, so it was easy to source this crystal. The crystal oscillator shown in the breadboard in Figure 15-8 was one of several that we had in the salvaged parts bin, and probably came from an older PC modem card.

Some programming languages such as C, Basic or Arduino have built in serial routines that use both the hardware USART on the microcontroller or simply bit bang the data stream out of an IO port. Keep in mind that these built in routines also suffer from errors if the master clock frequency is not chosen carefully. At 16 MHz, the Arduino

Figure 15-8 For high-speed serial communications, clock speed is important.

was unable to properly receive 38,400 baud the serial stream from the GPS module due to the resulting error in transmission speed. Using CodeVision C, we managed to get somewhat better communications at 20 MHz from the GPS module, but in reality you will probably need to choose a proper clock frequency to satisfy the high-speed serial USART if you want reliable communications from the GPS module or out to the PC through the serial port. These "magic" clock speeds are listed in the microcontroller data sheet and on some of the compilers (such as CodeVision AVR) when starting a new project. Consider this when working with serial data rates over 4800 baud.

Besides a few capacitors and a DC power supply, the MAX232 does not need any other components in order to operate as a serial level translator between your microcontroller and any PC (Figure 15-9). The USB counterpart (FT232) is also a single IC solution for interfacing the 3- or 5-V digital world to the PC through a virtual USB com port. If you only needed to send data to the microcontroller, then the level converter would not be necessary, but since the microcontroller cannot drive the PC serial port with enough voltage, it is needed to send data to the PC serial port.

Figure 15-9 The MAX232 will translate voltage levels for PC serial communication.

If you plan to work with serial communications, then the schematic shown in Figure 15-10 will be very handy to make on a small circuit board or perforated board. We have several boards that include the MAX232, as well as the FT232 so when we need to send or receive data between a microcontroller and PC, it's just a matter of plugging in the board to the project or breadboard. The schematic shown in Figure 15-10 is a one way system from the microcontroller to the PC, but to use it the other way (or both directions),

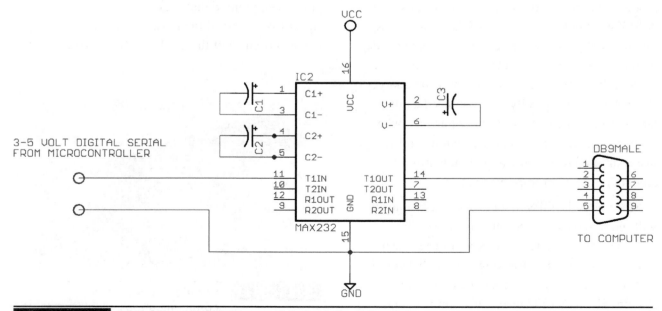

Figure 15-10 Using the MAX232 to send data to the PC serial port.

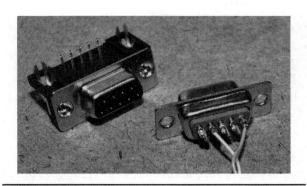

Figure 15-11 Connecting the 9-pin RS232 PC serial port connector.

just connect the input or outputs to the serial connector and MAX232 IC. The FT232 USB serial translator works the same way, allowing two way communications between any low-voltage logic and the PC USB virtual serial port.

To receive data on a PC serial port, you need to connect only two lines: ground (pin 5) and the serial receive line (pin 2). (See Figure 15-11.) This assumes that you are using a standard 9-pin serial port on your PC. Once these two lines are connected, the serial data stream can be sent to the PC from the level converter.

Once you have the circuit built on a breadboard and have the serial or USB connection to your PC ready, compile the simple GPS data logging program to your microcontroller. This program was done in CodeVision and is made for the AVR324p, but it would be very easy to convert it to just about any microcontroller. The real limitation is the amount of data that can be stored in the internal SRAM of the microcontroller, but by expanding the program to use a serial EEPROM or Flash memory, it could easily store hours or days worth of GPS data. For now, just get the basic system working so that you can verify that the microcontroller can decode, store, and then playback GPS data to your PC.

The operation of the microcontroller program is very simple. If you press the record button, the record indicator LED will stay on as data is stored into the internal SRAM. When the record LED is out, you can then press the playback button to stream out the data to the serial port on the PC. The data is formatted so that programs like Google Earth and Google Maps think that a standard GPS unit is connected. Having only a few kilobytes free, the AVR324 can store about 20 full NMEA strings before the SRAM is full. This is certainly not much data, but enough to test your hardware with a quick drive around the block. This GPS tracking test program is based on the source code from the GPS data reviver project, and is explained in much more detail there.

```
// ******************************************************************************
// **** PROGRAM = GPS TRACKER (C) 2010 LUCIDSCIENCE.COM
// **** TARGET = ATMEGA324P & FV-M8 GPS MODULE
// **** CLOCK SPEED = 14.7456 MHZ CRYSTAL
// ******************************************************************************
#include <mega324.h>
#include <delay.h>

// ******************************************************************************
// **** GLOBAL VARIABLES TO HOLD GPGGA DATA
// ******************************************************************************
char gpgga[80];
char track[1440];
```

```
// ***********************************************************************************
// **** SEND USART0 DATA TO SERIAL PORT @ 38400 BPS
// ***********************************************************************************
void DATSEND(char dat) {
while(!(UCSR0A & (1<<5)));
UDR0 = dat;
}



// ***********************************************************************************
// **** READ USART1 GPGGA DATA TO STRING @ 38400 BPS
// ***********************************************************************************
void GPSREAD() {
unsigned char ctr;
unsigned char data;

// $GPGGA, 223611.000, 4821.9234,N, 08916.4091,W, 1,   8,    1.17,   190.1,M,  -35.0,M, ,*6F
// $GPGGA, UTCTIME,    LATITUDE,    LONGITUDE,   FIX, SATS, HORIZ, ALTITUDE, HEIGHT   CHECKSUM

// CLEAR LAST STRING DATA
for (ctr = 0; ctr < 80; ctr++) {
gpgga[ctr] = 32;
}

// WAIT FOR $GPGGA HEADER
while (ctr != 6) {
ctr = 0;
while (!(UCSR1A & (1<<7)));
data = UDR1;
if (data == '$') ctr++;
while (!(UCSR1A & (1<<7)));
data = UDR1;
if (data == 'G') ctr++;
while (!(UCSR1A & (1<<7)));
data = UDR1;
if (data == 'P') ctr++;
while (!(UCSR1A & (1<<7)));
data = UDR1;
if (data == 'G') ctr++;
while (!(UCSR1A & (1<<7)));
data = UDR1;
if (data == 'G') ctr++;
while (!(UCSR1A & (1<<7)));
data = UDR1;
if (data == 'A') ctr++;
}
```

```
while (data != ',') {
while (!(UCSR1A & (1<<7)));
data = UDR1;
}


// READ GPGGA DATA UNTIL LINEFEED
data = 0;
ctr = 0;
while (data != 10) {
while (!(UCSR1A & (1<<7)));
data = UDR1;
gpgga[ctr] = data;
ctr++;
}
}



// ***************************************************************************
// **** PROGRAM VARIABLES AND PORT SETUP
// ***************************************************************************
void main(void) {

// MAIN PROGRAM VARIABLES
unsigned char ctr;
unsigned char mode;
unsigned char skip;
unsigned int trk;

// SET USART0 TO TX @ 38400,8,N,1
UCSR0A=0x00;
UCSR0B=0x08;
UCSR0C=0x06;
UBRR0H=0x00;
UBRR0L=0x17;

// SET USART1 TO RX @ 38400,8,N,1
UCSR1A=0x00;
UCSR1B=0x10;
UCSR1C=0x06;
UBRR1H=0x00;
UBRR1L=0x17;

// RECORD BUTTON PIN
DDRC.3 = 0;
PORTC.3 = 1;
```

```
// PLAYBACK BUTTON PIN
DDRC.2 = 0;
PORTC.2 = 1;

// RECORD LED PIN
DDRC.1 = 1;
PORTC.1 = 0;

// PLAYBACK LED PIN
DDRC.0 = 1;
PORTC.0 = 0;


// *********************************************************************************
// **** PROGRAM INITIALIZATION
// *********************************************************************************

// ERASE TRACK MEMORY
for (trk = 0; trk < 1440; trk++) {
track[trk] = 0;
}

// RESET VARIABLES
mode = 0;
skip = 0;


// *********************************************************************************
// **** MAIN PROGRAM LOOP
// *********************************************************************************
while (1) {


// *********************************************************************************
// **** CONTROL MODE BUTTONS
// *********************************************************************************

// RECORD MODE = 1
if (PINC.3 == 0) {
trk = 0;
mode = 1;
PORTC.1 = 1;
PORTC.0 = 0;
}
```

```
// PLAYBACK MODE = 2
if (PINC.2 == 0) {
trk = 0;
mode = 2;
PORTC.1 = 0;
PORTC.0 = 1;
}



// ****************************************************************************
// **** CONTROL STATUS LEDS
// ****************************************************************************

// IDLE MODE
if (mode == 0) {
PORTC.1 = 0;
PORTC.0 = 0;
}

// RECORD MODE
if (mode == 1) {
PORTC.1 = 1;
PORTC.0 = 0;
}

// PLAYBACK MODE
if (mode == 2) {
PORTC.1 = 0;
PORTC.0 = 1;
}



// ****************************************************************************
// **** MODE 1 : STORE (1440/80) = 18 POINTS TO TRACK MEMORY
// ****************************************************************************

if (mode == 1) {

// READ GPGGA DATA FROM GPS
GPSREAD();

// RECORD EVERY 20TH TRACK POINT
skip++;
if (skip == 20){
skip = 0;
```

```
// ECHO LIVE GPGGA DATA TO USART
DATSEND('$');
DATSEND('G');
DATSEND('P');
DATSEND('G');
DATSEND('G');
DATSEND('A');
DATSEND(',');
for (ctr = 0; ctr < 80; ctr++) {
DATSEND(gpgga[ctr]);
}
DATSEND(13);

// STORE GPGGA DATA TO TRACK
for (ctr = 0; ctr < 80; ctr++) {
track[ctr+trk] = gpgga[ctr];
}

// INCREMENT TRACK POINTER
trk = trk + 80;

// STOP RECORD AT SRAM FULL
if (trk > 1440) mode = 0;
}
}



// ******************************************************************************
// **** MODE 2 : SEND 18 TRACK POINTS TO USART
// ******************************************************************************

if (mode == 2) {

// SEND ALL 18 TRACKS
for (trk = 0; trk < 1440; trk = trk + 80) {

// CREATE GPGGA DATA STRING
DATSEND('$');
DATSEND('G');
DATSEND('P');
DATSEND('G');
DATSEND('G');
DATSEND('A');
DATSEND(',');

for (ctr = 0; ctr < 80; ctr++) {
DATSEND(track[ctr+trk]);
}
DATSEND(13);
}
```

```
// END OF TRACK MEMORY
mode = 0;
}

// END OF MAIN LOOP
}
}
```

Before taking your breadboard on a test journey, first make sure that your PC can receive the serial data stream by opening a terminal program such as Hyper Terminal, which is built in to Windows. Most compilers and IDEs also include a serial terminal, so connect your USB or 9-pin serial cable and then open the terminal of your choice for receive at the same baud rate that your GPS and microcontroller are set to. As shown in Figure 15-12, the serial communication settings are 38400,8,N,1 (38,400 baud, 8 bits, no parity, and 1 stop bit). "8,N,1" is a common setting from just about every device.

Figure 15-13 shows a screen full of various NMEA strings received directly from the GPS module. This data stream is taken directly from the GPS module by moving the TX pin from the GPS to the RX pin on the MAX232 chip. Basically, this takes the microcontroller temporarily out of the loop just so you can verify that your GPS module

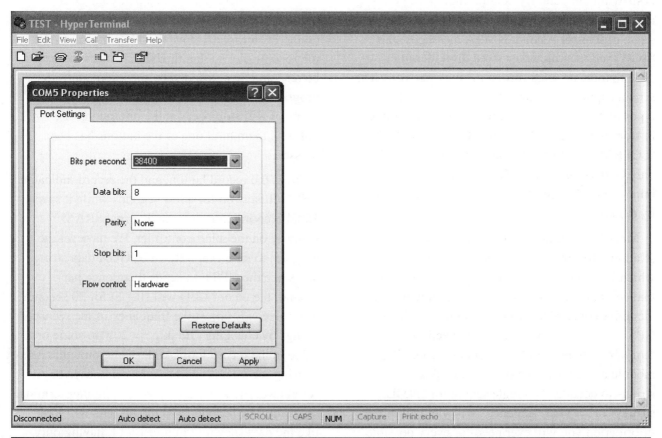

Figure 15-12   Testing the serial communications using a PC terminal program.

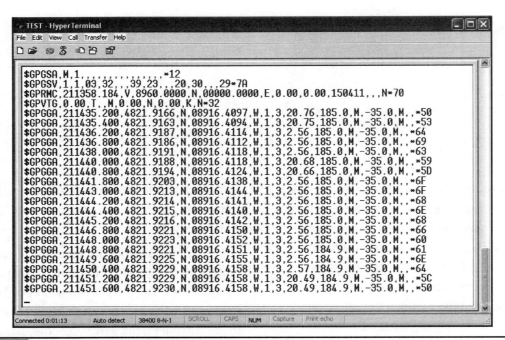

is spewing out data. By default, the FV-M8 GPS module was sending multiple NMEA messages at a rate of five per second. NMEA messages start "$GP" and then include a three digit identifier code to indicate what format the message is in. The screen capture in Figure 15-13 shows five different types of NMEA headers and messages streaming in from the GPS module; "$GPGSA," "$GPGSV," "$GPRMC," "$GPVTG," and "$GPGGA." We will only be decoding the GPGGA data string in the microcontroller, but the GPS module sends all five by default.

Most GPS modules output data in the NMEA (National Marine Electronics Association) format, which is a collection of standardized data formats that contain comma separated values flowing an identifier code. When you are coding your own GPS software, you will definitely need to get familiar with the data format sent by your GPS module and choose one of the many possible strings to decode. A Google search for "NMEA strings" or "NMEA sentence" will yield all of the information you will ever need for every possible NMEA string format.

Once you have determined that your GPS module is sending serial data by viewing it directly on the PC terminal, move the TX pin back to the microcontroller so that the NMEA messages can be read and then decoded by the microcontroller program. The data streaming in to the computer terminal will stop because the microcontroller will not send them until one of the buttons has been pressed.

Press the record button, and the record indicator LED will stay on for a few seconds while it saves 18 GPS coordinates into the internal SRAM memory on the microcontroller. We have set the program to save every 20th $GPGGA data string, and since the GPS unit sends them once per second, the record LED will stay lit for 20 seconds. You can easily adjust the frequency of the recorded strings by changing "if (skip == 20)" to some other value in the record loop. At the same time that data is being recorded, it will also be sent out to the serial level translator and to your computer terminal program. The decoded and rebuilt data will look like the data shown in Figure 15-14, having only the $GPGGA strings displayed. All other characters

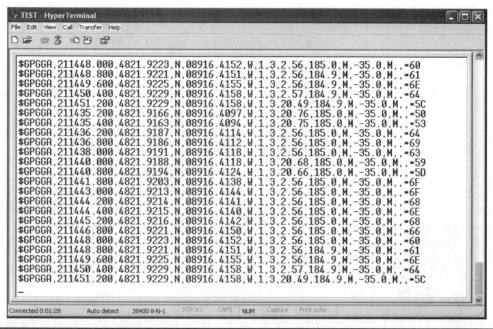

Figure 15-14   Displaying the decoded NMEA string from the microcontroller.

and strings have been stripped away, so the data is much more readable now.

This project assumes that you understand the workings of your GPS module, and if you are using a different GPS module from the FV-M8, you will likely need to adjust either the baud rate or the type of NMEA data being sent. If you can't set this on the GPS module, then you will need to change these parameters in the microcontroller program to match your GPS modules settings. More information on GPS modules and NMEA string formatting can be seen in the "GPS Data Receiver" project.

Now, let's use Google Earth or Google Maps to display our current location in real-time. Open Google Earth and then click on the "Tools" menu, followed by "GPS" in order to bring up the window shown in Figure 15-15. In this window, choose "NMEA" as the protocol, and then click on "Automatically follow the path" so that Google Earth will go directly to the location sent by the GPS. Once you press start, the window will disappear, and Google Earth will begin to track the real-time data received on the serial or USB port.

You can either move the GPS module TX line back to the RX pin like before or just press the record or playback button on the breadboard to get data to stream out to the computer. Either way, Google Earth will detect the data stream and then begin to decode whatever NMEA information is being sent. It may take a few seconds to get the globe to move, but once it starts, you will get to fly directly to the exact place on earth that you are currently located (give or take 20 feet [ft]).

Within a few seconds of sending the NEMA strings out to the PC, Google Earth jumps into action and rolls the globe around to zoom into our exact location as shown in Figure 15-16. Having a simple breadboard circuit able to know our whereabouts is both cool and a little creepy at the same time! As for accuracy, a GPS is usually accurate to about 20 ft, but the little location point on the Google Earth image seems to be right over our lab on the house roof exactly where expected. If we click on one of the "Street View" icons near our location, Google Earth zooms down to Street View and we are looking at a summer photo of the

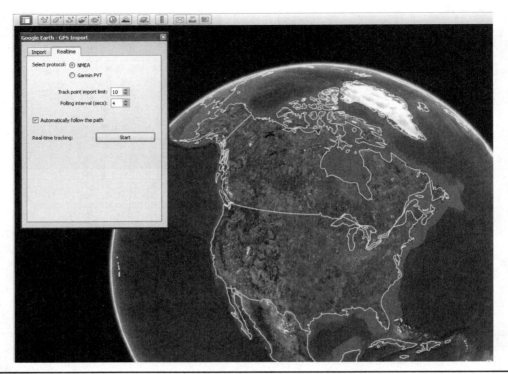

Figure 15-15   Setting Google Earth to display real-time NMEA data from a GPS.

Figure 15-16   Google Earth displays the exact location once the button has been pressed.

house, which is nice considering there is snow on the ground right now.

Once you have confirmed that the program can successfully receive, store, and then play back the NMEA data from the GPS module, it is time to take your prototype on a test journey. Since this version is limited due to the small amount of internal SRAM in the microcontroller, you will either need to make a short trip or adjust the record frequency to gain more time. Either way, you can only store 18 points in the AVR324 SRAM, but this is only a proof of concept model to expand on later.

We adjusted the record interval time so that we could record points as we drove around a two block radius. It took a few revisions to get the timing down to an even spread of 18 points, but in the end it worked out well. Figure 15-17 shows the breadboarded GPS tracking device ready to go on a trip around the block. As soon as we pulled out of the driveway, we pressed the record button and then made a loop around the next block over and back to the start location. At the end of the trip, the record LED was just turning off, so we knew the data would look like a round trip on the mapping program. Since the project was storing the points in the internal microcontroller's SRAM, we had to be careful not to disconnect the power as the data

Figure 15-18 En route around the neighborhood, recording GPS points.

would be lost. A large flash memory would be the optimal solution for a long-term GPS data logger.

This GPS module was very good at acquiring a lock in under a minute, which was actually a little bit faster than the handheld Garmin GPS unit. We did several trips around the neighborhood and had the handheld GPS along recording as well so that we could compare the tracks later on Google Earth (Figure 15-18). Both the GPS module and handheld GPS units showed an accuracy of about the width of the street, only wandering outside the street borders around corners. GPS technology is not accurate enough for indoor navigation, but an outdoor robot would do just fine in a large open area with some onboard obstacle avoidance.

Once the trip around the block was completed, the breadboard was carefully moved from the dashboard of the vehicle back to the lab for data downloading. In order to display the trip coordinates on the computer, a mapping program that can import and decode the NMEA strings saved by the program will be used. An easy method of analyzing the data is to simply capture the stream to a text file as it is played back from the microcontroller's memory. Figure 15-19 shows the GPS data being captured and saved as a text file from the Windows HyperTerminal program. To capture data from a terminal program, chose "Capture Text" and then select a location on your hard drive for the file. Once the file is open, press the playback button to

Figure 15-17 Getting set up for a quick trip around the block with the prototype.

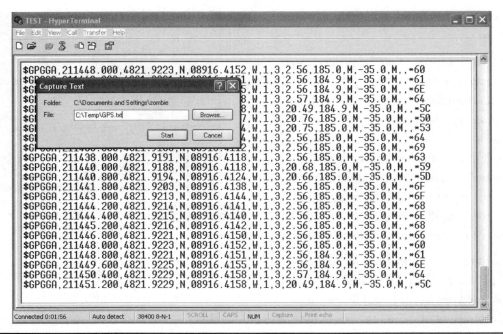

Figure 15-19 Capturing the GPS data for importing into a mapping program.

stream the data out to the text file. When you stop the capture, the file will contain the $GPGGA data saved by the microcontroller.

Since the Google mapping programs could not read the raw NMEA text file saved from the terminal program, it needed to be converted into a format the Google Earth or Google Maps could understand

and open (Figure 15-20). Searching Google for "NMEA to KML" yielded several GPS data format converters including one that worked directly online at http://www.h-schmidt.net/NMEA/. We clicked the box to open the file on the local hard drive after saving it from the Windows terminal program. There were a few other options available

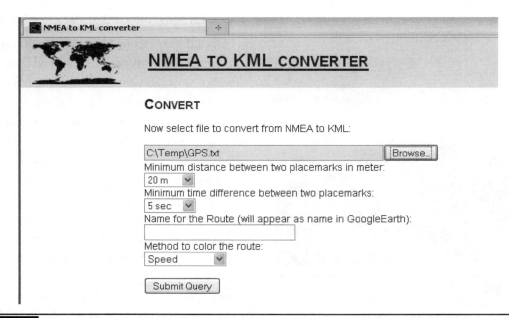

Figure 15-20 Converting the text file NMEA data into compatible KML data.

during conversion, but we just left them all at the default settings.

The online converter read practically any text file that contained valid NMEA strings and then converted them to a format (KML) that works with Google Maps or Google Earth. Having both programs installed gave the option to launch the application directly as shown in Figure 15-21. Hopefully, newer versions of Google Earth will be able to read NMEA text files directly, but for now the many online converters work just fine.

Once Google Earth opened, the KML file was read and the route was displayed as shown in Figure 15-22, with 16 points making up our entire journey. Comparing this to the data directly downloaded from the handheld GPS that was also along for the ride proved that this prototype was just as accurate as the commercially available GPS unit. For a future experiment, we will jam the works into a small black box, remove the indicator

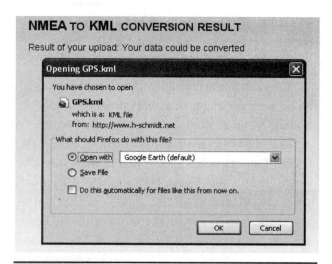

Figure 15-21 Once conversion was complete, the option to save or open appeared.

LEDs and include a strong magnet so the tracker can be stuck to a vehicle in some unnoticeable place. If the prototype passes this random "stick and run" test, then our next version will include

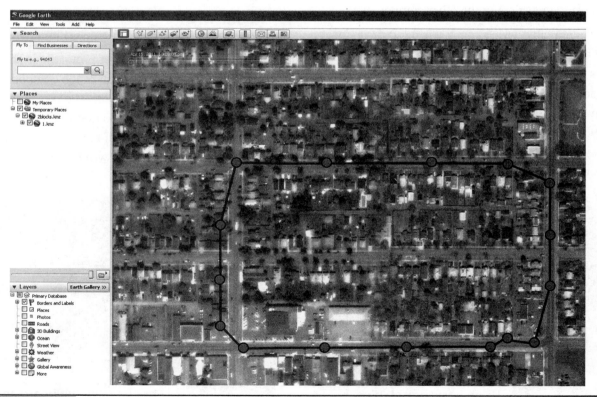

Figure 15-22 My prototype journey is now displayed as a path in Google Earth.

a much larger memory to store GPS data and possibly some more advanced NEMA string decoding and formatting.

Although a much smaller and cleaner prototype could be made using an actual circuit board and surface-mounted components, we just wanted a quick and easy-to-make boxed version of the GPS tracker for real-world testing. There wasn't much to this project besides the GPS module, a 40-pin AVR324 DIP, the MAX232 serial converter and a few capacitors so a small piece of perforated board was chosen to make the circuit board. The indicator LEDs were left out this time since they would be visible in the dark and use power for nothing during a long trip. The small black plastic box shown in Figure 15-23 has just enough room inside for the battery pack, the GPS module, and the small circuit board. Because the GPS has a built-in antenna, it cannot be installed in a steel or aluminum case as that would block the RF signal from the satellites. The black plastic box seemed to have no effect on the operation of the GPS as long as the antenna was against the plastic, not the other components.

There really wasn't much wiring to do under the small perforated circuit board since the circuit was very simple. Solder the components to the underside of the board using small wires as traces, leaving room between the pins so they can be bent at 90 degree angles to secure the components to the board. It is also a good idea to use a socket for any microcontroller that you may want to remove for

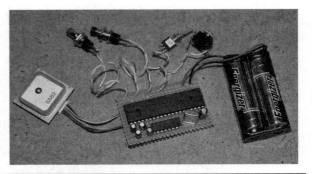

Figure 15-24   Creating the black box prototype circuit on a bit of perforated board.

reprogramming (if not using ISP programming), or for a future project. As for the 9-pin serial port, it was too large for the plastic box, so we changed it to a 1/8-in mono jack (shown next to the switch in Figure 15-24). Since only two wires are used for the serial port, any jack would work.

There was just enough room in the plastic box for the two switches and serial jack after installing the circuit board. This project could certainly be made much smaller with a little work, but it is still half the size of the pocket GPS, so it can be easily hidden under a car or in the box of a truck (Figure 15-25).

The completed GPS tracker prototype is shown in Figure 15-26 after fitting all of the components and hardware in the plastic box. The box has an aluminum lid, but that will be fine since the antenna on the GPS is pressed against the empty side of the box for optimal reception. It will be important to

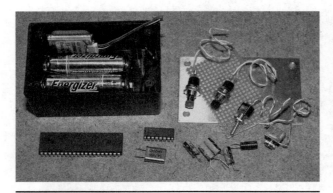

Figure 15-23   Using basic prototyping components for rapid development.

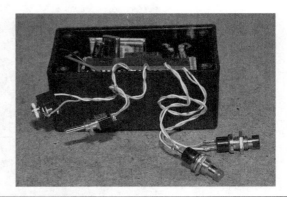

Figure 15-25   Fitting the circuit board and components into the small black box.

Figure 15-26 The completed GPS tracker prototype ready for action.

Figure 15-28 A ridiculously strong NIB magnet will hold the GPS to a steel object.

keep the plain side of the box facing outwards when sticking the unit to a vehicle so that the antenna is not blocked by any metal objects. The stealth installation of this tracking unit is very simple: turn on, wait 1 minute, affix to a vehicle on the outside or inside, press the red button. To get the data into Google Earth, just plug in the serial port and press the black button. All that is lacking in this proof of concept model is an external memory with much more room to store GPS coordinates.

Since the small plastic box did not have room for the DB9 serial connector, we just used a 1/8-in mono headphone jack inside the box. The adapter cable shown in Figure 15-27 has pin 2 of the DB9 serial connector going to the tip of the 1/8 connector and pin 5 of the DB9 going to the ground pin on the 1/8 connector. For two-way

communications between the PC and GPS module, a 1/8 stereo headphone connector could be used. We did not intend to change the default settings on the GPS module, so the two wire cable was fine.

To create a very secure mounting system, we fastened a 2-in long NIB (Neodymium) magnet (Figure 15-28) to the top of the plastic box using double-sided tape. This magnet was actually much too strong for this application, but we didn't want to lose the GPS on the road. When stuck to a metal body, the magnet holds on with such force that we need to pry it up with a plastic screwdriver as it would tear the lid right off the box if we tried to pull it free. A magnet of half the size would be just fine, but at least we know the GPS tracker unit wouldn't fall into enemy hands (or a storm sewer).

Installing the stealth GPS tracker is as easy as sticking it to any metal surface, but the antenna needs to have some room to receive the faint GPS signals. We found the unit to work well in most installations, including under a vehicle, but the license plate compartment was optimal if it included a small hiding place like the Chevy S10 shown in Figure 15-29. Having the super-strong NIB magnet meant that the GPS tracker could be mounted to the fiberglass parts of a vehicle as long as there was a metal object under the nonmetallic surface.

Besides much more storage memory, another improvement we will make in the next version is

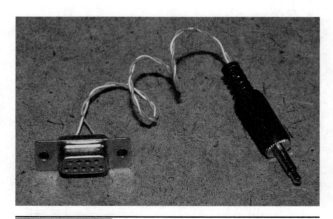

Figure 15-27 To retrieve the data, a 1/8 jack to 9-pin serial port cable is used.

Figure 15-29   Finding the perfect hiding place that would give the antenna a clear view.

until the recording starts would be best, so that there are no flashing lights on the box. Having a blinking red light would be an obvious giveaway at night, not to mention that the device would look like a small bomb! We are all for pranking our friends, but that would be too much!

The stealth GPS logger prototype performed as well as the breadboarded version and as well as the handheld GPS unit. Only once did the GPS lose the signal, but this was because the antenna was almost fully blocked by some metal objects in one of our test runs. To gain a bit more memory on this simplified prototype, we changed the AVR324p (4K internal SRAM) to an AVR1284p (16K internal SRAM) so that we could record four times as many NMEA strings, allowing about 20 minutes of decent resolution recording. This is still far from being much use in the real world, but did allow a much longer path to be shown on Google Earth (Figure 15-30).

to include the GPS Fix status LED once again so that the system can be started up before placement, saving time and guaranteeing that the unit will begin recording live coordinates as soon as the record button is pressed. Having the Fix LED blink

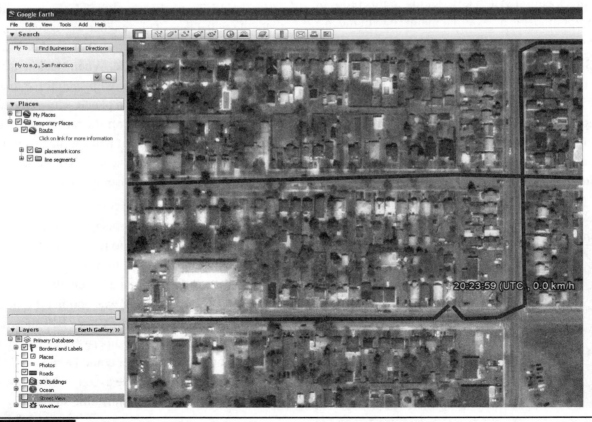

Figure 15-30   The stealth GPS logger prototype can track you around the globe.

The stealthy GPS tracker project was a great success. The simple microcontroller program can be much expanded on to offer greater control over recording frequency and operation. With the addition of an external EEPROM memory or Flash memory, the GPS tracker could be made to store hours or even days worth of information. By adding two-way communications between the microcontroller and PC, a complete terminal-based menuing system could easily be made that would allow the GPS tracker to be configured on the PC, adding features such as scheduling, motion detection, light detection, direction, and speed logging, and many other features that would only take minimal extra hardware. Using surface mount components and a lithium battery pack, the GPS tracker could be made as small as a matchbox, yet run for days using some power-saving microcontroller features. There is no limit to what can be done with this project, so warm up that soldering iron and put your ideas in motion

PART FIVE
Transmission and Interception

Laser Spy Device

THE LASER SPY SYSTEM is considered by many to be the Holy Grail of high-tech spy devices because it can give the user the ability to listen in on conversations that take place in a distant building without having to install a bug or transmitter at the location. The laser spy system was said to be invented in the Soviet Union by Leon Theremin in the late 1940s. Using a non–laser-based infrared light source, Theremin's system could detect sound from a nearby window by picking up the faint vibrations on the glass surface. The KGB later used this device to spy on the British, French, and U.S. embassies in Moscow. It is also interesting to note that Leon Theremin invented the world's first electronic instrument, a wand-operated synthesizer named "The Theremin" after him.

Figure 16-0 Build the long-range laser spy system.

PARTS LIST

- IC1: LM386 1-W audio amplifier IC
- Resistors: R1 = 1 K, R2 = 10 K, VR1 = 50 K potentiometer
- Capacitors: C1 = 0.01 μF, C2 = 10 μF, C3 = 470 μF
- Sensors: Q1 = NPN phototransistor or CDS cell
- Headphones: Mono or stereo headphones and jack
- Battery: 3- to 9-V battery or pack

The Laser Spy system goes by several names such as the Laser Microphone, Laser Listener, Laser Bug, Window Bounce Listener, and a few

similar names. The laser spy certainly works well under ideal conditions, but it has many strengths and weaknesses that will be discussed in this plan. Building your own laser spy is by far the best way to experiment with this technology as you can adjust the design to suit your needs, rather than forking over hundreds or thousands of dollars for an assembled kit that will likely be far inferior to one that you can build yourself. Many of the kits we have seen for sale over the Internet not only use dated technology, but they incorrectly state that the system uses a modulated laser beam to convert window vibrations into sound, which is simply not the case. Let's put the mysteries to rest once and for all and build a working laser spy system from the ground up and explore the functionality of each subsystem that makes a working unit.

We will be starting with an ultra-basic proof of concept test system that will show you how the laser spy converts vibration into sound and how careful alignment of both the laser and receiver are required for optimal performance. Ironically, the most basic configuration may prove to be the most useful, and the $20 you spend in parts could create a system that works as well (or better) than some of the ones that are for sale on the Internet for thousands of dollars. As you will find out, the key to spying with a laser beam is in the alignment and reception of the beam, not some magical black box full of fancy filters and optical components.

The obvious first component in the laser spy system is the laser, which will target a distant reflective object and send the beam back to your receiver for decoding. Before digging deep into this project, let us explain how this system works, and dispel some of the myths that are circulating on the Internet regarding the operation of this device. First of all, these laser spying devices do *not* work on modulation of the laser beam like some laser-communication devices. Modulation of the laser beam is impossible because it would take some type of circuitry installed in the laser driver to actually modulate the intensity of the beam, and the laser is going to be installed at your location, not the target location. The principle that is at work here is not modulation, but movement! As the laser reflects from the target window, the slight vibrations from conversations or noise that vibrate the windows cause a very slight change in position of the returning laser beam. This change in position is converted into voltage as the sensor in the receiver catches the returning laser beam. This is why the optimal operation of the receiver requires the laser beam to be slightly offset from the phototransistor as will be shown. So movement, not modulation is the principle on which this system operates.

You can use any laser you like for this project, and there will be no quality difference whatsoever between a state-of-the-art lab laser and a two-dollar pointer. The only disadvantage to using a cheap laser pointer is that you will have to modify it for an external battery pack if you plan to have it on for more than a few minutes at a time, but that is an easy task. Also, a visible red or green laser

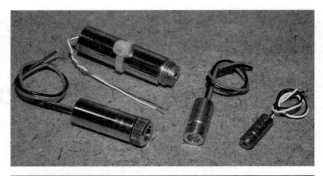

Figure 16-1 You will need some type of laser to bounce off of the target window.

will not allow for covert operation due to the fact that it is very easy to pinpoint the source of a laser beam, especially at night. For this reason, a visible laser is used for initial daytime targeting and then an invisible infrared laser is switched for optimal long-duration and night-time operation. Of course, this all depends on your intended use, and for demonstration purposes a visible red laser is best (Figure 16-1).

The first time we experimented with a laser spying device was in the 1980s when we found a DIY article in an electronics magazine. Lasers were huge, expensive beasts back then, but we were nerdy enough to have one to mess around with and followed the instructions in the article. In the end, the system was found to be 100% useless, and rumor at the time suggested it was all just a hoax. What happened was that the article failed to mention that as cool as this device was, it was extremely difficult to set up in the real world, especially when trying to bounce from a distant window. Believe us when we say that this device does indeed work, but using it to spy across the street will require a serious amount of setup, fine-tuning, and patience. To be perfectly honest, your chances of simply beaming toward your neighbor's window and hearing anything are about 1000-to-1 against you. So many factors have to be in your favor, such as the type of window, the alignment of the structure, the time of day, the level of sound, and mostly, your patience level. We have done

Figure 16-2   A small speaker will be used to simulate a vibrating window.

Figure 16-3   Add a reflective surface to the center of the speaker.

a successful window bounce from across a city street, but it was *not* an easy task, so keep that in mind. Any site selling this device in kit or plans form claiming that it is point-and-shoot should be deleted from your favorites in a hurry!

To create a "test window," to allow the deflection of the laser beam, a small speaker is connected to some audio source such as a radio or computer headphone port (Figure 16-2). Don't worry about how loud the audio source will be; as long as you can just barely hear the sound on the speaker, it will be good enough. Any small radio or portable music system will have a headphone jack that you can connect to your speaker. The size of the speaker is also not important as long as it is large enough that you can glue a small bit of mirror to the center cone to allow a surface to deflect the laser beam. Some speakers already have a chrome dome in the center, so if you can find one like that, then you will not need to use the mirror. Solder the appropriate jack to the speaker terminals; this will likely be a 1/8-inch headphone jack.

Any small piece of a highly reflective surface such as a mirror can be used to deflect the laser beam during these tests (Figure 16-3). A mirror works best, and a piece can be snapped from an old mirror using pliers or a small dental mirror can be

taken apart to remove the small round mirror from the plastic housing. A hot glue gun or even some double-sided tape can be used to glue the small mirror section to the center of the speaker. The size of the mirror is not important since the laser beam will only be a few millimeters across when it strikes the surface. If you intend to snap a bit off a larger mirror, use a cloth or paper towel to wrap the corner so that small slivers of glass do not fly from the mirror as you break it. A highly reflective plastic or metal surface will also work for this experiment, and even a shiny dime will do the job in a pinch.

The speaker needs to be driven by some audio source, but the level should be so low that you can only hear it when your ear is right next to the speaker (Figure 16-4). The goal is to recreate the same conditions that you will be dealing with during your covert spying operations, so the reflective surface should just barely be vibrating. A portable audio player is perfect for this test because it has a low-power amplifier and will run for hours at a time. Set your player to loop indefinitely and then adjust the volume as low as it will go until you can just barely hear the output from the speaker.

The alignment of the laser from the source to the target and back is not a trivial task, and as you

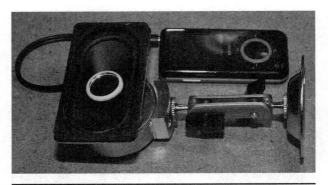

Figure 16-4   Feeding an audio source into the speaker.

Figure 16-5   Creating an easy-to-align speaker base.

increase the distance from the laser to the target, you also increase the error level. At 500 feet, the beam will become so sensitive to displacement that you will have to be careful when moving around the laser because the deflection of the floor inside your home will be enough to throw the beam out an inch or more. When we finally managed to set up a successful long-range configuration, we found that the alignment was so sensitive that even a passing car would create waves in the system due to the vibrations between the house and the road. The laser spy device is certainly capable of working from many hundreds of feet away (even across a few city blocks), but vibration would become so critical that you would probably need to have everything mounted on an extremely heavy concrete or metal base secured to the ground. These are things you will find out as you experiment with this project.

To make your life easy when experimenting with the basic indoor setup, create some kind of easily-adjustable speaker stand like mine using an old webcam base or adjustable bench vice (Figure 16-5). The ability to move the speaker to any angle and secure it will be key to testing the operation of your laser spy device as you build it. The targeting laser will also need some kind of adjustable base as well.

The adjustable window pane simulator will really help when creating and debugging this

project, as it would be next to impossible to try to target a distant window without first knowing what to expect in the captured audio signal or even if the unit is functioning properly (Figure 16-6). If the original 1980s magazine article had explained this, we might have had some success when we first tried this experiment!

The purpose of this initial experiment is to verify that you can indeed listen to sound that is vibrating a nearby reflective surface. In our case, it will be the small mirror glued to the speaker cone that will vibrate due to either sound from the radio feeding it, or by having a helper talk directly into the speaker as you listen for their voice at the receiver. This test receiver will be the most basic system possible, consisting of only a CDS cell, "cadmium

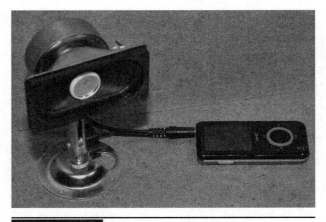

Figure 16-6   The adjustable window-pane simulator ready to use.

sulfide photocell," feeding a simple transistor amplifier that will convert the tiny changes in the laser beam position into changes in voltage that will be sent to your headphones as an audio signal. Think of the laser beam as being the "needle" on a record player and the vibrating window surface as the bumps in the record groove.

A CDS cell is basically a resistor that will change its impedance depending on how much light strikes the surface (Figure 16-7). By feeding it to a battery connected in series and then into an audio amplifier, the result is a light-sensitive audio system that will allow you to "listen" to light. Because this system has very little gain and no filtering, it will be a very minimal system, but will certainly let you hear the secret audio signal being sent along your test laser beam. This basic system would even work from hundreds of feet away if you could align the beam properly, which is part of the difficulty in using a laser spy system. A CDS cell can be purchased for a few dollars at most electronic suppliers, but if you don't want to wait for delivery or can't find one in stock, just purchase a night light and rip it apart to extract the CDS cell. The CDS cell will be the small disc with two leads and will have a wavy line on its surface. Just unsolder the two leads or bend the part back and forth until it frees itself from the tiny circuit board.

The photoresistor we are using in the receiver experiment was taken from a dollar store night light. The CDS cell is easily identified as the small

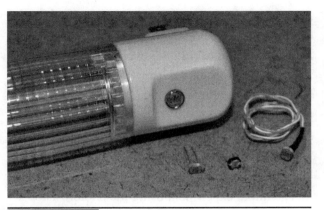

Figure 16-8   All night lights will contain a photoresistor.

disc behind the plastic lens. The internal circuitry of the nightlight is much like the circuit that will be used here to convert the light into sound. The change in resistance from the photoresistor is fed into the base of the transistor or switch that controls the current to the small light bulb. In our case, the light bulb will become headphones (Figure 16-8).

The laser will need to be installed on some type of adjustable stand or held in place at an approximately horizontal position so you can target the speaker and then catch the reflected beam at your breadboard. We clipped our red laser module into an adjustable bench vice so that it could be aimed at the speaker and then locked into position (Figure 16-9). When setting up the laser

Figure 16-7   A photoresistor will respond to a change in light.

Figure 16-9   Setting up the laser and speaker on a test bench.

and speaker, remember that according to the law of reflection, "the angle of incidence equals the angle of reflection." In other words, if your beam is lower than the speaker mirror, it will be bouncing back to you at a higher angle. As the distance from the target increases, this deflection becomes much greater. If you are feeling brave, try to bounce your laser back to your breadboard with the speaker at the other end of your lab—I dare you!

The initial experiment using the photoresistor will only require a single NPN transistor, resistor, and a battery in order to prove that the laser beam is definitely capable of picking up faint vibrations and changing them into audio. Any generic NPN transistor such as a 2N3904 or 2N2222 will work in this circuit. The battery voltage can be anywhere from 3 to 9 volts (V). If you really want to go basic, then just run a 9-V battery directly into the photoresistor and out to the headphones, although without the transistor to amplify the signal, the audio will be very faint.

This basic light-to-sound converter works because any change at the base of the transistor will amplify the current to the headphones (see Figure 16-10). Since the return laser beam will be bouncing around due to the vibration at the speaker, this will cause the headphones to respond to the beam as if it was an audio signal. Although the laser beam is moving,

not modulated, the principles at work are exactly the same with the exception that in this configuration, the beam needs to be slightly offset from the center of the photoresistor so that when it moves across the surface, there is a corresponding change in voltage. If the principle at work here was in fact modulation, then a direct hit onto the surface of the photoresistor would be optimal.

Turn on your audio source and adjust the volume so you can barely hear the output on your speaker, and then set up the laser and receiver so that the beam strikes the front face of the photoresistor (see Figure 16-11). As soon as the beam hits the photoresistor, the impedance will vary significantly, and you will hear a pop and probably a bunch of noise. Play around with the position of the beam to see how the position of the beam on the face of the photoresistor alters the reception of the audio signal. You will see that the optimal position for the beam is just touching the surface of the photocell so that any changes in the beam position from the vibrations will result in the most significant swing in voltage at the output of your headphones. If all you hear is a loud hum, then you probably have too much ambient light in your room. Incandescent light bulbs actually vibrate at 50 or 60 Hertz (Hz), and you will hear that in your system as a loud constant hum. As you have probably guessed, the laser spy system will not perform very well in the

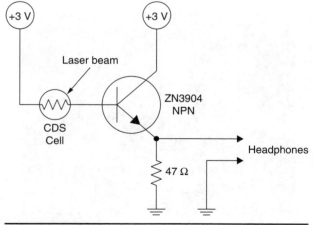

Figure 16-10 Schematic for an ultra-simple light-to-sound converter.

Figure 16-11 Targeting the photoresistor with the laser beam.

daytime due to ambient light sources competing with your laser beam, but this is fine since real spies usually operate in the darkness!

If you have your speaker and laser set up on the same workbench, then the process of targeting your photoresistor was probably only a 10-second job. Now, try to place your speaker at the other side of your room and see how long it takes to get the beam back to the target! We found that the distance across a room made the alignment significantly more difficult and even the deflection of the floor as we walked around made huge changes in the position of the beam. You will also have noticed that any slight vibrations of your desk or speaker stand resulted in all sorts of wild and wacky sound effects coming through your headphones. At one point, we were able to hear our own voices due to the thin surface of one of our tables vibrating the speaker stand.

These initial tests become very important so that you understand how well the laser spy system works, but at the same time how incredible finicky it will be to align at any real distance. Even though your test rig consists of nothing more than a cheap laser pointer and three semiconductors, it is actually a fully operational unit that could actually listen to a conversation a mile away if you could somehow capture the return beam (Figure 16-12). Seriously, this two-dollar unit is not much inferior to those "professional" laser spy devices you can find on the Internet being sold for hundreds or

thousands of dollars! All they have to offer is some built-in audio filtering and a more stable alignment hardware base. If you feed the output from your test rig into a real-time computer filtering software and mount everything to a solid base, you would have a system as capable as any available. Scary, huh? But we can actually improve the sensitivity of the receiver, so let's do that.

The photoresistor was able to convert the movement of the laser into a voltage change, but it is actually not the most optimal part for the job because it is slow to respond and has such a large reception surface. A phototransistor is a much better light receiver because it acts like an amplifier, has a much faster response time, and offers a smaller reception area to help catch the very slight changes in motion of the return laser beam (see Figure 16-13). A phototransistor is just a transistor in a clear case that has a photosensitive area feeding its base. In other words, it acts like a standard small-signal transistor but has no pin connected to its base.

Any NPN phototransistor will work in this project, and they come in a variety of styles and shapes. The most common variety will actually look identical to a clear LED, but may have a flat top to help focus the light onto the photosensitive base. This clear flat-top style will be your best bet. If you like to salvage your parts, then you can look inside an old roller ball computer mouse, as they

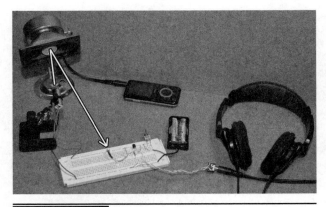

Figure 16-12 The test system is basic, but fully functional.

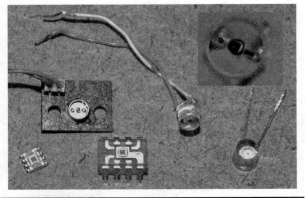

Figure 16-13 A phototransistor is faster and more sensitive to light.

will usually have several small phototransistors inside. Another source is any appliance that includes an infrared remote control.

Since laser light is so bright compared to any other type of light, the actual wavelength and lens type of the phototransistor is not all that important. Some phototransistors are more sensitive to infrared light than visible light, but they will still work just fine with any color of laser beam. If you are choosing a new phototransistor for this project, then look at the datasheet to determine the lens style, optimal wavelength, and make sure it is an NPN type not PNP (see Figure 16-14). There are hundreds of manufacturers and part numbers, so just go to an online electronics supplier like Digikey and enter "NPN phototransistor" to start your search.

If you have a huge selection to choose from, try to find one that has the best sensitivity to the wavelength of your laser, but keep in mind that for full covert operation of the laser spy, you will

probably want to switch to infrared light, which falls between 800 and 1000 nanometers (nm). Red lasers usually have a wavelength of around 650 nm, green lasers have a wavelength of 530 nm, and the newer blue lasers have a wavelength of around 470 nm. Again, don't be overly critical of the phototransistor wavelength chart as any light will create a response, especially laser light due to its brightness. Our phototransistor was rated for optimal performance in the infrared region but worked perfectly with the visible red laser as well as the green laser we tested.

The second version of the light-to-sound circuit is a bit more technical, but is still very basic as far as schematics go. The phototransistor is set up as an amplifier that feeds its output directly into the input of an LM386 audio amplifier integrated circuit (IC) in order to drive the headphones to a decent volume level. Oddly, the level of the output is never the issue; it's always the laser alignment

Silicon NPN Phototransistor, RoHS Compliant

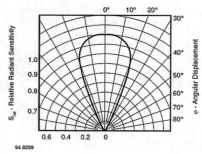

FEATURES
- Package type: leaded
- Package form: T-1¾
- Dimensions (in mm): Ø 5
- Leads with stand-off
- High photo sensitivity
- High radiant sensitivity
- Suitable for visible and near infrared radiation
- Fast response times
- Angle of half sensitivity: φ = ± 20°
- Lead (Pb)-free component in accordance with RoHS 2002/95/EC and WEEE 2002/96/EC

RoHS COMPLIANT

APPLICATIONS
- Detector in electronic control and drive circuits

DESCRIPTION
BPW96 is a silicon NPN phototransistor with high radiant sensitivity in clear, T-1¾ plastic package. It is sensitive to visible and near infrared radiation.

Fig. 8 - Relative Spectral Sensitivity vs. Wavelength

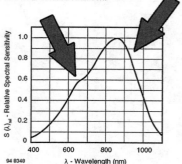

Fig. 9 - Relative Radiant Sensitivity vs. Angular Displacement

Figure 16-14   Choosing the best phototransistor for your receiver.

that causes the most effort, and we have yet to turn up the volume control more than about 10 percent of the way. Overpriced rigs from so-called Spy Stores will usually include some type of vocal pass filter or noise reduction circuit in the receiver as well, but in reality, these are pointless when you can simply feed the unfiltered output from this device into a computer sound card and process the audio using software that is going to be many times more effective than any filtering hardware. Your best bet is the recording and storage of the raw audio for post work on a computer later, so fancy front-end filter hardware is really a waste of time.

The LM386 is a very common audio amplifier IC that requires very few external components to operate. It has more than enough output power to drive a set of headphones to a very loud level, so the variable resistor that controls the volume level is often set very low when in use. You could certainly feed the output from the phototransistor to any audio amplifier or even a computer sound card as well. If you do plan to include some type of advance real-time audio filtering, then take the audio from the output of the phototransistor so that no extra noise is included in the signal as it is amplified. Many audio editing programs include support for real-time filters and audio processing,

so for very little money you can create a real-time post-processing system that will rival anything that was available to even the most well-funded spy agencies just a few years ago. There are audio-processing programs available for your PC that can restore the faintest conversation from the noisiest background (see Figure 16-15).

Before committing any design to hardware, it is a good idea to test your circuit on a solderless breadboard so that you can work out any bugs and add whatever modifications may be necessary (Figure 16-16). We found the new phototransistor-based light-to-sound circuit performed much better than the previous photoresistor version and had more than enough amplification to hurt our ears when turned up to the maximum. There comes a point where a louder signal is pointless as you will only be amplifying noise, and that is where more advanced audio processing may come in handy. If you do plan to add some post-processing audio filters into your schematic, then the most useful of these would be a 50- or 60-Hz notch filter to reduce the buzz from ambient AC light sources.

Try your new system using various levels of sound from your audio source, and even try to catch the beam from a reflective object across your room. As you will see, this is no trivial task! We were able

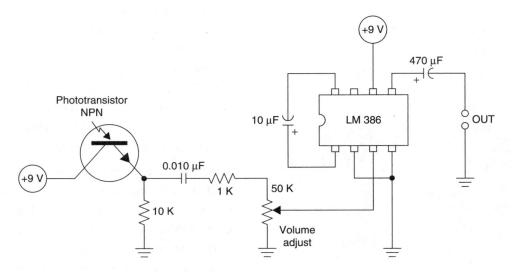

Figure 16-15   The laser spy light-to-sound receiver schematic.

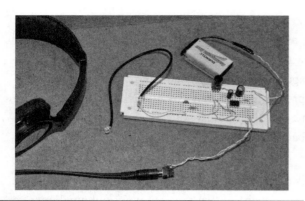

Figure 16-16 Testing the phototransistor-based circuit on the breadboard.

Figure 16-18 Perforated board (perfboard) is all you need for this circuit.

to tape the mirror to a lampshade and hear music from across the room, but this task involved very careful adjustment of the beam and the lampshade, a luxury you will not have when lurking in the shadows on a real covert spy operation.

Once you are happy with the performance of your light-to-sound receiver, it will need to be installed in some kind of cabinet so that you can mount it on a tripod for real-world use (Figure 16-17). The optimal installation will place the lens of the phototransistor on the front of the cabinet and place the headphone jack and volume control on the rear. Any square metal or plastic box that has enough room for the battery, controls, and the small circuit board will do the trick. We found a 3- by 2-inch (in) plastic box that had just enough room.

Since the LM386-based light-to-sound circuit is so simple, a tiny bit of perforated board is

all you need to create a circuit board for the device (Figure 16-18). Actually, there are so few components that you could just solder the pins together and hand them right off the variable resistor if you wanted to. This type of installation is called a "dead bug" circuit because the IC will look like a bug that has died and turned over. We prefer to use the perfboard method as it makes it easier to alter the circuit later. The LM386 only has to share the board with two capacitors and a resistor since the other components are placed right on the variable resistor terminals. All wiring is made on the underside of the small perfboard.

Rather than running a wire from the phototransistor to the perfboard and back, a few of the parts were soldered right to the variable resistor. When dealing with what might be a noisy signal, you need to create the most noise-free amplifier possible so that problem signals have a chance at being restored on an audio-processing program at a later date (see Figure 16-19). It is amazing what can be done to a noisy signal on decent audio-processing software.

Once your small circuit has been completed, install all of the hardware onto the project box such as the power switch, volume control, headphone jack, and the phototransistor. Hopefully, you will still have room for the battery as well as the small circuit board! We came very close to running out of room, but managed to squeeze all of the parts into the small plastic box (Figure 16-20).

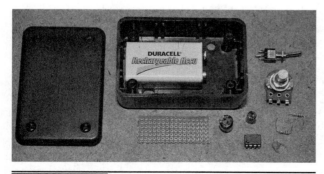

Figure 16-17 Finding a suitable project box for the light receiver.

Figure 16-19 Keeping components close to the source reduces noise.

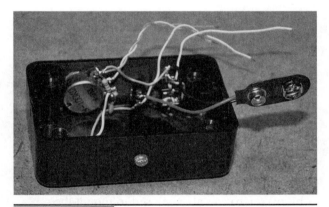

Figure 16-20 Installing all of the hardware into the cabinet.

Here is an "official" schematic for those who think a hand-drawn schematic is just too damn old-school! Personally, we like the look of the coffee-stained, hand-drawn schematics with the ink smudges and spelling mistakes…there is a certain character to them (see Figure 16-21).

The perfboard we used had solder pads on the underside so that it would be easy to solder the leads as well as the connecting wires. Some perfboards are just fiberglass wafers with holes, so you will have to use wires on the underside or bend the leads of the semiconductors to create traces. With a circuit as simple as this, any method would be just fine (see Figure 16-22). The 9-V battery connector and all of the wires that connect to the switch, headphone jack, and volume control were also added.

If your project box has no slots to fasten the battery or circuit board, a bit of double-sided tape, Velcro, or hot glue can be used to hold all of the parts from bouncing around inside the box (see Figure 16-23). Make sure that the underside of the small circuit board does not come into contact with any of the other components or the metal case on the battery, or you might cause a short circuit. When installing all of the components, consider

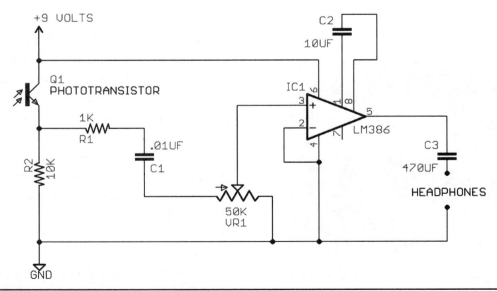

Figure 16-21 The final laser spy light-to-sound receiver schematic.

the fact that eventually you will have to change the battery, so place it in a position that makes it easy to remove or insert. It's also bad luck to put the lid on the box before you test the device, as the law of Murphy will most certainly make your unit fail if you assume everything is done properly! Put on the headphones, and just pass the laser across the surface of the phototransistor, and if everything is working, you will hear a loud thump as the amplifier is saturated with voltage. You can also "listen" to other light sources such as incandescent light bulbs and even your TV remote, which will make a *blip-blip-blip* sound as you point it at the unit and press any button.

| Figure 16-23 | Fitting all of the hardware into the project box. |

| Figure 16-24 | The completed receiver with a tripod mount. |

The laser spy system will require a tripod mount for both the laser and the receiver so that it can be used in the real world (Figure 16-24). To affix a plastic project box to any tripod, drill a hole that is slightly smaller than the size of the tripod bolt, and then thread it into the hole. Keep in mind that the bolt cannot press into any of the components, so choose the placement of the bolt hole carefully if space is tight in the project box. A bit of Velcro can also be used to mount the devices to a tripod, but remember that any slight change in position can easily throw your beam out of alignment, so ensure that all parts are mounted securely.

Since the only real obstacle between you and some distant eavesdropping mission is alignment, you must ensure that nothing moves once you have managed to lock onto the return beam. Since most camera tripods are fairly lightweight and easy to move, the headphone cord should be affixed to the base of the tripod leg so that it does not vibrate the tripod when you move around while wearing the headphones (Figure 16-25). Just use a zip tie or tape to hold the headphone cable to the bottom of one of the tripod legs, but allow enough slack so you can aim the unit. As you will soon find out, using the laser spy device is only a matter of alignment, and is very capable if you can learn

Figure 16-25 The tripod-mounted light-to-sound receiver.

the black art of catching the return beam. Do not expect success on every mission especially if the target window is above your line of sight.

If you are serious about your covert spy operations, then you can't go around pointing a visible laser at your target, or you will instantly compromise your position at night. Even if you are a mile away, people looking out the window can easily pinpoint your exact location if you point a laser in their direction, so your only option for truly covert operation is to use a laser beam that is invisible to the human eyes. A 3-5 milliwatt (mW) infrared laser module is no more expensive than a typical visible red unit and will work just as well, although you certainly increase the complexity of catching the return beam.

So how do you send an invisible beam across the street and attempt to have it land exactly on the 1 millimeter phototransistor when the beam is completely invisible? Well, you can't, so you will need two lasers—one to do the initial targeting, and the other to do the covert surveillance work. Now, this may sound completely impossible, but we can assure you that it does work, and we have managed to target a window using the infrared laser as well. It is *much* more difficult, but it can be done if all of the conditions are in your favor. As for the

performance of the receiver, it actually works better with the infrared laser, as the collimating lens on a real module is usually better than on a pointer, and most phototransistors are centered around the infrared band. So if you can aim the thing, it will perform!

The key to using two lasers is to have them pointing in the exact same direction so that you don't have to mess around too much once you have a visible target acquired. You could try fastening the bodies of the two lasers together using some type of clamp, or drill two very accurate holes through a half-inch thick piece of aluminum as we have done. The distance between the two lasers is negligible, but the angle of the two lasers must be as precise as possible. Even a degree of error could mean 20 feet of error on a beam that is returning from a 500-foot round trip! You will need to use a drill press to make the holes so that they go through the material at exactly the same angle. If you have an unlimited budget, have a machine shop do this drilling and explain to them that alignment is ultra-important (Figure 16-26).

The two lasers will also need to be mounted to a box that will allow room for a battery pack and a switch to turn each laser on or off (Figure 16-27). Once you have found a way to mount both lasers so that they are pointing in the exact same direction,

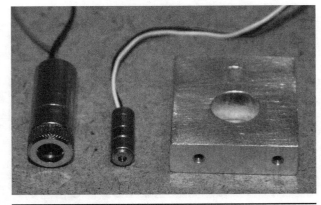

Figure 16-26 Making a dual-laser alignment block.

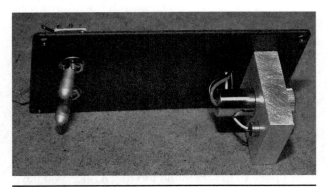

Figure 16-27   Mounting the lasers to a box with switches.

affix them to a plastic or metal box that can also be attached to a tripod just like the receiver box. Since you will have no control over the position of the target, you will certainly have to move both the laser and the receiver in order to get a lock on the beam. Don't forget to label the infrared laser's switch as you will not be able to see the beam when it is operational!

With the two units completely boxed and operational, do a few basic tests using your speaker and mirror arrangement, starting with only a few feet of distance and working your way up to the length of your room (see Figure 16-28).We can easily get our system aimed and working from any distance across a room and even across a street if we have total control over the placement of all three components. Alignment in the real world

Figure 16-28   The completed laser spy components.

using a distant target is another story altogether, but the system does work if you can just get that laser beam back to the same room it came from. We would have to say that you have a 1-in-10 chance of acquiring a target from a window that visually seems to be almost at the same angle and horizontal position of your own window. Remember that every degree of misalignment adds up to a large distance when the target is more than a few feet away from your location. Alignment really is a black art.

If you are still on good terms with your neighbors across the street, and they get a kick out of your evil genius experiments, ask them if you can set up your mirrored speaker system at their house for a long-range test. You will need to send the beam through the window and then set up the speaker so that the beam returns to your own location. All of these tests need to be done at night as well, as you will have little chance of seeing the low-power laser beam outdoors in the daytime. Once you can get the beam back to your receiver, verify that you can hear the output from the audio system, and then switch off the output so that only the sounds from the room are vibrations from the speaker. The speaker will act much like the window pane, allowing you to see the difference between a controlled audio source and one that is being driven by vibration alone. As you will see, the tonal qualities are very different due to the way the object vibrates. Larger objects will produce a sound with much more bass, and this is where your computer can be used to work magic on the received signal.

When you are working with your laser spy system in the real world, you may get a lock on the beam yet fail to understand the received audio due to the effects of vibration and ambient room noises (see Figure 16-29). If there is a furnace or fan running in the room, huge waves will be produced with extremely low frequencies, or maybe a bright light is creating a loud AC hum in your audio signal. A good audio equalizer or some type of

Figure 16-29 The completed laser spy system ready to use.

real-time computer filtering software can really do a lot to clean up these types of ambient background noises, so learn how to use notch, band pass, or noise-filtering software, and you will be able to work magic on some noisy signals.

Once you get used to setting up your laser spy system to acquire an audio signal, you can then decide if some kind of front-end or post filtering will be necessary. We actually found that for the most part, filtering was not necessary, especially when working with the unit at night. A simple 8-band equalizer was a good method of removing 60-Hz AC hum or booming bass, but these noise sources were mostly present at the

source location, and easy to control. If you are recording your captured audio, then a simple 60-Hz notch filter will almost eliminate most of the AC hum picked up by ambient light sources, and the vocal frequencies can be further enhanced using a band pass filter that is set to pass only frequencies between 80 and 100 Hz (Figure 16-30). We also found that vehicles passing by our location would create a bounce in the return beam, adding a thump into the signal, but most of this was removed by cutting the bass frequencies on the equalizer. Again, you will have to try these tests for yourself as you will likely encounter very different sources of noise at your location.

As the laser beam travels further, the beam spread becomes greater (Figure 16-31). Some laser pointers have very cheap plastic collimating lenses and will end up spreading the beam so wide that it will cover your entire wall by the time it bounces off a window across the street. For this reason, a decent laser module with an adjustable lens will always be better for long-range operation. Of course, you only need to catch the very edge of the return beam, and as long as it strikes the photosensor, there will be sound. Sometimes, the dull, spread-out beam gave much better results than a highly focused point.

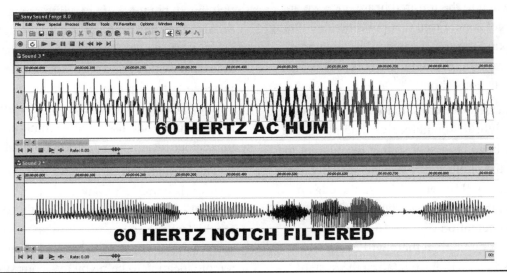

Figure 16-30 Computer software will do wonders for a noisy audio signal.

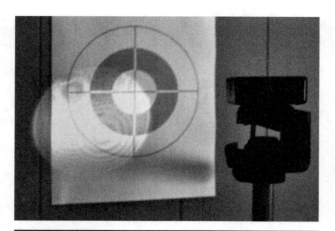

Figure 16-31   The return laser beam will no longer be a tiny spot on the wall.

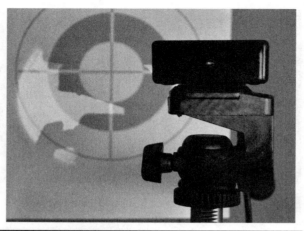

Figure 16-32   The edge of the beam is the optimal point for clear audio.

For alignment, we usually tape a paper target on the wall behind the area where we will place the receiver so that we can carefully adjust the laser beam tripod while looking at the wall, rather than waiting to catch the return beam in our eyes. A 5-mW laser will not have very much focused intensity by the time it travels across a street and back, but that does not mean you can ignore laser safety. This target is also the only way to verify the alignment of an infrared laser if you are using one because you will need to view the laser spot using a camcorder as the human eyes cannot see infrared light. We do all targeting using the visible red beam and then switch to the infrared beam once we are near the target zone. From that point the fine tuning is done looking through a camcorder viewfinder or security camera on a monitor. This makes alignment a tedious chore!

Once you have the beam onto the target area, turn on the light-to-sound receiver and place it in front of the large laser spot so that the edge of the beam is on the photosensor surface. Usually, the best reception will be received with the edge of the beam on the photosensor, but this will require some experimentation (Figure 16-32). If the target room is quiet, then there will be no window vibrations so you will have to align the system based on the *pops and clicks* you hear as the beam

saturates the photosensor. When the beam is not striking the photosensor properly, there will be a crackling sound similar to an FM radio that has lost its station and when you oversaturate the photosensor, there will be a loud *wump* followed by complete silence. The optimal "noise" you want to hear when setting up from a quiet target is a slight hiss.

Unless you are on a ground floor or a concrete slab, the laser will probably jump all over the place as you walk around the room or when nearby vehicles drive past. We found that this laser bouncing did not degrade the received audio all that much, but there are certainly things you can do to reduce this problem. Adding weight near the top of the tripod will reduce any oscillations or at least slow them down. A 5-pound steel weight fastened to the top center of the tripod worked well on certain occasions. Of course, you will just need to experiment and use creative thinking in order to make your installation perform optimally.

The ultimate test of the laser spy system will involve listening to a conversation in a room that is remotely located from the device. Of course, you would "never" try to eavesdrop on an unsuspecting party, so you will need to have a helper at the remote location to help you test your system. We originally built this unit to dispel the many myths about its "ease of use," but were seriously surprised at how well it does work if you can actually get the return beam aligned properly. In our test, we set up the unit and then went to the target location and spoke loudly right at the window, expecting that the received audio would be barely understandable (see Figure 16-33). Well, if you've seen the video then you can see that the reception was very loud and clear, to the point that it almost sounded like a radio transmission! We honestly did not expect it to work that well.

Of course, make no mistake—alignment from across the street is no easy task, and every single factor has to be completely in your favor to have any chance at all of success. The target window must be at a very precise angle to allow the beam to come back to your location, and the conversations must be loud enough to create a decent vibration on the window. Gas-filled high-efficiency windows further reduce your chances, and any noise in the room may completely overpower the conversations. Remember, we were in the totally silent room almost yelling directly at the window. Further testing proved that lower level conversations were also possible, but computer filtering would be necessary. In our tests we also found that the

Figure 16-34   **The laser spy system is a great source of entertainment.**

infrared laser did better than the visible red laser, but it was a magnitude more difficult to get proper alignment. So, our conclusions are that the laser spy device is definitely a powerful tool, but your chances of pulling up in a darkened vehicle to point and shoot are almost zero. Set-up and alignment require extremely careful and tedious planning, often resulting in failure. It's not the hardware that fails; it's the path of the laser beam.

The completed laser spy system was a great demonstration platform, proving that a person with enough motivation could potentially eavesdrop on a conversation a mile away without access to the location (Figure 16-34). The chances of this actually happening are extremely rare, but nonetheless possible! So, the next time you catch the flicker of a laser beam out of the corner of your eye, ask yourself, "Was that some kid with a keychain pointer, or a highly motivated individual who just read this article?" Yes, indeed, a laser beam can steal your secrets, so your best bet is to be on the receiving end of the beam, not the transmitting end.

Figure 16-33   **Beaming a window across the street in a controlled test.**

Basic Spy Transmitter

RADIO FREQUENCY PROJECTS can seem more difficult than most electronics projects because most of the time you cannot build them on a solderless breadboard, and they may use parts that are not easy to source, such as coils and adjustable capacitors. This project is focused toward those who have not yet attempted to build any kind of radio frequency (RF) project, and it is laid out in such a way as to make it easy to explore the basic principles of RF circuitry and ensure a successful final product.

Figure 17-0 Build the two-transistor spy transmitter.

PARTS LIST

- Resistors: R1 = 2.2 K, R2 = 22 K, R3 = 22 K, R4 = 4.7 K, R5 = 1 K, R6 = 100 Ω

- Capacitors: C1 = 0.047 μF, C2 = 10 μF, C3 = 0.22 μF, C4 = 0.47 μF, C5 = 10-50 pF variable, C6 = 5 pF, C7 = 0.022 μF

- Transistors: Q1 = 2N3904 or 2N2222A, Q2 = 2N3904 or 2N2222A

- Inductors: L1 = 5 turns of #18 copper wire at ¼-in diameter

- Microphone: Electret 2 wire style microphone

- Antenna: Antenna is 6- to 12-in insulated copper wire

- Battery: 3- to 9-V battery or pack

This simple two-transistor audio transmitter will send the sounds picked up in a room to any FM radio tuned to the same frequency as the transmitter, somewhere between 80 and 100 megahertz (MHz). The expected range will be at least 100 feet (ft) and could be substantially longer depending on the parts used and the quality of your final product. This circuit is based on one that has been around since the 1960s and published thousands of times, so it is tried, tested, and guaranteed to work if you follow the instructions. Performance is "okay," but since this is the one of the most basic transmitter circuits possible, don't expect high quality or rock-solid performance.

Since many of the parts only need to be "close enough," you will probably be able to salvage all that you need from any old radio, TV, or RF-based circuit board (Figure 17-1). Even the two transistors used are generic, and practically any small-signal NPN transistor will work here. Looking up datasheets on the various transistors pulled from old circuit boards is a great way to learn the important parameters such as VCEO (Collector Emitter Voltage), VCBO (Collector-Base Voltage), VEBO (Emitter-Base Voltage), and IC

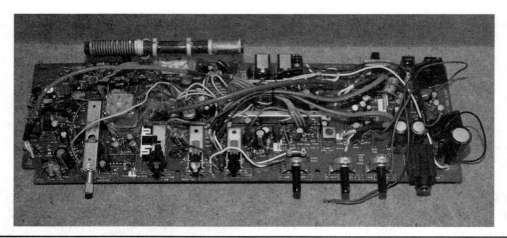

Figure 17-1 You can salvage most of the parts needed from an old radio.

(Continuous Collector Current). Anything close to the specs shown in the parts we are using for this project will be good enough for this transmitter. Out of the 10 or more random NPN transistors, we tried, only two of them would not transmit onto the FM radio band.

Resistors, capacitors, and even the wire needed to wind the small coil can also be salvaged from an old circuit board if you hunt around for the correct values. The only component you may have to order from a supplier will be the small 10 to 50 picofarads (pF) adjustable capacitor, but we will offer an alternative solution to the circuit that will not require the trimmer capacitor at all. Read over the entire project before heating up your soldering iron so you know what parts you may need and how the circuit can be modified to suit your own junk box inventory.

This transmitter can be thought of as two independent stages that work together: an audio preamplifier stage that feeds a radio frequency and oscillator stage. The audio preamplifier stage consists of most of the semiconductors in the circuit, including the small electret microphone, which is basically a tiny microphone in a can with its own built-in amplification circuit. Because of this built-in amplifier, the electret microphone is able to drive the single-transistor preamplifier circuit to a very decent level, enough to hear just about every whisper in a room (Figure 17-2).

Figure 17-2 The electret microphone will drive the audio preamplifier stage.

The output from the audio preamplifier is then sent to the radio frequency stage in order to create the needed frequency modulation. To make it easy to build and test this project, the audio stage will be built first and then tested before any of the radio frequency components are added to the circuit.

Electret microphones can be salvaged from most small consumer electronic devices that records audio. Answering machines, old tape decks, dictation machines, and even kids' toys will have one inside. The electret microphone is very easy to identify—it will be the pencil eraser-sized metal can with a felt pad on one side and two wires or terminals on the other. Sometimes the microphone will be wrapped in a rubber casing, which can easily be removed.

Figure 17-3   The radio frequency coil is extremely easy to make.

Figure 17-4   Wrap the bolt five times and you now have your RF coil.

Coils are usually the most difficult part of any radio frequency circuit, as most hardware hackers do not own equipment to measure and test coils. The good news is that there is only a single coil in this circuit, and it is so easy to make that it is almost impossible to do it wrong! (Figure 17-3) All you need is some small enameled copper wire and a ¼-inch (in) diameter bolt or dowel to wind it on. This copper wire can be pulled from an old transformer, toy motor, relay, solenoid, or purchased new at most electronics supply outlets. As for the wire gage (thickness), don't worry too much about it—1 millimeter (mm) or somewhere near 1 mm is close enough. We have built many versions of this transmitter using all kinds of varying scrap parts, and it usually works. Most times, errors are due to wiring, not the parts used in the circuit.

The making of the RF coil is as easy as wrapping the enameled wire around the bolt five times and then cutting off the ends (Figure 17-4). Leave about ½ in of wire at each end so you can scrape the enamel off of the wire and solder the coil to your circuit board. That's all there is to making the necessary coil. Practically any wire close to 1 mm in diameter will work as long as it has the protective enamel coating. You could actually get away with bare wire, but the coils would have to be extremely close together but not touching each other or the coil would fail. Enabled copper wire is much easier to use.

At one point we even tried making the coil with four and also six turns, and were still able to make

the transmitter send the RF signal to an FM radio. Five turns seems best for reaching the center of the FM radio band, though. Try to get the coil loops as close to each other as possible by keeping the wire tight as you wrap it around the bolt. Once the five turns are made, just unscrew the bolt from the coil.

Once you unscrew the forming bolt from the coil, cut off each end of the wire to about a ¼ in in length and then gently press the coil together to take up any gaps left between each loop of wire. The completed coil will perform just as well as any manufactured open-air coil of similar size, and it only took 5 minutes to make. Some radio coils are much more complex, containing a ferrite bead or even some internal circuitry, but for these simple, low-power, low-frequency open-air coils, you can almost always roll your own if you know how many turns are going to be needed (Figure 17-5).

Figure 17-5   The completed RF coil after removing the forming bolt.

Figure 17-6 The enamel needs to be scraped from the end of the coil.

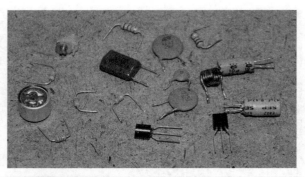

Figure 17-7 Sourcing the rest of the needed components for the transmitter.

In order to use the coil in your circuit, the enamel needs to be scraped from a small portion of each end so that you can solder the coil into your circuit (Figure 17-6). A small razor knife or blade can be used to remove the insulating coating by scraping it along the edge. You don't have to get all of the coating off the wire, as the heat from the soldering process will likely melt away what is left. Just make sure that you can see some of the copper that is underneath the coating, and you will be able to complete the circuit between the solder joint and the copper wire.

If you have a good supply of junk circuit boards in your scrap bin, then you will be able to find all of the parts needed for this transmitter project, or at least all of the parts minus the small

trimmer capacitor (Figure 17-7). Use the following schematic as a guide to source the parts, trying to keep the values as close as you can. Remember that just about any small-signal NPN transistors will work, but this will require some searching of online datasheets while you scrounge around the old circuit boards. If you want to work with all new components, you could try an electronics shop or online supplier if they deal in small quantities. We usually always salvage our parts as it seems crazy to pay 20 bucks in shipping for a 10-cent component!

The schematic for the two-transistor spy transmitter is shown with the audio-preamplifier stage in green and the radio-frequency stage in red (Figure 17-8). As you can see, the audio-preamplifier

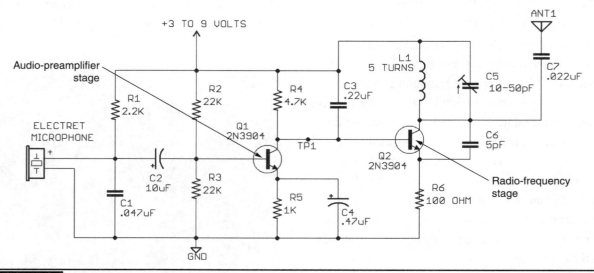

Figure 17-8 The basic two-transistor spy transmitter schematic.

stage takes up most of the schematic real estate, and parts count. Transistor Q1 forms a simple audio amplifier that is fed directly from the output of the electret microphone. Since the electret microphone also contains its own built-in amplifier, the system becomes very sensitive to any nearby sounds, often able to hear a whisper in a large room. Most of the capacitors in the audio preamplifier stage are there to stabilize the circuit. Because the transmit frequency is so dependent on voltage and loading, changes in current consumption from the amplifier would make the RF stage unstable without the capacitors to act as a buffer. You will see this as you tune up the circuit for the first time.

The radio frequency stage consists of the transistor Q2 and a "tank circuit" made from the hand-wound coil (L1) and the variable capacitor (C5). The coil and capacitor form a tuned circuit, which will oscillate at a frequency somewhere in the FM radio band, dependant on the setting of the variable capacitor. Since the tuned circuit is switched by transistor Q2, which is in turn switched by transistor Q1, changes in the audio preamplifier result in modulation of the RF stage. This is about as simple as transmitters get!

Commercial transmitters must include a lot more circuitry and quality components in order to achieve any kind of real stability, and as you will see when experimenting with the finished product, this unit is very prone to drifting if the voltage changes or if anything comes close to the antenna or RF stage. Our little room bug certainly works well if left undisturbed, but simplicity is exchanged for quality here.

Laying out all of the parts inventory right onto the printed schematic is a system we have used for many years, and makes it easy when scavenging for parts or possible replacement values (Figure 17-9). A resistor color chart will be handy unless you already have all of the color codes memorized since you can't just drop your ohm meter across any resistor that is still soldered into a circuit board and expect a valid reading. A small magnifier will help as well, since resistors and capacitors are so small on most modern boards.

As far as capacitor codes go, the numbers state the value in microfarads (μF) and include the number of places the decimal is moved to the left. We know…why not just put the actual value??? Hey, that would be too logical and use less ink, so

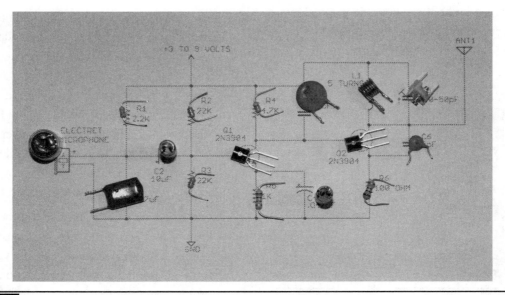

Figure 17-9   Laying out the parts inventory before the build.

let's not go there! The capacitor codes will be as follows: C1 is 0.047 µF (#473), C2 is electrolytic and will be marked as 10 µF, C3 is 0.22 µF (#224), C4 is electrolytic and will be marked as 0.47 µF, C6 is 5 pF and will be simply marked as 5. C5 is an adjustable trimmer capacitor (approximately 10–50 pF), and will probably have no markings at all. You may have to order one from a supplier or just take one from some old RF circuit board and hope for the best. The good news is that most of them will work anyway, as they all have very small values.

As an option to avoid having to source an adjustable capacitor for C5, you could just use a six-turn coil and insert a ferrite slug into it to allow the tuning to be done by the coil instead. This will also require that you replace C5 with a 10-pF fixed capacitor instead of the adjustable capacitor. Try to find the adjustable capacitor first though, as slug coil tuning is even more finicky than capacitor tuning.

Before we show how the completed circuit is created, let us take a small diversion and demonstrate how "not" to build an RF prototype (Figure 17-10)! When we first started electronics, we made many successful and extremely complex projects on large solderless breadboards, sometimes reaching digital speeds of 100 MHz in circuits containing hundreds of semiconductors. Success is almost guaranteed with digital circuitry and slow analog projects on a breadboard, so much that we often leave out the bypass capacitors altogether.

Figure 17-10 This is how "not" to build an RF circuit.

This is *not* the case when working with radio frequency circuits!

We had to learn this lesson the hard way after having so much easy success with our non–RF prototyping work. We experimented with RF many years later into the hobby and simply did not believe that we could not breadboard such a simple circuit. Hell, we made a 100-MHz computer on a breadboard with no bypass caps! Well, now we know that it is basically a waste of time, and we will prove it here just for fun.

Before we waste our time on this endeavor, let us explain why you have a 1 in 100 chance of ever getting any RF circuit to function on a solderless breadboard. The reason is capacitance. A look at the underside of a solderless breadboard shows that it is made up of multiple metal strips that join the holes together in rows. These metal strips actually act like small-value capacitors, often with more capacitance than some of the capacitors you may actually need in your circuit. So, imagine taking a working circuit and then randomly dropping 20 or more randomly placed 5-pF capacitors all over the place! As you can imagine, you will either radically alter the frequency or get nothing at all as a result. Don't get us wrong…sometimes it may work, but usually it's not worth the attempt!

We added all of the components to a tiny solderless breadboard, taking care to isolate the RF stage as much as possible and then hooked it all up for a test using an FM radio to receive the signal. Well, this time we were very lucky as the circuit actually made a tiny bit of crackle on the FM band if we placed our hand in the right place over the circuit! In other words, the stability was so awful that the RF circuit was oscillating all over the place and hardly sending out any RF energy at all (Figure 17-11).

Even when we finally managed to hear a faint crackle on the FM radio receiver, the output was barely registering, and no audio was being transmitted, just the carrier wave. It took many attempts to even get that far with the breadboard circuit, and required that we have a hand almost

Figure 17-11   This circuit almost worked...but barely!

Figure 17-13   The final circuit will be built on perforated board.

right over the entire circuit to add even more capacitance to the already confused oscillator! (See Figure 17-12.) So the moral of this waste of time is to not expect any RF circuit to work on a solderless breadboard no matter how much of a breadboard Jedi you may be. Of course, don't let us stop you from trying, after all—that's how we hardware hackers learn—from real-world trial and error!

To avoid as much stray capacitance as possible, the final circuit must be built on a bit of perforated board (Figure 17-13). There are several types of perf board, but you must choose one that does not have any soldering pads or strips, or you will be in the same boat as with the solderless breadboard— stray capacitance all over the place. This circuit is so simple that it only needs a piece of perf board

about 1 by 2 in in size, and you could even just build it on some cardboard by punching holes for the component leads if you wanted to. Plain perf board is your best bet, and if you are doing a lot of prototyping, these boards can be ordered in large squares for very little cost at most electronics suppliers.

The perf board needed for this circuit is about the same size as the side of a 9-volt (V) battery, although you could certainly make it larger or try to make it much smaller. Remember that the closer you jam all of the components, the more stray capacitance you will introduce to your circuit. If you follow the general size and layout shown in the next few steps, then your circuit is almost guaranteed to work unless you have bad wiring or components far out of specification (Figure 17-14).

Figure 17-12   Tuning was next to impossible due to stray capacitance.

Figure 17-14   Getting ready to lay out the small circuit board.

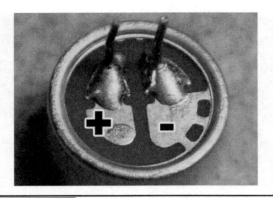

Figure 17-15   Start by soldering leads to the electret microphone.

Figure 17-16   Insert the electret microphone at the end of the perf board.

Most likely, the electret microphone will have no leads if it was salvaged from some appliance. A pair of leads can be made by cutting the legs off of another component and then soldering them to the pads on the underside of the electret body (Figure 17-15). Also take note that one of the leads has a connection to the outer ring on the microphone body; this pad will be the negative lead. Polarity is important on an electret mike as it needs to have power supplied properly in order to power up the small built-in amplifier circuit contained in the can. In the schematic, resistor R1 supplies this power to the positive lead on the electret microphone. If the microphone is inserted in reverse, it won't be damaged, but there will be no audio sent to the preamplifier stage. It's a good idea to mark the outside of the can with a marker or by scratching a point for the positive lead side.

The first component to install on the perf board is the electret microphone, once it has leads (Figure 17-16). Insert it near the end of the board so that it will not be close to the RF stage or have its opening blocked by any of the other components. If you want to make the microphone a bit more sensitive to higher-frequency sounds, you can peel away the top felt cover as it is only there as a "wind screen" to block out puffs of air if a person was to speak directly into the opening. In a spy transmitter, you are hoping that the speaker will not see the device, so a wind barrier is not really much use.

Also, take note of which lead is positive so that R1 can feed the required current to the internal amplifier.

Making a simple circuit on the perforated board is just a matter of inserting the components and then bending the leads to hold them in place (Figure 17-17). If the leads are long enough (as with new parts) then you can form the traces directly with the leads. Very complex circuits can actually be formed this way on perforated board, although when your part count rises, the number of wires will also rise.

The best way to build this circuit successfully is to first build up the audio preamplifier stage and then test it using a pair of headphones or an oscilloscope (Figure 17-18). Once all of the components on the green side of the schematic are

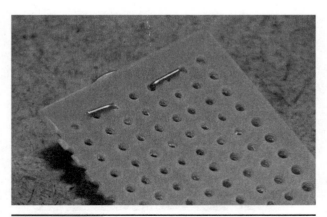

Figure 17-17   Components are held to the perf board by their bent leads.

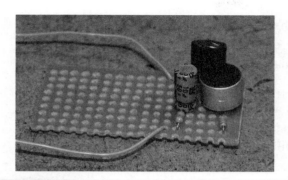

Figure 17-18 Adding the audio preamplifier stage components.

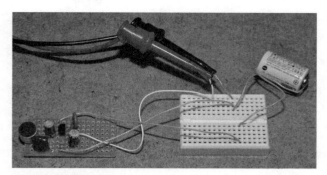

Figure 17-20 Tapping the test point to test the audio output.

inserted and wired together, you will hear or see an audio output by taking a reading or by connecting headphones between test point 1 (TP1) and ground. Once you know the audio stage is working properly, any dubbing will be limited to the new components that make up the RF oscillator stage.

Use as few jumper wires as possible as they will add extra capacitance to your completed circuit. By connecting the bent leads of each component together, you can make traces on the underside of the perforated board (Figure 17-19). Small wires or even cut-off leads from other components can also be added to make solder bridges and traces on the underside of the board. The results won't be pretty, but they will be as functional as any professionally manufactured board—if wired correctly.

Once all of the components that make up the audio preamplifier stage have been wired and

checked, solder a temporary wire to the test point (TP1) and add the power and ground wires to allow connection to some battery source with a voltage between 3 and 9 V. If you hook up the connection between TP1 and GND to a set of headphones, you should be able to hear a rumbling sound when you blow into the electret microphone. An oscilloscope will also register the analog waveform from the preamplifier stage (Figure 17-20).

We like to use two different oscilloscopes in our evil genius lab—a nice new digital scope with all of the bells and whistles, and this beat-up, old clanker from 1970s that does not even have a front panel overlay! The old (cathode ray tube) CRT-based scope is great for looking at real-time low-frequency data from audio sources as it just clips on and works every time (Figure 17-21). You can often purchase an older scope like this for a

Figure 17-19 Making traces in the underside of the perforated board.

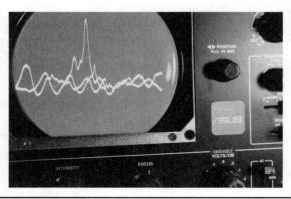

Figure 17-21 The oscilloscope shows the audio wave in real time.

few bucks at a surplus supply store, and for many projects it will be more than adequate. We actually prefer the beater scope to our brand-new 400-MHz digital scope for looking at live sound data.

If you are not seeing any change on your scope or hearing any noise in your headphones when you blow into the microphone, recheck all of your wiring. Almost 99 percent of the time, a failure will be a wiring problem, so check and recheck before claiming that a circuit is no good in a forum! Nobody wants to look like a noob when they finally realize that their component is reversed, right? If you want to experiment a bit with the audio preamplifier, changing the value of R1 will affect the output from the electret microphone internal amplifier. A value of 1 K will push the amp to its max, while a value of more than 5 K will reduce the output slightly.

Once you have verified that the audio preamplifier stage is actually sending out an audio signal, complete the rest of the circuit by adding the components that make up the RF output stage (Figure 17-22). If you have the room on your perf board, try to keep the coil and the tuning capacitor slightly isolated from the other components by giving it a small gap. This isn't absolutely critical, but remember that anything that comes near the coil or tuning capacitor will add capacitance and

offset the frequency slightly. If you offset the frequency too much, you will be transmitting outside the FM radio band or not at all.

The actual antenna is just a bit of wire, and is not critical since it will not be tuned to the transmit frequency anyhow. A 2-in long wire will probably give you a range of across your house and a 6-in antenna may cover the entire yard. Again, don't expect any kind of amazing performance out of this device as falsely claimed on one too many websites trying to sell them off as high-tech law enforcement–grade spy transmitters! If you can pick up the audio on your receiver from one room to the other, then you have done well, but expect static and frequency variances as people move around the room. Without a more complex stabilizing circuit such as a SAW oscillator or crystal-driven oscillator, the output will vary when anything like capacitance or battery level change.

The completed circuit board looks like a random trail of soldering blobs, but it is perfectly functional and will work as well as a professionally manufactured circuit board providing the wiring is all correct (Figure 17-23). Before connecting any power to the circuit, check the polarity of transistor Q2, as this is about the only component that could suffer damage from a reversal. There are not many components to add in the RF stage, so it should be an easy task.

Figure 17-22   Adding the RF output stage to the circuit.

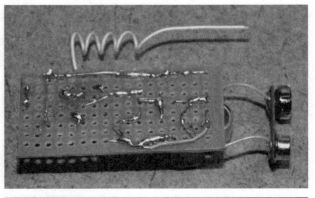

Figure 17-23   The completed circuit board on the underside.

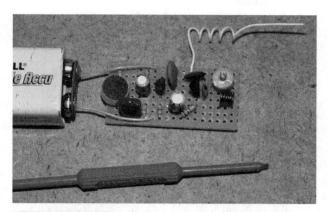

Figure 17-24 The tuning must be done with a nonferrous tool.

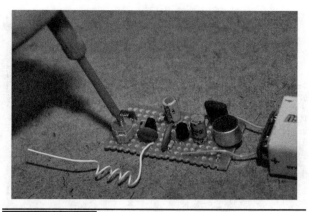

Figure 17-25 Tuning the transmitter might seem like black magic.

Tuning of the transmitter is simple…as long as it works! You will need some kind of nonferrous (nonmetal) tool to adjust the trimmer capacitor as any metal objects near the capacitor will throw off the frequency, making it highly annoying to adjust (Figure 17-24). What will happen is that you will get the transmitter adjusted where you want, and then removal of the tool will alter the transmitter frequency. Actually, this even happens slightly with only your hands being near the circuit board, so be prepared for a little black magic here! Also, when you change the battery voltage, retuning will be necessary unless you luck out and end up on a totally different dead spot on the FM radio band.

To start the tuning process, find a dead spot on the FM radio band near 100 MHz. Place your transmitter close to the radio and slowly turn the trimmer capacitor with the plastic screwdriver all the way around. If you are lucky, your radio will go from a smooth hiss to a loud thump as you hit the exact receiver frequency. Since the tuning bandwidth cannot cover the entire FM radio band (from 88 to 108 MHz), you will most likely have to repeat this process in steps of 10 MHz or so until you get luck and catch the correct frequency (Figure 17-25). An auto-scanning radio tuner may also find the transmitter, but it will probably be easy to tune the transmitter to the radio.

We kid you not when we say this may seem like black magic. If your transmitter is working,

it will likely be transmitting somewhere between 80 and 110 MHz, but depending on parts used, board layout and stray capacitance, who knows where it may be. We built two versions of this project for this write-up—this one and a smaller version shown next. Although we used the same exact parts, the output frequency was wildly different. The larger version hit about 100 MHz, and the smaller one was down to 88 MHz. Oh, and remember that as soon as you move your arm away from the transmitter, you can expect some kind of frequency shift. Hey, that's why radios have dials! Just tune your transmitter back in.

Although we have been warning you about the many pitfalls with this ultra-basic transmitter, we will say that once it is set and running on a fresh battery, it will perform quite nicely. You have to get used to all of the variables that can affect the output frequency, such as large objects near the unit, battery voltage changes, and even people moving around a room close to the unit (Figure 17-26). Get the transmitter working and then grab hold of the end of the antenna and watch the output frequency dive to the lower regions of the FM band. Once we had our transmitter set up and placed on a table, we were able to clearly hear conversations in the room from two rooms away, so it does work once it is set up and left undisturbed.

To make this project more spy-like, we added some double-sided tape to the underside of the

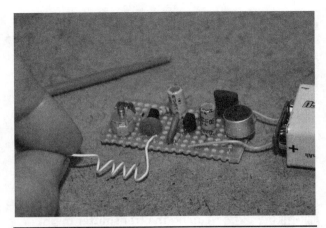

Figure 17-26   Many factors can cause the frequency to drift.

Figure 17-28   The completed stick and run spy transmitter.

circuit board in order to fasten the circuit board to the battery casing and then added the tape to the other side of the battery casing so we could just walk by our target area and "stick" the unit to something the way they do in the spy movies (Figure 17-27). Of course, the real spooks have frequency hopping, digitally encrypted burst transmitters running in the gigahertz bands, but this thing was cool nonetheless because it was built from a dead radio in one afternoon!

After sticking the circuit board to the battery casing, we had to retune the unit again to place it back on a nice high dead band on the FM radio (Figure 17-28). Higher frequencies are better because there is less crackle on the receiver,

allowing for a cleaner reception. The position of the tiny antenna was also important, as any changes would throw off the transmit frequency. You may actually find better performance using a shorter and more rigid antenna that is not easily bent around.

The simple transmitter worked fairly well considering how few components it has, and from a fresh 9-V battery has been running for many hours. In fact, we have not run down a battery yet in all of our tests, including the time we forgot to unplug the battery (Figure 17-29). The transmitter uses such a small amount of current that even a 3-V coin cell will run it for many hours, which is what prompted us to make the second size-reduced version.

Figure 17-27   Making the transmitter more spy-like.

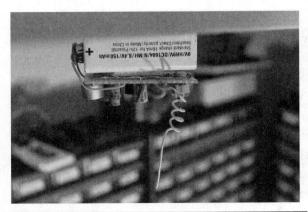

Figure 17-29   Bugging our own stock room to protect it from pirates.

Figure 17-30 How small can you make this project?

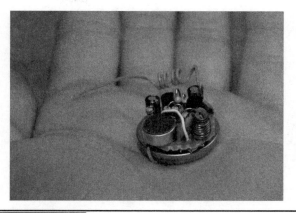

Figure 17-31 The completed miniature version of the transmitter.

Having so few components, the project lends itself well to miniaturization (Figure 17-30). Of course, the tighter you pack the components, the more chances of stray capacitance, so beware that size reduction may also mean a performance reduction—or complete elimination! We managed to cram all of the parts on a bit of perforated board no larger than the diameter of the nickel-sized 3-V coin battery and it worked just as well as the larger version with a slight frequency shift. Actually, this is version 2 of the mini transmitter as version 1 seemed extremely unstable. We just moved the component around a bit, and the instability was magically cured. Welcome to the dark art of RF circuit building!

If you have some skill with a soldering iron and a steady hand, this pop-top sized version of the spy transmitter would look gigantic compared to one built using surface mount components. It would be possible to make a performing unit that would be

no larger than a pencil eraser, running from a pair of watch batteries (Figure 17-31). Of course, for that kind of miniaturization, some kind of copper-clad board will need to be made—using a Dremel tool as perf board will not be of much use for surface-mounted components. We will leave you to send us a photo of this project made smaller than our current 1-in diameter version!

Thanks for tuning in. We hope this project was an enjoyable pathway to many more interesting and complex radio frequency gadgets. We will be adding many more transmitter projects in the future as well, including those that can transmit video and encoded information. Of course, all of these devices are for fun only, and never to be used to intrude on anyone's privacy or operate in places where prohibited. Now, we must run, as that unmarked van has once again pulled up in front of our secret lab, and the men in black are heading to the door… (click)!

PROJECT 18

Remote Control Hijacker

THIS FUN PROJECT lets you take control away from the person holding the remote control by intercepting the invisible signals as they travel through the air so you can play them back to the TV or video machine. You can also "train" your remote hijacker by recording certain button presses directly from the remote so that you can play them back later on, taking total control over the target appliance. Because this project records the remote control pulse stream directly, it will work on any infrared-based remote control, able to learn a few button presses.

| Figure 18-0 | This hidden device can intercept remote control signals and play them back! |

PARTS LIST

- IRMOD: rc5 infrared remote control decoder module
- IC1: ATMEL ATMEGA88P or comparable microcontroller
- Q1: 2N3904 or 2N2222A NPN transistor
- LED 1: 940-nm infrared LED
- LED 2: Red or green visible LED
- S1, S2, S3: Normally open pushbutton switches
- Resistors: R1 = 100 Ω, R2 = 10 K, R3 = 1 K
- Battery: 3-V AA or AAA pack

This project uses a very simple microcontroller program that just times the pulses coming into the infrared decoder and then stores them in the internal SRAM for later playback. The source code is made as simple as possible, allowing for plenty of room for modifications and alterations to suit your evil genius agenda. Because no interrupts are used, the C program could be ported to just about any microcontroller, and will work on all of the Atmel microcontrollers as is. Larger internal memory allows more button presses to be stored, with the ATMega88 (1K SRAM) allowing about three button presses to be recorded and played back.

Almost every electronic appliance that includes an infrared remote control will use a standard method of communication over the invisible beam called the "RC5 Protocol" (see Figure 18-1). This simple protocol works by sending a series of 1.5 to 2.5 millisecond (ms) long pulses that are modulated by a carrier frequency of 36 to 45 kilohertz (KHz). The pulses make up a frame of data, which is usually 12-bits long, encoded using a system of inversion called Manchester Encoding. Of course, we won't have to dig all that deep into any of this stuff because this project just records the length

Figure 18-1   The remote control signals are first decoded by the RC5 module.

of pulses and stores them as byte values into the microcontroller's internal memory for later playback.

Of course, you could actually decode the data and store a lot more, but this would require some crafty programming to measure the exact pulse rate and then understand the stream that it is seeing at the input. We just wanted a quick and dirty hack that would allow us to prank the remote control user, so we opted to just measure the time between pulses and store that value. This allows any remote to be recorded and played back as the program does not care what the exact frequency or command being sent really is.

To deal with the very fast 40-KHz modulation, a readymade solution is used that will strip out the modulation and leave only the millisecond pulse train. These remote-control decoder modules are very common as they are used in most of the appliances that we are going to hijack. These tiny 3-pin blocks have a power, ground, and output, and do nothing more than look for RC5 pulses in order to strip them of their modulation. We have collected many of these remote-control modules from various dead appliances and electronics

suppliers, and all of them do basically the same thing. Some of them are contained in a metal can, while others look like transistors with a bubble on one side to input the infrared light. All that matters is that you can figure out which pins are power, ground, and output on the device.

In our version of the remote control hijacker, we used the Sharp GP1UM26X decoder module, as we had a few of them that were salvaged from old DVD players and VCRs (see Figure 18-2). If you are going to salvage your decoder module, make sure to take a look on the board before unsoldering so that you can identify the power and ground connections as most OEM modules will have no markings. The ground wire will be easy to locate, but you may have to do a bit of tracing to identify the signal pin. If you have to take a wild guess, then just connect the unit to a 3-volt (V) coin cell with a piezo buzzer and try random combinations until you hear a *blip-blip-blip* on the buzzer when you aim a remote control at the device and press a button. A coin cell has very low amperage, so it will not kill the module if you hook it up completely reversed while brute force hacking to find the pinout.

■ Electro-optical Characteristics

(T_a=25°C, V_{CC}=5V)

Parameter	Symbol	Conditions	MIN.	TYP.	MAX.	Unit
Dissipation current	I_{CC}	No input light	–	0.95	1.5	mA
High level output voltage	V_{OH}	*3	V_{CC} –0.5	–	–	V
Low level output voltage	V_{OL}	*3 I_{OL}=1.6mA	–	–	0.45	V
High level pulse width	T_1	*3	600	–	1 200	μs
Low level pulse width	T_2	*3	400	–	1 000	μs
B.P.F. center frequency	f_0	–	–	*4	–	kHz
Output pull-up resistance	R_L	–	70	100	130	kΩ

*3 The burst wave as shown in the following figure shall be transmitted by the transmitter shown in Fig. 1
  The carrier frequency of the transmitter, however, shall be same as *4, and measurement shall be from just after starting the transmission until 50 pulse
*4 The B.P.F. center frequency f_0 varies with model, as shown in ■ **Model Line-up**

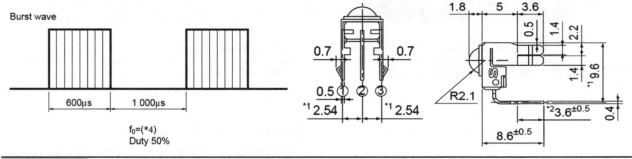

Burst wave

600μs   1 000μs

f_0=(*4)
Duty 50%

Figure 18-2   The datasheet for a typical RC5 remote-control decoder module.

As for the playback portion of this project, it will utilize the same exact system that is found in the remote control, using an infrared light-emitting diode (LED) to send out the modulated pulse stream (Figure 18-3). You can use practically any type or number of infrared LEDs, as the decoder module is highly sensitive to the modulated pulse stream. In our testing, it was even found that visible LEDs would work! You only need a single

Figure 18-3   The pulses are sent to the appliance using an infrared LED.

infrared LED unless you plan to wreak havoc on the entire neighborhood or the TVs in a large area, but we will leave that modification up to you. A single infrared LED will have more than enough reach to control any TV or video machine in a living room, and since stealth is the goal, size matters.

Infrared light has a longer wavelength than any of the colors on the visible spectrum, crossing into invisibility around 800 nanometers (nm), and continuing well beyond 1000 nm (Figure 18-4). Most remote controls use 940-nm infrared LEDs, which is why those types are most common and inexpensive. Of course, you could use any infrared LED, even the 800 to 880 nm types often used for night vision illuminations as well. Either way, your buddy (victim) will not see the invisible light so you can hijack the TV or video player in complete darkness without compromising your funny prank.

If you like to mull over datasheets, then the important factors of any infrared LED are the

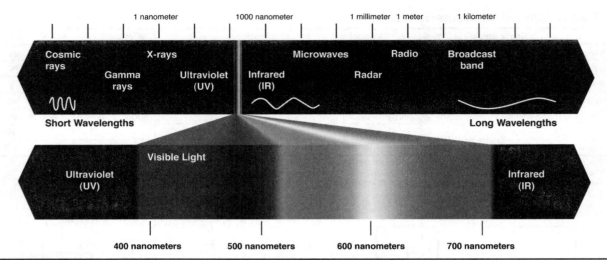

Figure 18-4 The light spectrum, showing the invisible infrared region.

center wavelength and the forward continuous current, as this will determine what kind of light and how much light you can expect (Figure 18-5). Of course, for this project, an infrared LED will work, so ratings are not important as the current is low enough due to the supply as well as the short pulses. If you really wanted to get long range from this project, then look at the datasheet for the value of pulsed current rating, which is a much higher value than continuous current, allowing the LED to be pushed into extremely bright pulsed-mode operation.

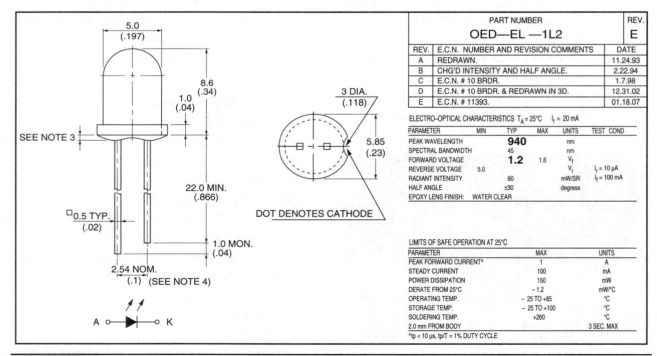

Figure 18-5 An infrared LED datasheet, showing the optimal wavelength.

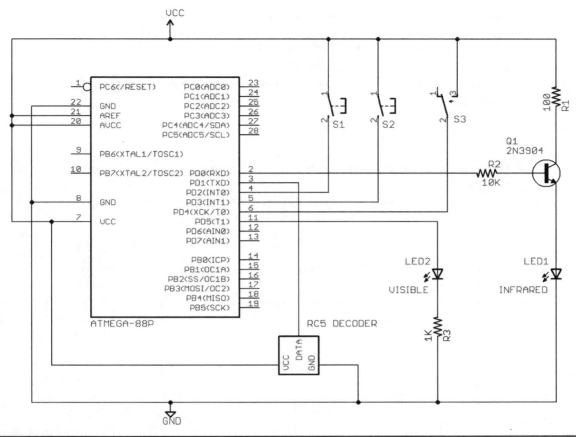

Figure 18-6 The Remote Control Hijacker Schematic.

Having a very basic schematic and using a microcontroller programmed in C allows the use of just about any microcontroller with a bit of internal SRAM and a few input/output (IO) lines (Figure 18-6). Only five IO lines are needed for the basic version: a digital input for the RC5 decoder, an output for the infrared LED, an output for the visible indicator LED, and two pushbutton inputs. The output to the infrared LED is amplified though an NPN transistor to offer a bit more range, but it will also work if you just push the LED directly with the output pin of the microcontroller through a 500-ohm to 1-K resistor. The timing has been calculated using an 8-MHz clock source so that the ATMega88 that we are using can run from its internal oscillator, further reducing the parts count.

The push button switches put the microcontroller into record or playback mode, allowing the interception of a few button presses that are stored in the internal SRAM for later playback. The visible LED flashes to show that the system is receiving pulses when in record mode, but it can also be omitted if you want a smaller and stealthier project.

The source code is written in CodeVision C for the AVR, and is simple enough to be easily ported to just about any microcontroller with an internal memory to store the incoming signals. One kilobyte (K) of memory will store approximately 3 button presses, and a 4 K of memory will store about 16 button presses. Open the source code in your favorite IDE and we will go through it one section at a time.

```
// ***************************************************************************
// **** PROGRAM = REMOTE CONTROL HIJACKER (C) 2010 LUCIDSCIENCE.COM
// **** TARGET = ATMEGA88P
// **** CLOCK SPEED = 8 MHZ INTERNAL
// ***************************************************************************
#include <mega88p.h>
#include <delay.h>


// ***************************************************************************
// **** PROGRAM VARIABLES
// ***************************************************************************
unsigned char RC5[750];


// ***************************************************************************
// **** RECORD FUNCTION
// ***************************************************************************
void RECORD(){
unsigned int CTR;
unsigned char TMR;
unsigned char REC;
unsigned char DIV;

// STATUS LED ON
PORTD.5 = 1;
REC = 0;

// ERASE STREAM ARRAY
for (CTR = 0; CTR < 750; CTR++) RC5[CTR] = 0;

// WAIT FOR START BITS
while (PIND.1 == 1) {}

// RECORD RC5 PULSE STREAM
for (CTR = 0; CTR < 375; CTR++) {

// RECORD MODULATED PULSE
TMR = 0;
while (PIND.1 == 0) {

// DIVIDE TIME BY 10
for (DIV = 0; DIV < 10; DIV++) { //$$$

// 40 KHZ DELAY
#asm
ldi r23,66 ;1
DL4:
dec r23 ;1
```

```
brne dl4 ;1/2
nop ;1
nop ;1
#endasm
}
TMR++;
}
// STORE TIME VALUE
RC5[CTR] = TMR;

// INCREMENT COUNTER
CTR++;

// RECORD UNMODULATED PULSE
TMR = 0;
while (PIND.1 == 1) {

// DIVIDE TIME BY 10
for (DIV = 0; DIV < 10; DIV++) { //$$$

// 40 KHZ DELAY
#asm
ldi r23,66 ;1
DL5:
dec r23 ;1
brne dl5 ;1/2
nop ;1
nop ;1
#endasm
}
TMR++;
}
// STORE TIME VALUE
RC5[CTR] = TMR;

// STATUS LED FLASH
REC++;
if (REC == 40) REC = 0;
if (REC == 0 ) PORTD.5 = 1;
if (REC == 20 ) PORTD.5 = 0;
}

// STATUS LED OFF
PORTD.5 = 0;

// WAIT FOR BUTTON RELEASE
while (PIND.2 == 0){}
delay_ms(500);
}
```

```
// ******************************************************************************
// **** PLAYBACK FUNCTION
// ******************************************************************************
void PLAYBACK(){
unsigned int CTR;
unsigned char TMR;
unsigned char DIV;

// STATUS LED ON
PORTD.5 = 1;

// PLAYBACK RC5 PULSE STREAM
for (CTR = 0; CTR < 375; CTR++) {

// SEND 40KHZ MODULATION
TMR = RC5[CTR];
while (TMR > 0) {

// DIVIDE TIME BY 10
for (DIV = 0; DIV < 10; DIV++) {

// 40 KHZ MODULATION CYCLE
PORTD.0 = 1;
#asm
ldi r23,33 ;1
DL1:
dec r23 ;1
brne dl1 ;1/2
nop ;1
#endasm
PORTD.0 = 0;
#asm
ldi r23,33 ;1
DL2:
dec r23 ;1
brne dl2 ;1/2
nop ;1
#endasm
}
TMR--;
}

// INCREMENT COUNTER
CTR++;

// SEND NO MODULATION
TMR = RC5[CTR];
while (TMR > 0) {
```

```
// DIVIDE TIME BY 10
for (DIV = 0; DIV < 10; DIV++) { //$$$

// NO MODULATION CYCLE
#asm
ldi r23,66 ;1
DL3:
dec r23 ;1
brne dl3 ;1/2
nop ;1
nop ;1
#endasm
}
TMR--;
}
}


// STATUS LED OFF
PORTD.5 = 0;

// REPEAT DELAY
delay_ms(100);
}



// ***********************************************************************************
// **** IO PORT SETUP
// ***********************************************************************************
void main(void)
{
DDRD.0 = 1; // IR LED OUTPUT
DDRD.1 = 0; // IR SENSOR INPUT
DDRD.2 = 0; // RECORD BUTTON INPUT
DDRD.3 = 0; // PLAYBACK BUTTON INPUT
DDRD.4 = 0; // MODE SWITCH INPUT (NOT USED)
DDRD.5 = 1; // STATUS LED OUTPUT

// ENABLE INPUT PULLUPS
PORTD.1 = 1;
PORTD.2 = 1;
PORTD.3 = 1;
PORTD.4 = 1;



// ***********************************************************************************
// **** MAIN PROGRAM LOOP
// ***********************************************************************************
while (1)
{
```

```
// CHECK RECORD AND PLAYBACK BUTTONS
if (PIND.2 == 0) RECORD();
if (PIND.3 == 0) PLAYBACK();

// END OF MAIN LOOP
};
}
```

Under "*program variables*," a large 8-bit array is created that will store all of the pulse duration times into the internal memory. The line "unsigned char RC5[750];" sets the largest possible array for the microcontroller, which is based on the amount of internal SRAM, the compiler needs, and the stack. In our case, we have 750 bytes available in the 1024 bytes of the ATMega88 available. You will have to mess around with this number if you are using a different microcontroller, making use of all of the free internal memory space.

The "*record function*" is called when the record button is pressed, and it takes care of measuring the time between received pulses from the infrared decoder module. On entering the routine, the status LED is turned on to alert the user that the record function is armed. The entire array is then erased. To avoid false recording, the routine first waits for a start pulse from the remote control, sitting in a loop until one is received. Remember that the pulse stream from the decoder module is stripped of all 40-KHz modulation, leaving only the inverted pulses, so this is what will be recorded as a timed value into the storage array.

Once a pulse stream begins, the recording routine then resets a counter and stays in a time-controlled loop, incrementing the counter until the pulse changes polarity. The resulting time is the length of the first negative-going pulse (modulated pulse), which is then stored into the array. The following timed loop does much the same, but is now recording what will be the duration of the high pulse (none-modulated pulse). An inline assembly routine is used to mimic the exact timing of the

40-KHz cycle since this will be needed for the playback mode later on.

This recording of low- and high-pulse times continues until the entire array is full, flashing the status LED so that the user can see that valid pulses are being received and stored into the array. To avoid wrapping the memory, the record routine does not exit until the user has let go of the record button. The control is the returned to the main loop, which just sits around waiting for buttons to be pressed.

The "*playback function*" is much like the record function except that instead of recording pulse times to the array, it reads them in series and sends back the missing 40-KHz modulation during the low pulses, recreating the signal that was originally coming into the remote control decoder module. The status LED is turned on during playback to show the user that the unit is transmitting, and continues to read the array contents and send the signal until the end of the array. To create a stable 40-KHz modulation cycle, an inline assembly routine is used again during the low-pulse part of the routine. At the end of the playback routine, the status LED is turned back off, and the control returns to the main loop to wait for another button press.

"*IO port setup*" is easily adjusted to suit your microcontroller or board layout, and the pull-up resistors are enabled for the switches so they are active low. In the "*main program loop*," the routine just sits in a tight loop waiting for either the record or playback button to be pressed so it can call the corresponding routine.

This source code was made as minimal and as easy to understand as possible, so there is plenty of room for improvement and modification. Much better use of the array could be made by first timing the actual start bits so that pulses could be stored as single bits (0:off 1:on) instead of time values. This would allow many more button presses to be recorded into the memory, even on small microcontrollers. This type of efficient pulse decoder would either have to take advantage of interrupts or be done in assembly to achieve maximum accuracy. We wanted this to be a simple project, so we will leave you to make the improvements!

It's always a good idea to first build your projects on a solderless breadboard so that modifications or debugging is easy (Figure 18-7). This project lends itself well to further improvement, since only a small portion of the program memory and IO space is used. You will also want to test the range, and consider increasing the pulse current to the LED for longer range or simply removing the driver transistor altogether for one room operation. Also shown in the breadboard is the 7805 regulator to allow operation from a 9-V battery, although we used one that was much too large only because that's what we found in our junk bin. You could probably run the system from 3 V as well if your microcontroller supports it, allowing the use of a tiny 3-V button cell for a highly reduced footprint.

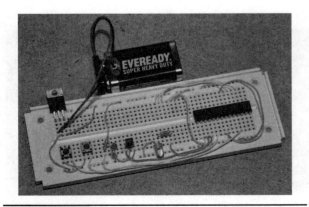

Figure 18-7   The remote control hijacker on a breadboard.

Figure 18-8   The infrared input module and output LED.

Notice that the infrared input module and output LED are facing the same direction (Figure 18-8). This is done so that you always know which side of your project should face the TV when trying to be sneaky. Since the infrared module is extremely sensitive to the RC pulses, it does not need to actually face the remote control during recording. Infrared pulses reflected from walls and other objects will be strong enough to allow the record circuit to function, but the infrared output needs to be at least pointing in the general direction of the appliance to be hijacked. When you are testing the breadboard circuit, notice how sensitive the infrared input module is, even working with your finger covering the input.

Once you have your project on the breadboard with the program running in the microcontroller, place the infrared section close to the TV so that the output from the infrared LED is fed almost directly into the sensor on the TV or appliance. The infrared (IR) input is often hidden behind a dark-black round or square lens near the base of the unit. With the TV on, press the record button on the hijacker and then use the remote control to operate some obvious function like volume or power. As soon as the TV begins to respond, the visible LED on the hijacker should start to flicker, showing you that pulses are being recognized, recoded, and stored into the internal memory.

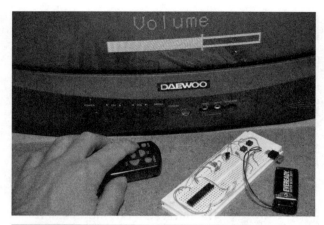

Figure 18-9   Initial testing of the prototype project (record).

The indicator LED will stay solid as soon as the record button is pressed, but will flash as pulses are received to show you that it is working. Once the SRAM inside the microcontroller is filled, the LED will stop flashing, indicating that recording is now completed. If the LED fails to flash, then either the wiring is wrong, or your IR module is not sending out pulses properly (see Figure 18-9). You can either probe the output pin with an oscilloscope to verify pulses are being sent, or just drop a piezo buzzer on the output to listen for the *blip-blip-blip* sound from the output of the sensor.

Once you have recorded a button press, aim the infrared LED toward your TV and then press the playback button (Figure 18-10). The visible LED will light for a moment as the RC5 pulse

stream is sent back to the TV, which will respond to the signal just as it did when you used the actual remote control. If your TV fails to respond, check the output at the infrared LED using a scope or with a piezo buzzer to verify that pulses are being sent. If you hear pulses but the TV fails to respond, compare the time base or sound of the pulses coming in to those coming out to ensure that it is not a clocking or timing problem. If you are modifying this code to work at another frequency, then you will have to alter the timing in the recording loop.

If you keep holding down the playback button, the pulses will continue to loop, which is good when hijacking commands like volume, as you can keep cranking up or lowering the volume by simply holding down the playback button. When you want to drive your remote-control-wielding pals crazy, adjusting the volume is a good method as they will never be able to regain control with you secretly working the remote hijacker. The on and off command is also a good one to prerecord ahead of time as it can be used to turn on or shut off the TV at any time, just like that fun "TV-B-Gone" device.

Once you have verified the operation of the solderless breadboard prototype, it will need to move to a more permanent home on a circuit board or perforated board (Figure 18-11). There are many options to create a stealthy installation, ranging from small and unnoticeable to hidden in plain sight. If you are good with a soldering iron,

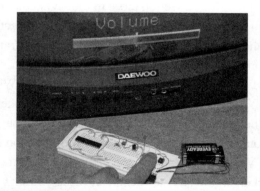

Figure 18-10   Initial testing of the prototype project (playback).

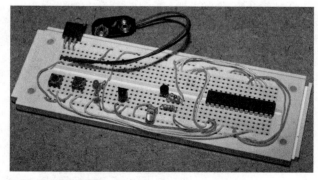

Figure 18-11   Completed prototype ready for a permanent installation.

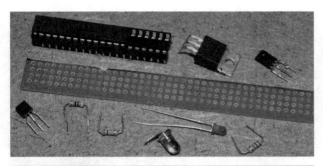

Figure 18-12 Perforated board is good for small component count projects.

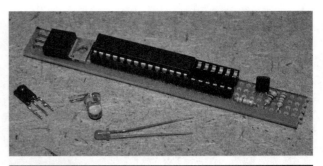

Figure 18-13 Laying out the parts to create the smallest board.

you could try a surface-mount circuit board using a small pin microcontroller package and then power the unit from a coin battery, making a final product no larger than a pop lid. We opted for the hidden-in-plain sight version using a pop can, as there is nothing suspicions about holding a can while you are sitting there in front of the TV.

The easiest method to take a small component count project off the solderless breadboard is with a bit of perforated board (Figure 18-12). The most basic perf board is just a wafer with holes, but there are also protoboards available with copper pads and even interconnected strips just like the breadboard. Having only nine parts makes it easy to use any method of installation, so we opted for a bit of perfboard. Also shown is a pair of smaller sockets connected together to make one large enough for the 28-pin AVR. We did not want to wait around for days to order a new socket, and we always mount microcontrollers in sockets in case we need to remove them for reprogramming or use in another project later on.

Depending upon your installation, the perfboard could be square or long to allow for mounting into an item such as a marker case along with a few coin batteries. We decided to make the board only as wide as the AVR socket and then as long as needed to install the other few components. If we had had a small regulator, the board would be about half an inch shorter, but we usually work with what we can salvage at the time (Figure 18-13).

If your components are new, then the leads will be long enough so that you can simply bend them around on the underside of the perf board to create your traces (Figure 18-14). We also bent the four corner socket pins so that it would be held securely to the perf board. If you consider your component layout ahead of time, you will be able to create the connections with minimal wiring, using the leads as traces.

Since ground and power connections will make up the bulk of the wiring, they are all done first using color-coded wires (red for VCC and green for ground). If there are any mistakes, they will likely be made in these wiring paths, so it makes sense to do them first so that the wiring can be rechecked before applying power (Figure 18-15). Don't forget to tie down the microcontroller reset pin as well or your system may behave erratically.

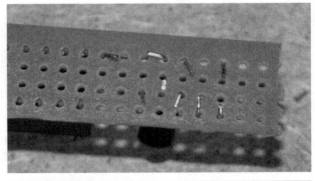

Figure 18-14 The pins are bent to form traces and hold the components.

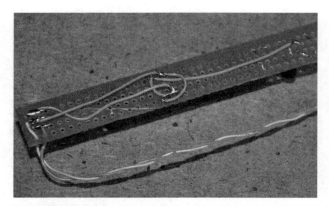

Figure 18-15 Starting the wiring with ground and power connections.

Figure 18-17 Wiring all of the external components and switches.

There are quite a few connections that need to be made off of the main circuit such as the infrared module, infrared LED, two switches, visible LED, and battery, so there will be a mess of wires coming from the small board (Figure 18-16). If you already know what kind of enclosure you are going to use, then many of the external parts could be mounted right on the perf board, but we wanted flexibility in our final enclosure so we opted to place all of the parts at the end of the wiring.

We added 10-inch (in) long wiring to all of the external components and switches so that we could first mount them in our enclosure (pop can) and then complete the wiring to the circuit board while working outside of the enclosure (Figure 18-17). It always helps to use some kind of color coordination so you know which pin is ground, VCC, or output for each external component.

What is less suspicions than sitting by the TV with a can of pop? We figured that the installation would be perfect as there is plenty of room for the board and battery and all of the switches could be operated away from our victims by holding the can strategically to conceal its true nature. Also, since the infrared signals reflect off of walls, the sensor and infrared LED could also be hidden from those in the room by aiming it away from view. Using a pair of scissors, we popped the necessary holes and then carefully turned the blade until they were the correct size for each external component (see Figure 18-18).

The underside of the can is also cut open to allow the insertion of the battery and circuit board

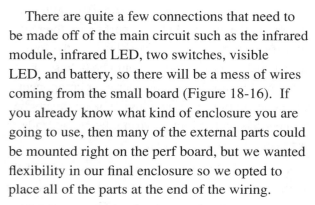

Figure 18-16 The completed perfboard is a mess of wiring.

Figure 18-18 Punching holes in the pop can for the external components.

Figure 18-19   Making an opening for the battery and circuit board.

Figure 18-21   Installing the circuit board after the wiring is completed.

after the switches and infrared components have been installed (Figure 18-19). To feed the switches into the holes, they are guided in place by holding onto the wires.

Once we had the switches installed, we found the best placement for the infrared components so that when we were holding the can with the switches hidden from the person next to us, the infrared components would have a clear line of sight to the front of the room where the TV was located (Figure 18-20).

The external components are now wired to the circuit board after being stalled into the can (Figure 18-21). It is much easier to test and trouble-shoot the installation with the circuit board hanging out of the can, which is why the wires to the external components were made longer than necessary. This also makes future updates to the microcontroller firmware easier as well.

To ensure that there are no short circuits, all of the components are held inside of the can using a bit of double-sided tape (Figure 18-22). The tape keeps parts from bouncing around, yet allows the battery to be pulled free if it ever needs to be replaced. The novelty of your evil prank will probably wear out long before the new 9-V alkaline ever will.

Figure 18-20   Installation of the infrared components.

Figure 18-22   Securing all of the internals into the can.

Figure 18-23   The completed remote control hijacker.

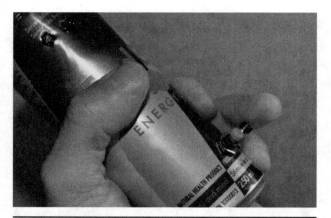

Figure 18-24   The controls are conveniently mounted for hidden operation.

The completed remote control hijacker is now ready to use, and with this stealthy hidden installation, it can be used in plain sight without detection (Figure 18-23). Of course, if your buddies are used to your evil prankster ways, then you might be found out sooner or later, but no doubt you will have some good fun before anyone suspects that your refreshment is stealing the invisible signals from the remote control right out of the air! If you really want to make it look good, open a real can in front of your unsuspecting victim and then do a switch when they are not looking so that any suspicion will not be towards the can you are now holding.

Give your new toy a few test runs so you can get used to using the controls in a nonsuspect manner before unleashing your chaos onto the world (Figure 18-24). It may even work best if you get a hold of the remote beforehand and record the power button or mute button so you can enter the room already drinking your favorite refreshment and then just start hijacking the TV without suspicion. Once your victim begins to fiddle with the remote in frustration, you can then intercept new commands just to mess with them even more soon after!

You might have noticed that in our schematic there is an extra IO pin connected to a switch that is not used in the source code. Our intent was to

add a timer mode that would just count down for a minute or two and then issue the last recorded command. This way you could be out of the room and just listen to the cursing from a distance, taking away all suspicion from you or your evil contraption! (See Figure 18-25.) Another idea for modification is to just send random pulses on a button press to effectively jam the remote control so that the user can't do anything at all after you mess with the TV or VCR. Again, there is plenty of room for modification of the remote control hijacker, so feel free to let your evil mind wander and send us a few photos of your completed work for our gallery!

Figure 18-25   Playing back the mute command to the TV after a successful intercept.

PART SIX
Personal Protection

PROJECT 19

Camera Flash Taser

THIS PROJECT SHOWS you how to turn 1.5 volts (V) into almost 400 V, creating a hand-held device capable of charging high-voltage capacitors or delivering a low-current high-voltage shock. Using only the tiny circuit board from a cheap disposable camera, you can step up the voltage from a single AA battery to a level that is three times higher than the voltage coming out of your AC wall outlet! Of course, the output current is extremely low, but make no mistake—the high-voltage output from this handheld taser/zapper is painful enough to make anyone jump up to the ceiling.

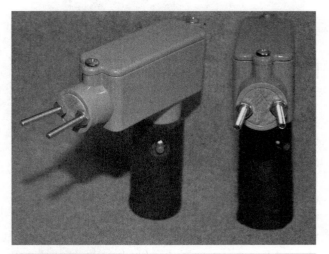

Figure 19-0 Build the disposable camera flash taser.

PARTS LIST

- Camera: Any disposable camera with a built-in flash
- Battery: Single 1.5-V AA or dual 3-V AA pack
- Terminals: 2- to 4-in long 1/8-in diameter bolts
- Enclosure: Must be large enough for battery, PCB and switches

This circuit can also be used to make a small florescent light inverter, front-end charger for a Marx generator, or any other device needing several hundred volts of high-frequency power. Voltages over 1000 V can also be generated from this circuit by increasing the DC power supply or by changing the single transistor to one that can handle a higher current. Of course, the circuit shown here will deliver more than enough voltage

to make you afraid to test the device on yourself more than once!

Any type of disposable camera that has a built-in flash will work for this project as they are all based on the same type of charged-capacitor operation (Figure 19-1). A quality camera or flash unit will not be the best choice though as they have a lot more circuitry, unlike the disposable camera, which is made as simple as possible to keep the costs of manufacture low. A new or used camera will be fine, but if the camera has taken many flashes, you will probably need a new alkaline AA battery to get the full potential out of the high-voltage charging circuit.

If you have actual photos that you intend to keep then let the store remove the film canister and then ask for the camera body back. Opening the camera in the following steps will destroy the film, if it is still inside the camera. If you don't want to destroy

Figure 19-1 You will need a typical disposable camera with a flash.

a new disposable camera for this project, then ask your local camera shop if they would give you a few used camera shells as they normally throw them out. Sometimes, they will hand over an entire bag of camera shells for you to use in your high-voltage experiments, and sometimes they will look at you like you are some kind of lunatic. Hey, there is always the dumpster out back as well for the hard-core junk collector!

Many of the disposable cameras are the exact same under the paper or plastic branding wrapper, so just cut or rip it away to expose the main plastic camera shell body (Figure 19-2). At this point, you are still safe from an accidental shock you

will never forget, so don't worry about that evil high-voltage capacitor lurking inside as you peel away the cover. During the next few steps, you may want to consider wearing gloves if you have never experienced a 400-V shock or have some fear of electrical jolts.

Let us warn you before you precede that there is a real nasty surprise waiting for you under that innocent-looking plastic shell in the form of a 360-V capacitor that can hold a charge for weeks or even months since the last time the flash was charged. The voltage and amperage from the capacitor can be a real risk if you somehow managed to get your hands across both small terminals while it was fully charged, so work carefully or wear gloves until the capacitor is removed and discharged. We are not using the capacitor in the project, so the risk will be removed in the next few steps.

A small flat-head screwdriver is used to pry open the camera shell by placing the head into the plastic clips to snap them apart (Figure 19-3). Work carefully and try not to dig around in the circuit board area with the screwdriver blade as you may damage the circuit or scare yourself out of your chair if you sort out a charged capacitor. The end of a small screwdriver will also get mangled if it has to endure a full capacitor blast, so don't use your good tools as a discharging device! Remove the

Figure 19-2 Remove the plastic or paper branding cover.

Figure 19-3 The camera shell will pry open into two halves.

Figure 19-4   Identify the capacitor and its two output terminals.

top half of the cover, and set the lower half on your bench to inspect the newly exposed circuit board.

This is where the real fun begins! After you have exposed the small circuit board and camera mechanics, look for the photo flash capacitor and locate the two output terminals. The flash capacitor will be about the length of an AA battery and about the same width, usually with a black covering (Figure 19-4). There will be two leads that exit the flash capacitor and enter the circuit board. This is the high-voltage input and output for the flash charging circuit and the place we will be adding our high-voltage leads later.

Make no mistake, the flash capacitor can carry a 350- to 400-V charge for months after the last use, and there is enough power stored there to become a health hazard if you found a way to get your hands across both terminals. Most likely, the only risk will be a bruised elbow after you jump from shocking your finger, but don't treat any fully charged capacity lightly, as you are dealing with more than enough power to be dangerous. The shock from the charging circuit is just as painful, but because it carries such low amperage, it is not a risk like the capacitor, which takes time to charge and can store real energy.

Now that you have been warned, it is time to drop a screwdriver or some metal object across the high-voltage terminals to remove any residual

charge that may be lurking in that flash capacitor. If you like to make loud sparks, then charge up the flash first, but don't use a good screwdriver to blow the charge! A good rule to follow when dealing with high voltages is to keep one hand away from the danger zone so any shock you may accidentally incur will be across your hand, not your entire body. A shock to your finger from the charged flash capacitor will feel like a heat burn as compared to a *zinging* blast from the charger circuit since the capacitor is pure direct current (DC). You might also want to wear ear plugs.

It is recommended that you discharge the flash capacitor before lifting out the circuit board. We have shorted the fully charged capacitor while the circuit board is exposed so that it can be clearly seen where the high-voltage terminals are located and what happens to the tip of a small screwdriver (Figure 19-5). Okay, we admit, we enjoy making loud, angry sparks and have spent a good part of the day charging up the capacitor to fry various conductive objects with it! A single charge could blow several holes in a tinfoil sheet or create two small craters in the surface of a coin. After our ears stopped ringing from the blasts (ear plugs would have been a good idea), it was back to the cause.

Once the flash capacitor has been discharged, pull out the circuit board from the camera shell.

Figure 19-5   Another screwdriver tip is sacrificed for the cause!

It will either lift right out, or there will be a few plastic clips that need to be pushed in order to release the board. The camera flash circuit board is amazingly simple considering what it does, and the entire high-voltage circuit can be further narrowed down to only four components, if you want to dig in and reverse-engineer the circuit. The charging circuit is a simple voltage inverter.

The important parts of the high-voltage generator circuit are as follows: Q1 is a high gain NPN power transistor (see Figure 19-6). Some typical part numbers of this transistor are; 2SD965, 2SD1960, 2SD601A, 2SD879, FXT617 or XN4601. T1 is a tiny step up transformer with a primary winding count of 5 or 6 turns and a secondary winding count of 1500 to 1800 turns with a flyback tap at 15 to 20 turns. One possible part number for this transformer would be T-14-013 (Tokyo Coil). R1 is a resistor of some varying value that controls the feedback from the flyback tap on the transformer, creating a high-frequency oscillator circuit. This value can be changed to create different frequencies as well. D1 is a power diode that converts the high-voltage AC into chopped DC to charge the capacitor. Any diode capable of 400 V or more will work.

You will probably want to leave the camera flash circuit unmodified with the exception of the removal of the high-voltage capacitor. It is easy to

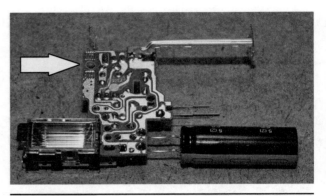

Figure 19-7   Identifying the flash charging switch.

just replace the capacitor with your high-voltage output wires and then solder to the charge switch in order to use the circuit board for your taser project. If you like to reverse-engineer, then you can actually reduce the circuit down to the size of a small marble as will be shown later on. A 400-V shock from a device the size of a marker top is always a great demonstration of one's hardware hacking skills to an unsuspecting buddy!

To engage the high-voltage circuit, the camera had a "charge" or "flash" button on the front (Figure 19-7). This will be seen as two solder or copper pads on the underside of the circuit board. This is where you will solder the wires that lead to your trigger switch, as this will turn in the high-voltage inverter. The other points of interest are the two capacitor high-voltage points, as well as the battery connection points.

To remove the flash capacitor from the circuit board, unsolder the two points and then pull out the two high-voltage leads (Figure 19-8). But, before doing so, identify the polarity of each point in case you want to use your completed project to charge up high-voltage capacitors. The negative side of the capacitor case is clearly marked by the white-and-black dashed stripe. Use a marker to mark the polarity on the small circuit board so you can identify the polarity later on. If you attempt to charge the capacitor in reverse, it may damage the driver transistor or burn out the inverter coil.

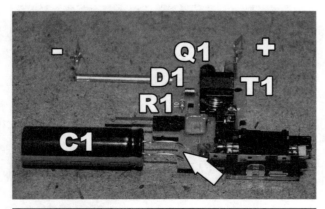

Figure 19-6   Identifying the important parts of the circuit board.

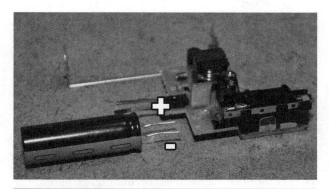

Figure 19-8   Removal of the flash capacitor.

Before moving on to the actual taser project build, we had to take some more time to mess around with the fully charged 360-V flash capacitor as it was much fun blowing holes in things! (See Figure 19-9.) It was easy to charge the capacitor by soldering a temporary wire across the "charge" switch pads and then holding the capacitor in place to allow it to charge. The faint whizzing sound would slowly climb in frequency until the charge light-emitting diode (LED) came on, indicating that there were at least 250 V now in the flash capacitor. To verify this, we used our multimeter set on the 1000 VDC setting, which showed anywhere between 250 and 350 V, depending on charge time and battery condition.

Handle the fully charged capacitor as if you were holding a small stick of unstable dynamite!

We have had plenty of shocks from these things, and they do burn a little on the fingers, but across both hands it is a painful experience that you will never want to repeat, so keep one hand to your side when dealing with high voltages. There isn't enough energy to cause any metal shrapnel when blasting things, but the sound is certainly loud enough to make your ears ring. Avoid shorting into an inductive load as well or you may end up making a high-energy radio frequency (HERF) weapon (Yes, a good idea for a future project here).

When fully charged, the flash capacitor will easily blow two holes through some tinfoil with a load crack and an angry spark (Figure 19-10). We were able to blow holes through the foil three times on a single charge, which is a good reminder that you should double- or triple-check your capacitors after a discharge to ensure that they are not holding a surprise for you later. Tiny wires were also turned into instant vapor if dropped across the charged capacitor terminals. Salt water would sizzle, steel wool would make huge sparks, and various semiconductors would pop or explode violently, sending silicon shrapnel in all directions. We would be lying if we claimed that blowing stuff up in the lab wasn't a damn fun time waster!

The charged photo flash capacitor had enough juice to actually spot weld small bits of metal together and make decent craters in the surface of many metal

Figure 19-9   The little capacitor has more voltage than your wall outlet!

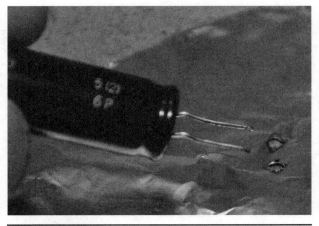

Figure 19-10   Tinfoil is no match for the capacitor's 360-angry volts (V).

Figure 19-11   Spot welding the surface of a quarter.

surfaces such as a quarter (Figure 19-11). No doubt, this dead shorting of the charged capacitor is probably not good for it, but when you can get them for free it is not such an issue. We have an entire bag full of disposable cameras after asking a few of the developing places if we could have them. For some serious sparks, 10 charged capacitors in series (like a Marx generator) will offer an output of 3600 V, but at that power level, you really better know what you are doing! Okay, enough destruction…it's time to get back on track!

After removing the capacitor from the circuit board, collect all of the other goodies from the camera body for use in some later project (Figure 19-12). There are a few lenses, springs, and

various mechanical bits that might be of some use, and a true inventor can never have enough junk to mess around with. Also, test the battery to make sure it has at least 1.5 V left after being used in the camera. A fresh alkaline battery will run your completed taser for a very long time.

Out of the 12 or more semiconductors on the small camera circuit board, you only need the 5 components that make up the high-voltage inverter as shown in the schematic (Figure 19-13). There are many variations of this circuit, but they are all based on the same flyback transformer oscillator circuit. The value of R1 can actually be experimented with to obtain a much more painful shock by lowering the output frequency unto the 100 to 200 Hertz (Hz) range. The charging frequency is much higher in the nonmodified circuit in order to charge the capacitor quicker.

If you just plan on putting the circuit board in a box as it is, then the schematic is not of much importance, as the system will work just fine by simply soldering a pair of wires to the charge switch to turn it on and then a pair of wires where

Figure 19-12   Salvaging all of the parts in the disposable camera.

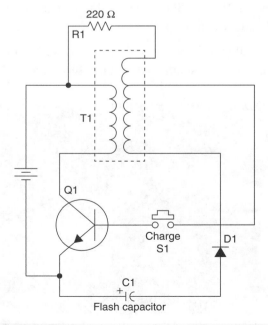

Figure 19-13   The high-voltage inverter schematic.

the flash capacitor used to live to get access to the high-voltage output. We decided to trace our board and reduce it to the minimal needed parts just to learn about the inverter circuit.

The small circuit board is very easy to reverse-engineer since it is a fairly simple circuit with few semiconductors. You won't need any of the flash tube circuitry, which includes another small transformer, a few resistors, and a capacitor, and you won't need the charge indicator LED and its voltage dropping resistor. By holding the circuit board up to a bright light, you can easily see the traces under each component and follow the path through the inverter section (Figure 19-14). If you are removing components, check the high-voltage output after each change in case you pull a needed part from the inverter circuit. We call this method brute force hardware hacking.

The absolute minimal schematic has only five semiconductors: the transistor, transformer, resistor, diode, and capacitor, although the capacitor is not used in the taser project (Figure 19-15). The high-voltage flash capacitor is removed, and the outputs are fed to the "probes" on the outside of the taser case. If you really wanted a minimal unit, the diode can also be removed, leaving on the three components that make up the high-voltage inverter. With a very small battery, the shocker could actually fit into a matchbox. We left the diode

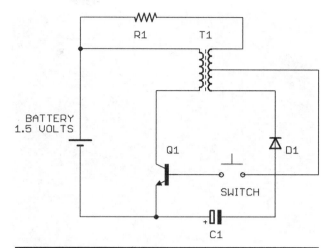

Figure 19-15 The absolute basic high-voltage schematic.

so that we could charge up capacitors when we felt like making more sparks!

Once you have figured out which components make up the high-voltage inverter circuit, you can unsolder the other semiconductors for use in some other project (Figure 19-16). There will be a few resistors, another transformer, a flash tube, and an indicator LED or neon bulb left over.

To use the small high-voltage circuit board in your taser project, it will need to have wiring added for a battery pack, trigger switch, and the high-voltage output probes (Figure 19-17). If you know your circuit, then you only need one common

Figure 19-14 Hold the circuit board up to a light to see the traces.

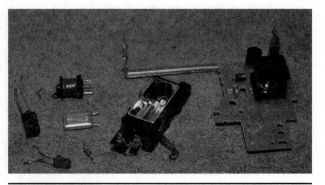

Figure 19-16 Removing any components that are not needed.

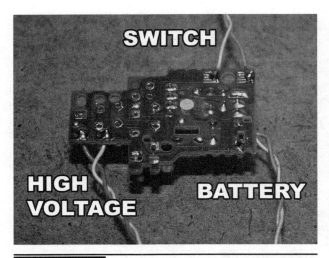

Figure 19-17 Adding the needed wiring to the circuit board.

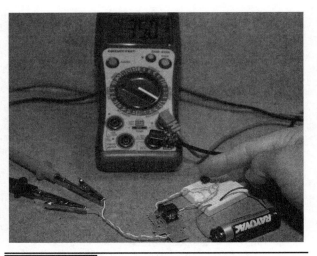

Figure 19-19 The typical output will be around 350 V from a battery.

ground wire and the positive battery, high voltage, and switch wires, but it is probably just as easy to solder pairs of wires to the three connections points as we have done. Remember to mark the polarity of the battery, as well as the high-voltage output points if you plan to charge up capacitors using the output probes later on.

There isn't much left of the camera circuit board after removing the unnecessary components from the high-voltage section. With the switch, battery, and probe wires installed on the board, you can now test the circuit to ensure that it is working properly, outputting between 250 and 400 V from a single 1.5-V battery (Figure 19-18).

Figure 19-18 The minimal high-voltage generator ready to use.

Depending on the transformer and battery quality, the voltage should register between 250 and 350 V using a multimeter set at the 1000 VDC scale (Figure 19-19). An alkaline battery must be used as well since a dry cell battery will not have enough available current to drive the inverter properly when the high-voltage probes are in contact with a semiconductive surface such as your unsuspecting friend's hand. If you want to experiment with even more output voltage, try a pair of batteries to run the high-voltage generator from 3 V instead of 1.5 V. At 3 V, we registered over 600 V on our meter, and the frequency was much higher.

Be careful when upping the voltage from 1.5 V as you may actually blow the transistor or the small flyback transformer. We have not hurt one using 3 V, but some of the circuits we tested did not survive a 9-V battery. The ones that did spit out a crazy 1500 V, more than 10 times the voltage that comes out of the wall outlet! Needless to say, those units hurt like crazy, could actually light wet paper on fire, and cause tiny burns on your fingers. Don't say we didn't warn you!

When making a shocking device, remember that old golden rule: "Don't do on to others something you are too afraid to do unto yourself!" Seriously,

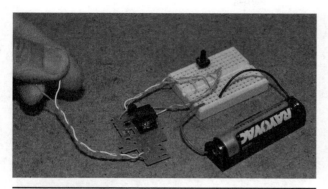

Figure 19-20   If you can't handle the shock, then don't zap your friends!

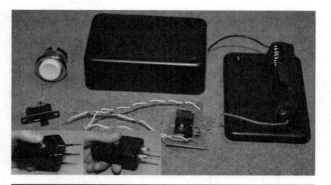

Figure 19-21   Several different versions in various plastic boxes.

you need to zap yourself a few times to not only test your unit, but to understand the wrath you plan to unleash on your unsuspecting buddies (Figure 19-20). Go ahead and take a good grip on the high-voltage wires with your fingers and press down on that trigger. Go ahead, we will wait for you to return…

So? How did 350 V feel as it blasted through your knuckles? Yes, sir, it's like 100,000 carpet shocks in a row, and now your finger will tingle for the rest of the day. If you are really brave, or have some unnatural tolerance for pain, grab a wire in each hand and have someone push the trigger so you can feel the full glory of 350 V though your body. We guarantee that you will not be able to hold on to those wires for more than a fraction of a second! Now, do you "really" want to up the voltage and try that with 1500 V? We didn't think so.

We have made many versions of this disposable camera taser, with varying voltages and enclosures, and it is fun to find creative ways to mask the device as a joke so that an unsuspecting friend will come along and pick it up (Figure 19-21). Using a motion switch or a sensitive switch on the underside of a curious-looking enclosure can make for a really fun gag. Since this project was designed to look dangerous, we wanted a hand-held pistol-type enclosure with a set of ominous-looking probes sticking out front. A simple pistol

grip enclosure will be shown later that can be made from a standard PVC box and some bicycle handlebar tubing.

If you have been an evil genius for some time, then it is going to be difficult to find a buddy who will be gullible enough to let you press some kind of black box with probes up to his/her body and hit the trigger. Hey, they probably know you by now! Of course, if the device is not much larger than a matchbox or fits inside a large marker, then most people will not assume that it could generate any kind of real voltage, so reducing the circuit to the bare minimum may increase your chances of getting one over on your lab buddies (Figure 19-22).

Figure 19-22   The ultimate stealth zapper reduced to the bare minimum.

Using the transformer as a mounting base, solder the transistor, resistor, and diode right to the pins to make the final high-voltage circuit no larger than a marble. This type of installation is called "dead bugging" since you are using the legs of each component to create a circuit board, and often the results look like a bug that has died and turned onto its back. Now how many geeks would be afraid of something this small? Little do they know, the tiny block is capable of pushing out 600-V or more on a tiny 3-V camera battery. Seriously.

One possible installation for the miniaturized taser circuit is installation into a single AA battery flashlight body, replacing the bulb with your high-voltage probes (Figure 19-23). The probes need to be insulated from each other or any conductive mounting surfaces, and should be placed as far apart as possible for maximum shocking pain. If the probes are too close together, then the shock will not disturb as many nerves, creating less of an effect. The most painful effect is a probe in each hand followed by a probe on two different fingers, but you already know this from your "self-tests," right?

The tiny flashlight taser system looked so innocent that we could often trick our victims into zapping themselves just by claiming that this was an all-powerful shocker (Figure 19-24). The miniaturized unit was so unintimidating that most who knew something about electronics promptly

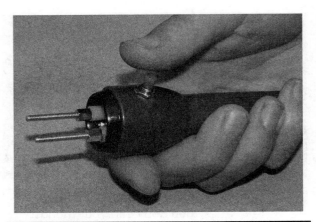

Figure 19-24   Come on, how much could this little thing really hurt?

grabbed the unit to self-administer a painful shock, usually dropping the unit instantly after pressing the trigger! Because of this "drop effect," you should consider a rugged construction if the device is to be picked up.

A nice pistol-grip enclosure can be made using a standard PVC box from the electrical department of the hardware store (Figure 19-25). These enclosures can withstand much abuse and offer a place to mount a handle as well as the high-voltage contact probes. There are several sizes available, so you can experiment with a larger battery pack if higher output voltages are wanted. You will also need some kind of trigger switch, which will be a normally open (NO) pushbutton switch.

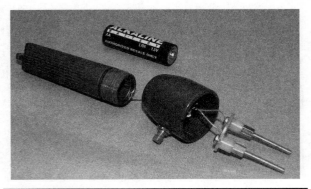

Figure 19-23   The reduced circuit can easily fit in a small flashlight.

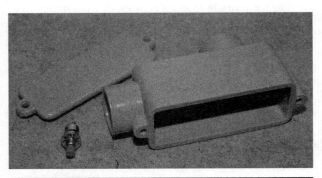

Figure 19-25   A PVC wiring box makes a for a strong enclosure.

Figure 19-26   Making a pistol grip from some bicycle handlebar parts.

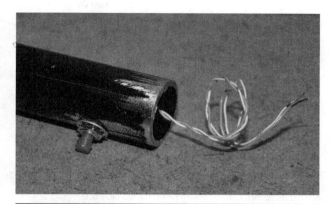

Figure 19-28   The pushbutton trigger switch installed.

The round opening on the PVC box is just the right size to jam in a bit of bicycle handlebar tubing or plastic PVC pipe. You can use this tubing to make a pistol grip by adding a rubber handle-grip and a pushbutton switch (Figure 19-26). A ski pole grip should also work nicely. We cut off a small length from an old mountain bike handlebar tube using a pipe cutter.

A small normally open pushbutton trigger switch is inserted into the piece of handlebar tuning, which will serve as the pistol grip handle (Figure 19-27). The tubing fits snugly into the round opening on the PVC box one it has been filed a bit with a round file. Epoxy glue will be used to seal the tubing into the PCV box.

A pair of wires soldered to the pistol-grip trigger switch will run to the two contact pads on the camera circuit board to engage the high-voltage generator to start. It may be a bit of a task to get the switch fed into the tubing and out through the hole, so solder the wires to the switch first and then use them as a way to guide the switch to the hole opening (Figure 19-28).

A rubber handle-grip is slid over the metal tubing and switch to create the completed trigger handle (Figure 19-29). Make a hole for the switch that is about three times larger than the switch button so it will be easier to stretch the rubber grip over the switch into the correct position. The tubing is then glued into the PVC opening after it has been filed to allow a snug fit.

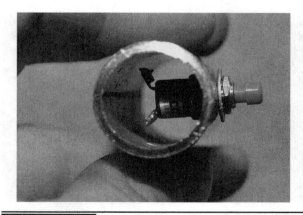

Figure 19-27   Fitting the trigger switch into the hand grip tubing.

Figure 19-29   The handle grip is placed over the metal tubing.

Even though the hand-held taser is more of a novelty than a self-defense device, ergonomics are always important. The position of the trigger switch should allow easy operation while your hand has a firm grip on the unit (Figure 19-30). You might also want to consider a secondary "safety switch" on the box in case you want to carry the unit in a bag or your pocket. We can tell you from painful personal experience that you should never carry a high-voltage device with a single trigger switch in your pocket! Keep in mind that carrying this device around in public may be considered a concealed weapon.

On a real taser gun, an explosive charge fires a tethered double pointed dart into the victim's skin, allowing the high voltage to bypass the "skin effect" and attack the central nervous system directly. Our project simply buzzes the skin with a very high-voltage, low-current shock through a pair of blunt-end conductors (probes). These probes are easily made by fastening a pair of 2- or 3-inch-long bolts to the end of the device to make contact with your target (Figure 19-31). These probes can also be used to connect capacitors for charging, as long as you have clearly marked the polarity of each probe.

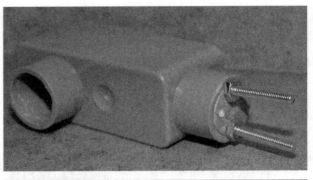

Figure 19-31   The high-voltage probes are just long bolts.

For maximum voltage delivery, place the probes as far apart from each other as your enclosure will allow. The plug that is used to fill the open gap is just a bit of broom handle cut and sanded to fit the opening.

To connect the bolts or probes to the high-voltage circuit, just wrap the exposed end of the wires around the end of the conductors so that there is no slack (Figure 19-32). You can solder the connection if you want, but it does not have to be perfect as high voltage seems to find its way a lot better than lower voltage. Another nut and washer can also be used to squash the end of the wire onto the bold head.

When installing the circuit board and battery into the cabinet, ensure that nothing is left to

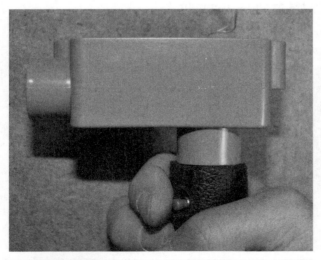

Figure 19-30   The trigger is placed in an easy-to-operate position.

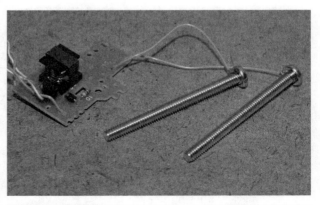

Figure 19-32   Connecting the high-voltage circuit to the output probes.

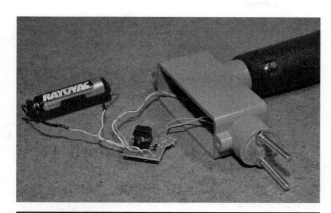

Figure 19-33 Fitting all of the parts into the box.

bounce around, or you may short out your circuit (Figure 19-33). A bit of double-sided tape will easily hold the circuit board to the cabinet, as well as the battery pack. Also, install your battery pack so that it is easy to change the battery without the need to remove the circuit board. If your unit might be dropped to the floor or ground, adding cotton balls around the components can help save them from damage.

Before putting on the cover screws, test the voltage output. It is always bad luck to assume you have made no mistakes before putting the lid on any project, as Murphy's Law will come and give you trouble (Figure 19-34). Hook up your high-voltage meter and check the output, or grab a hold

of the probes and let your fingers do a real test if you think you have the nerve!

The completed unit looks and performs great considering it is built from a throw-away camera and a two-dollar PVC box. You won't be using this device to save your hide in an emergency situation, but it does a great job of showing off your command of the screaming electron to fellow hardware hackers! Charging the photo flash capacitor and then welding it to a coin is also an impressive demonstration of the device and how the power from a single 1.5-V battery can be converted into a real force to be reckoned with (Figure 19-35).

Now that your taser project is completed, it is mandatory that you zap yourself a few times before unleashing the angry electrons on any of your unsuspecting fellow hardware hackers (Figure 19-36). Don't be a wimp, get a serious grip on those two probes, and give that trigger a good push. If you can take it for even a microsecond, then either you are the living dead, or your battery is no good! If you think it hurts across the fingers, then we dare you to take a probe in each hand and feel the lightning!

Here are a few warnings and suggestions now that you own a new toy. Never zap an unsuspecting person, and never point it at anyone you don't know

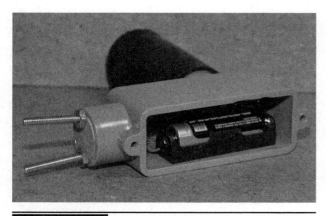

Figure 19-34 The completed circuit board and battery installation.

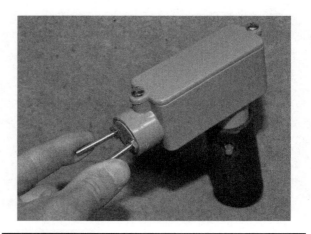

Figure 19-35 Here is our completed pistol style taser.

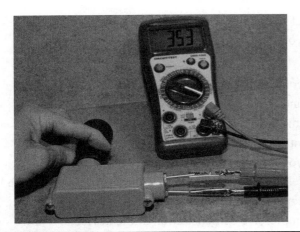

Figure 19-36   Zap yourself before you zap your fellow nerds!

Figure 19-37   Yes, this is three times the voltage available at your AC outlets!

very well. It may be considered a weapon, and you could end up in legal troubles or worse. Never bring this device to any public place, especially an airport, school, or your workplace. Don't try to use this device for personal protection as it will just piss off your attacker; it's not nearly as effective as tasers used in law enforcement. Never mention our name if you get expelled from school or in trouble with The Man due to the misuse of this device! This project is designed to help you learn about electronics, not commit felonies.

Well, we hope you had fun building this project and that by the end of the night that

tingling sensation in your knuckles goes away. The hand-held taser device is a great educational unit that demonstrates how the voltage of a single 1.5-V battery can be stepped up to almost 400 V (Figure 19-37). The taser device can also be used to charge high-voltage capacitors or serve as a handy power supply to drive various high-voltage experiments. By adding a second battery, you can push the output well beyond 600 V, and by altering the value of the flyback resistor, you can adjust the output frequency of the high-voltage oscillator. Have fun, and remember to respect the screaming electron!

Portable Alarm System

THIS PROJECT DETAILS a simple yet effective security system that is perfect for temporarily protecting an area or building. This portable alarm system functions much the same way as a typical hard-wired home security system, allowing for a timed entry and deactivation via numerical keypad. The alarm system will respond to practically any type of sensor input such as a motion switch, magnetic door switch, window break tape, infrared motions sensor, and any other sensor that acts like a simple switch. Because the main unit contains all of the electronics, alarm, keypad, and battery, the alarm trigger sensor can be remotely located, making it difficult to disarm the system once it has been activated by an intruder.

Figure 20-0 This security system works as both a permanent or portable alarm.

PARTS LIST

- IC1: ATMEL ATMega88P or similar microcontroller
- Keypad: Sparkfun 12 button keypad or similar
- Resistors: R1 = 10 K, R2 = 1 K, R3 = 1 K
- Transistors: Q1 = 2N3904 or 2N2222A NPN
- LEDs: LED 1 = red LED, LED 2 = green LED
- Diodes: D1 = 1N4001 or similar
- Relay: Any small 3- to 5-V relay
- Buzzer: Small 2 wire piezo element
- Alarm: Battery operated pocket alarm
- Battery: 3- to 9-V battery or pack

The goal of this project is to demonstrate how to connect a keypad matrix to a microcontroller to scan the rows and columns for key presses and to create a simple security system that can be used as the framework to create your own security system, leaving a lot of room for modifications and improvements. Practically any type of matrix keypad can be used, and the alarm siren used is a common and inexpensive pocket alarm that has been modified to allow switching through a relay. The microcontroller used is an 8-bit ATMega88, but just about any microcontroller can be used, as only a few IO pins are needed. The programming is done in C for portability and clarity.

The alarm is activated or deactivated by entering a secret code on a 12-digit keypad like the one shown in Figure 20-1. This keypad is from SparkFun.com, part number COM-08653, but any keypad will work as long as you know the pinout. You could even make this alarm system with no

Figure 20-1   Arming and disarming the alarm is done by entering a code on a keypad.

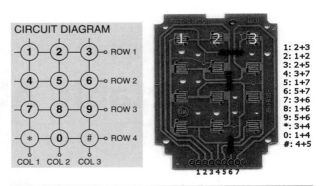

Figure 20-2   Most keypads are wired as a matrix of rows and columns.

keypad at all and just use the microcontroller as a simple time delay to allow arming and disarming the system, using the power switch to turn the alarm on or off.

Unlike a permanent home security system where the keypad is right at the door and the alarm is hidden, this system places the keypad and alarm in the same box, so the goal is to locate it deep inside a building and run a wire to the actual trigger sensor. In portable mode, the sensor can be affixed to the alarm box, allowing it to be propped up against a door or placed in a drawer or even a suitcase. Having the alarm, battery, and main electronics located in the same box means that this particular alarm system design is probably not the best solution for a permanent installation, but it does work very well as a portable system or when you just need temporary security for a building.

Although it may look confusing at first glance, the wiring diagram shown in Figure 20-2 is commonly referred to as a "matrix," as the button switches are made of common row-and-column connections. It certainly takes a little more work to decode a single button press into a microcontroller on a matrix system, but if each key had its own wire, then this 12 button keypad would need 12 input/output (IO) lines on the microcontroller. Using a matrix, the keypad now only requires 7 IO lines, a significant savings.

In a matrix-connected keypad, there is no common ground to all button switches since they all share a row or column. To decode a matrix keypad, you have to turn on a single column at a time and then scan the rows for the current button press. By doing this, you basically turn the keypad into a commonly-connected row of switches momentarily. This may seem like a lot of work, but it happens in a few microseconds in the microcontroller with just a few lines of program code as will be seen later.

The SparkFun keypad has a rounded rectangular shape, so mounting it into a box will require some fancy cutting to make a snug fit. There are several methods that can be used to mount this keypad in a box: cut the exact rounded rectangular shape as we are going to do, cut a smaller 90 degree rectangle so that only the keys are visible, or mount the key pad on the top of the box, leaving the PCB header exposed. The rounded rectangle opening will certainly look most professional if it can be made accurately, so that is how we decided to proceed. Take the basic length and width measurements from the key panel as shown in Figure 20-3. If you have the same keypad that we do, then the dimensions are 46-millimeters (mm) wide and 57-mm tall with 4-mm radius corners.

When choosing a project case for the numerical keypad, consider the width of the main panel circuit board because it has to fit between the edges

Figure 20-3 Measuring the dimensions of the keypad for mounting.

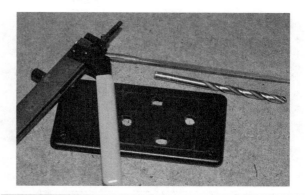

Figure 20-5 The notching tool is a great addition to your hacking toolkit.

of the box lid or sides. If you have the SparkFun keypad, there is a very detailed drawing and printable template on their Web site. Since our hand cutting would be the weak link, we just decided to trace out the dimensions as shown in Figure 20-4 and then hand-bombed the rounded edges by drawing them freehand. Hand cutting the rounded rectangle will be okay using a notching tool and a small round file, so perfect accuracy won't be possible.

Whenever you are faced with the task of cutting a square or rectangular opening in a thin plastic or metal project box, the notching tool shown in Figure 20-5 will make the job easy. This tool basically nibbles a tiny square bit out of the surface each time you close the pliers, so you can notch away along a straight line, creating any

size opening with straight edges. The tool is no replacement for a CNC mill, but it does a decent job for prototyping. You will have to drill starter holes at each corner and then work the notching tool along the edge until all four corners have been connected. A fine pitch flat file can be used to clean up the edge later if necessary.

Cutting along the edge of the lines, the notching tool takes out one small nibble at a time, eating a straight line out of the plastic enclosure as shown in Figure 20-6. The key to making a smooth cut is to keep the tool aligned along the edge of the line so that no jagged edges are formed. The cutout can be made fairly accurately using this tool, but don't expect perfection.

Figure 20-7 shows the results of using the notching tool to cut the opening and then spending

Figure 20-4 Marking the dimensions of the keypad for cutting.

Figure 20-6 Cutting the rectangular opening one nibble at a time.

Figure 20-7 The keypad fits like a glove after cutting the opening.

Figure 20-9 Soldering the keypad matrix wires to the circuit board.

some time with a fine pitch file to clean up the edges and create the rounded corners. The opening is not perfect, but once the keypad is inserted, it looks like a factory job. We wouldn't want to do 10 of these this way, but for this prototype, it did the job.

Once inserted, the keypad circuit board has just enough room between the box lid edges and the edges of the board as shown in Figure 20-8. This is why an enclosure with an inside diameter slightly larger than the edges of the keypad circuit board was necessary. An external mounting of the keypad would not be so critical.

You will need to solder a wire for each row and column needed on your keypad as shown in Figure 20-9. Having four rows and three columns, our keypad needed seven wires in total. If your

keypad has digits you don't plan to use, it may be possible to leave an entire row or column unconnected, but you will need to look at the datasheet to see how many digits you will lose by leaving a row or column unconnected. If we dropped row four from the panel, we would lose keys "*," "0," and "#" according to the matrix mapping shown in Figure 20-2. We didn't care about the symbol characters, but wanted to use the zero, so we had to connect all of the rows and columns.

To scare trespassers away, a very loud, ear-piercing siren will be needed. The high-pitch sound will travel a great distance and make it almost impossible to locate the actual alarm system, making it impossible to disable in a hurry. You have many choices for alarm hardware, with battery voltage being the only consideration. The pocket alarm shown in Figure 20-10 is a perfect choice for

Figure 20-8 The inside edges of the circuit board have just enough clearance.

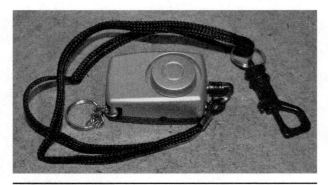

Figure 20-10 This tiny pocket alarm is amazingly loud for its size.

Figure 20-11 Converting the pocket alarm for relay operation.

Figure 20-12 Replacing the internal batteries with a pair of wires.

this project because it will run from any voltage between 3 and 5 volts (V), and can be switched on using a small relay or even transistor. The pocket alarm is inexpensive (normally retails for under $10), even less than a basic DC-operated piezo buzzer, so consider it for this project.

Inside the pocket alarm (Figure 20-11) you will find a few coin batteries, a small circuit board with an oscillator and inductor, and the piezo transducer that makes the high-pitch siren sound. This thing is so loud that it makes our ears ring, which is quite impressive given its size and the fact that it runs from the low current 4.5-V coin battery source. At 3 or 5 V, the siren worked the same, so conversion to a microcontroller-based project is easy. The pocket alarm operation is simple: you pull a rip cord pin, which closes a switch and turns on the alarm. To convert this alarm for triggering through a relay, remove the rip cord pin and replace the battery pack with the switched output from a relay.

With the rip cord pin removed, the alarm will now be in the on position permanently. To control it externally, solder a pair of wires to the battery terminals as shown in Figure 20-12, and whenever a voltage between 3 and 5 V is applied, the alarm will begin to shriek. Use a pair of color-coded wires that will make it easy to determine polarity so that you don't accidentally reverse the power to the alarm circuit.

Once you have the wire pair soldered to the battery terminals, apply power to verify that the alarm is working properly. Drill a small hole in the back or side of the alarm casing as shown in Figure 20-13 so that you can connect them to the relay that will be installed in the alarm system circuit.

This alarm system will work in "self-contained" mode where the sensor is affixed to the alarm box or in "remote mode" where the intruder sensor is installed via wire to another location. In self-contained mode, this alarm system is great for protecting drawers, luggage, and doorways as it is as simple as arming the alarm and placing it in an area that would be triggered by an intrusion. In self-contained mode, the goal of the alarm is to

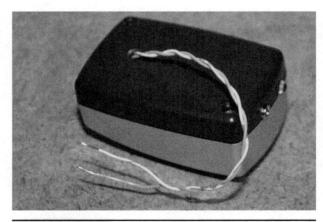

Figure 20-13 The completed externally powered pocket alarm.

instantly alert you of an intrusion into your space. In remote mode, the alarm can protect several windows or doors at the same time by connecting magnetic or break switches in a series loop back to the alarm unit. Remote mode allows the system to act as a semipermanent building alarm that can be installed in a few minutes.

This project will be completed by testing the alarm in self-contained mode, using a motion-sensitive switch as the alarm trigger. Motion switches come in the form of ball switches, mercury switches, or electronic motion detection IC packages. We just happened to have a few old mercury switches salvaged from some 1980s thermostats, so we will use one of them as a motion switch. Mercury switches are rather rare these days, but the ball bearing switches shown in Figure 20-14 next to the mercury switches are a fairly decent replacement that does not contain toxic liquid metal. Warning: Mercury is poisonous. Any motion switch will work as long as it acts like a real switch, opening or closing when motion is detected.

A passive motion sensor does not require a power supply as it is mechanical in nature, opening or closing a switch in response to motion on its body. You can see in Figure 20-15 that the mercury switch is just a blob of liquid metal that can be at either end of the sealed tube, making contact with

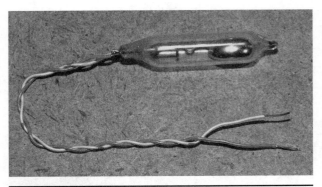

Figure 20-15   Passive switches like this mercury switch are optimal for this project.

a pair of pins at one end. Ball-bearing switches and magnetic switches also work the same way. Since this project puts the microcontroller into deep power saving mode when activated, a passive switch is used so that the alarm can run for a very long time off a battery pack. An electronic switch would slowly drain the power pack, even if it only used a small amount of current, so a passive switch is the best choice here.

The schematic for the portable alarm system shown in Figure 20-16 is fairly straightforward. The rows and columns of the keypad are connected directly to the microcontrollers IO pins so that the program can switch on one column at a time and then scan for row button presses. The input from the alarm sensor (switch) is connected directly to an input pin, and the relay can be turned on by sending voltage from the IO pin connected to the base of the driver transistor. The two light-emitting diodes (LEDs) and piezo buzzer give an audio and visual indication of the various functions of the alarm system.

Also note that the wiring shown in the keypad matrix may not match your own keypad, so you will need to determine which pins are rows and which pins are columns, and then make the appropriate changes in the code dealing with the IO pins on PORT-D that connect to the keypad rows and columns. As for VCC, either 5 or 3 V will work fine with the ATMega88, but when using 3 V, you

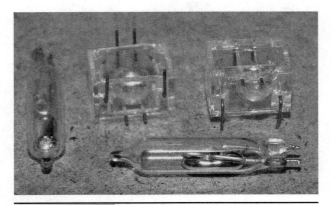

Figure 20-14   A few different motions switches for self-contained mode.

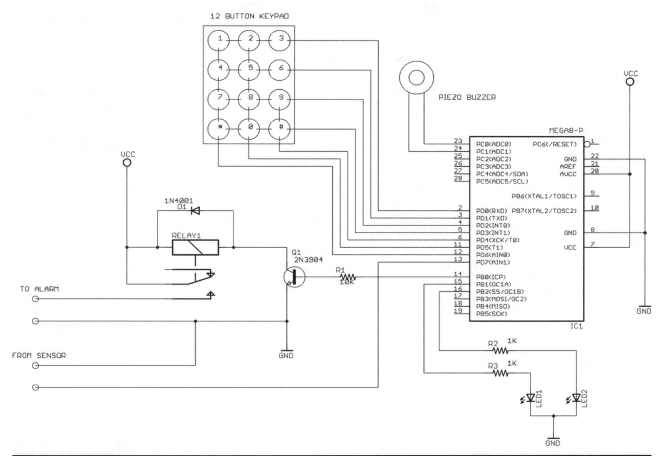

Figure 20-16 The portable alarm system schematic using an ATMega88 microcontroller.

will have to ensure that the relay chosen will close on this low voltage. Most relays rated for 5 V will work just fine on 3 V.

Before committing any permanent circuit board wiring, build the circuit on a solderless breadboard as shown in Figure 20-17 so that you can easily program the microcontroller, make source code changes, and alter the circuit to suit your needs. Some of the things you will want to alter in the source code will include the secret password (found under *program initialization*), which is currently set to "1234," as well as the times between activation and triggering. Some of the program code is specific to the ATMega-88, allowing it to be sent into deep power-saving mode once the alarm has been set to armed.

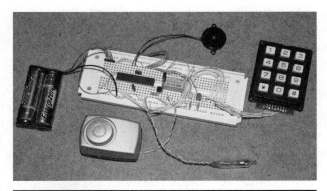

Figure 20-17 Building the portable alarm system on a solderless breadboard.

Power-saving mode is deactivated by any change on the alarm sensor pin via pin change interruption, allowing this system to run for a very long time once set to armed.

```
// ********************************************************************************
// **** PROGRAM = HOME ALARM SYSTEM (C) 2010 LUCIDSCIENCE.COM
// **** TARGET = ATMEGA88P & 12 DIGIT KEYPAD
// **** CLOCK SPEED = 8 MHZ INTERNAL
// ********************************************************************************
#include <mega88p.h>
#include <delay.h>
#include <sleep.h>



// ********************************************************************************
// **** SEND SHORT BEEP TO PIEZO BUZZER
// ********************************************************************************
void BEEP1() {
unsigned char ctr1;

// SINGLE SHORT BEEP
for (ctr1 = 0; ctr1 < 50; ctr1++) {
PORTC.0 = 0;
PORTC.1 = 1;
delay_us(400);
PORTC.0 = 1;
PORTC.1 = 0;
delay_us(400);
}
}



// ********************************************************************************
// **** COMMAND CONFIRM BEEP
// ********************************************************************************
void BEEP2() {
unsigned char ctr1;

// LEDS ON
PORTB.1 = 1;
PORTB.2 = 1;

// BEEP FREQUENCY 1
delay_ms(250);
for (ctr1 = 0; ctr1 < 100; ctr1++) {
PORTC.0 = 0;
PORTC.1 = 1;
delay_us(500);
PORTC.0 = 1;
PORTC.1 = 0;
delay_us(500);
}
delay_ms(100);
```

```
// BEEP FREQUENCY 2
for (ctr1 = 0; ctr1 < 200; ctr1++) {
PORTC.0 = 0;
PORTC.1 = 1;
delay_us(200);
PORTC.0 = 1;
PORTC.1 = 0;
delay_us(200);
}
delay_ms(100);

// BEEP FREQUENCY 3
for (ctr1 = 0; ctr1 < 150; ctr1++) {
PORTC.0 = 0;
PORTC.1 = 1;
delay_us(300);
PORTC.0 = 1;
PORTC.1 = 0;
delay_us(300);
}
delay_ms(250);

// LEDS OFF
PORTB.1 = 0;
PORTB.2 = 0;
}



// ****************************************************************************
// **** SEND AVR INTO POWERDOWN MODE
// ****************************************************************************
void SLEEP() {

// TURN OF LEDS AND BEEP
PORTB.1 = 0;
PORTB.2 = 0;
BEEP2();

// ENTER POWERDOWN MODE
#asm("sei")
sleep_enable();
powerdown();
}
```

```
// ***********************************************************************
// **** WAKE FROM SLEEP ON ALARM PIN CHANGE
// ***********************************************************************
interrupt [PCINT2] void pin_change_isr2(void) {
sleep_disable();
#asm("cli")
}



// ***********************************************************************
// **** READ KEYPAD AND RETURN ASCII VALUE
// ***********************************************************************
unsigned char KEYPAD() {
unsigned char key;
unsigned char scn;
key = 0;

// SCAN KEYS IN COLUMN 1
PORTD.4 = 0;
delay_ms(1);
scn = (PIND & 15);
PORTD.4 = 1;
// CONVERT KEY TO ASCII
if (scn == 14) key = '1';
if (scn == 13) key = '4';
if (scn == 11) key = '7';
if (scn == 7) key = '*';

// SCAN KEYS IN COLUMN 2
PORTD.5 = 0;
delay_ms(1);
scn = (PIND & 15);
PORTD.5 = 1;
// CONVERT KEY TO ASCII
if (scn == 14) key = '2';
if (scn == 13) key = '5';
if (scn == 11) key = '8';
if (scn == 7) key = '0';

// SCAN KEYS IN COLUMN 3
PORTD.6 = 0;
delay_ms(1);
scn = (PIND & 15);
PORTD.6 = 1;
// CONVERT KEY TO ASCII
if (scn == 14) key = '3';
if (scn == 13) key = '6';
if (scn == 11) key = '9';
if (scn == 7) key = '#';
```

```
// SEND KEY BEEP
if (key != 0){
for (scn = 0; scn < 50; scn++) {
PORTC.0 = 0;
PORTC.1 = 1;
delay_us(200);
PORTC.0 = 1;
PORTC.1 = 0;
delay_us(200);
}
}

// HOLD WHILE KEYDOWN
scn = 0;
while (scn != 45) {
scn = 0;
PORTD.4 = 0;
delay_ms(1);
scn = scn + (PIND & 15);
PORTD.4 = 1;
PORTD.5 = 0;
delay_ms(1);
scn = scn + (PIND & 15);
PORTD.5 = 1;
PORTD.6 = 0;
delay_ms(1);
scn = scn + (PIND & 15);
PORTD.6 = 1;
}

// RETURN KEY VALUE
delay_ms(100);
return(key);
}



// ******************************************************************************
// **** PROGRAM VARIABLES AND PORT SETUP
// ******************************************************************************
void main(void) {

// MAIN PROGRAM VARIABLES
unsigned char ctr1;
unsigned char ctr2;
unsigned int ctr3;
unsigned char mode;
unsigned char code[4];
```

```
// 3X4 KEYPAD ROW INPUTS
DDRD.0 = 0;
DDRD.1 = 0;
DDRD.2 = 0;
DDRD.3 = 0;
PORTD.0 = 1;
PORTD.1 = 1;
PORTD.2 = 1;
PORTD.3 = 1;

// 3X4 KEYPAD COL OUTPUTS
DDRD.4 = 1;
DDRD.5 = 1;
DDRD.6 = 1;
PORTD.4 = 1;
PORTD.5 = 1;
PORTD.6 = 1;

// GREEN LED OUTPUT
DDRB.1 = 1;

// RED LED OUTPUT
DDRB.2 = 1;

// PIEZO BUZZER OUTPUTS
DDRC.1 = 0;
DDRC.1 = 1;

// ALARM TRIGGER INPUT
DDRD.7 = 0;
PORTD.7 = 1;

// ALARM SIREN OUTPUT
DDRB.0 = 1;


// ****************************************************************************
// **** PROGRAM INITIALIZATION
// ****************************************************************************

// ENABLE PIN CHANGE INTERRUPT
EICRA = 0x00;
EIMSK = 0x00;
PCICR = 0x04;
PCMSK2 = 0x80;
PCIFR = 0x04;
```

```
// SEND STARTUP BEEP
delay_ms(1000);
for (ctr1 = 0; ctr1 < 4; ctr1++) {
BEEP1();
PORTB.1 = 1;
PORTB.2 = 1;
delay_ms(100);
PORTB.1 = 0;
PORTB.2 = 0;
delay_ms(100);
}

// SET SYSTEM TO IDLE MODE
mode = 0;
ctr1 = 0;
ctr2 = 0;
ctr3 = 0;

// SET ACCESS KEYCODE
code[0] = '1';
code[1] = '2';
code[2] = '3';
code[3] = '4';


// ****************************************************************************
// **** MAIN PROGRAM LOOP
// ****************************************************************************
while (1) {


// ****************************************************************************
// **** MODE 0 : IDLE AND UNARMED
// ****************************************************************************
while (mode == 0) {

// TURN OFF EXTERNAL ALARM
PORTB.0 = 0;

// TURN ON GREEN LED
PORTB.1 = 1;
PORTB.2 = 0;

// SEARCH FOR KEYCODE
if (ctr1 == 0 && KEYPAD() == code[0]) ctr1 = 1;
if (ctr1 == 1 && KEYPAD() == code[1]) ctr1 = 2;
if (ctr1 == 2 && KEYPAD() == code[2]) ctr1 = 3;
if (ctr1 == 3 && KEYPAD() == code[3]) ctr1 = 4;
```

```
// ARM SYSTEM ON KEYCODE
if (ctr1 == 4) {
ctr1 = 0;
ctr2 = 0;
ctr3 = 0;
mode = 1;
}
}



// ***************************************************************************
// **** MODE 1 : ALARM ARMED WITH 60 SECOND COUNTDOWN
// ***************************************************************************
while (mode == 1) {

// BEEPS FOR 60 SECONDS
for (ctr1 = 0; ctr1 < 60; ctr1++) {
BEEP1();
PORTB.2 = 1;
delay_ms(500);
PORTB.2 = 0;
delay_ms(500);
}

// ARM SYSTEM AND POWER DOWN
SLEEP();
ctr1 = 0;
ctr2 = 0;
ctr3 = 0;
mode = 2;
}



// ***************************************************************************
// **** MODE 2 : ALARM TRIGGERED WITH 60 SECOND COUNTDOWN
// ***************************************************************************
while (mode == 2) {

// ALARM WARNING BLIP
for (ctr1 = 0; ctr1 < 50; ctr1++) {
PORTC.0 = 0;
PORTC.1 = 1;
PORTB.2 = 1;
delay_us(150);
PORTC.0 = 1;
PORTC.1 = 0;
PORTB.2 = 0;
delay_us(150);
}
```

```
// ACTIVATE SIREN WITHIN 60 SECONDS
ctr3++;
if (ctr3 == 400) mode = 3;

// SEARCH FOR KEYCODE
if (ctr2 == 0 && KEYPAD() == code[0]) ctr2 = 1;
if (ctr2 == 1 && KEYPAD() == code[1]) ctr2 = 2;
if (ctr2 == 2 && KEYPAD() == code[2]) ctr2 = 3;
if (ctr2 == 3 && KEYPAD() == code[3]) ctr2 = 4;

// ENTER IDLE MODE ON KEYCODE
if (ctr2 == 4) {
PORTB.1 = 0;
PORTB.2 = 0;
ctr1 = 0;
ctr2 = 0;
ctr3 = 0;
mode = 0;
BEEP2();
}
}



// ***********************************************************************************
// **** MODE 3 : ALARM SIREN ACTIVATED
// ***********************************************************************************
while (mode == 3) {

// TURN ON EXTERNAL ALARM
PORTB.0 = 1;

// POWER DOWN AND WAIT FOR CODE
SLEEP();
ctr1 = 0;
ctr2 = 0;
ctr3 = 0;
mode = 2;
}


// END OF MAIN LOOP
}
}
```

If you intend to use the alarm system for personal protection where you want to be alerted instantly to any intrusion, then change the counter check value under the section "*Mode 2: Alarm triggered with 60 second countdown*" and in the line "if (ctr3 == 400) mode = 3;" from 400 to some smaller value like 50. You will of course have to experiment with the code to change it to suit your needs, but it is well-commented and very easy to modify or improve.

Without altering the source code, the operation of the portable alarm system is as follows: Once powered on, the LEDs will flash to indicate that everything is working, and then the green LED will stay lit to indicate the system is idle and waiting for the secret code to be entered. Once you enter the secret code of "1234," the red LED will flash for 60-seconds, giving you time to exit the building in order to set the alarm trigger switch to the armed position. After 60 seconds, the alarm will enter a deep power-saving mode, and both LEDs will be off. Any change on the alarm trigger switch or sensor will wake up the alarm and begin another 60-second countdown. If you do not enter the secret code within 60 seconds, the relay will close, sending power to the loud alarm until the entire system is powered off. If you do enter the secret code within 60 seconds, the system will enter the idle state again. This operation is much like a typical house alarm system.

Once you have verified the operation of the alarm system and made whatever source code changes may be necessary, collect all of your required parts and find a suitable perforated circuit board to hold the six or seven semiconductors, including the microcontroller. As shown in Figure 20-18, most of the inventory will be hardware, and the actual circuit board will only take up as much room as needed for the microcontroller, relay, and a few resistors. When working with microcontrollers, it is always a good idea to use a socket on your circuit board so that you can remove the microcontroller for

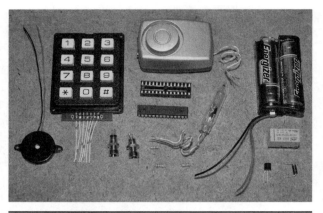

Figure 20-18 Getting ready to build the final version on a circuit board.

reprogramming if you don't intend to add an in-circuit programming header.

Perforated board is available with solder pads like the board shown in Figure 20-19, or with unclad holes. Both types will be fine for this project since there are so few components to wire. Parts are placed on the perforated board, and then small wires are soldered on the underside, between the component pins to create traces. Since the microcontroller and relay will take up most of the room, the circuit board does not have to be much larger than both parts. The wiring on the keypad has also been trimmed so that there is just enough slack that the circuit board can be folded around and stuck to the back of the keypad, making for a compact final product.

Figure 20-19 Adding the components to a small perforated circuit board.

Figure 20-20 Creating circuit board traces using small wires.

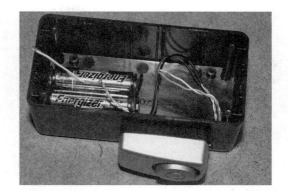

Figure 20-22 Mounting the alarm siren and the battery pack to the enclosure.

To make the perforated board design easier, start with the VCC and ground connections, using red and black or red and green wires so that there will be no polarity reversal errors. The rest of the connections are easy enough, with the seven IO lines from the keypad taking up most of the wiring. Figure 20-20 shows our completed circuit board, with various colored wires used to make debugging easier.

Once your circuit board has been completed and tested, some space can be conserved by folding the circuit board to the back of the keypad as shown in Figure 20-21, holding there with a bit of double-sided tape. We also added the piezo buzzer to the board, making the entire system not much larger than the keypad. The 1/8-in jack shown in Figure 20-21 is to allow the external alarm trigger

switch to be removed and swapped with various different trigger systems for more versatility.

To get the most volume out of the alarm siren, it should be mounted to the outside of the box, or at least in a way that places the speaker portion outside the enclosure. Figure 20-22 shows the alarm siren held to the side of the enclosure with a few tiny screws as well as the 3-V battery pack, which is held to the bottom of the box with a small bit of double sided tape. Keep all wiring long enough so that when you remove the lid (and circuit board), you can get at the underside of the board easily.

Figure 20-23 shows the completed and working system before placing all of the components into the enclosure. Having the wiring long enough to spread out the components like this means that debugging a bad connection will be much easier.

Figure 20-21 Affixing the circuit board to the back of the keypad.

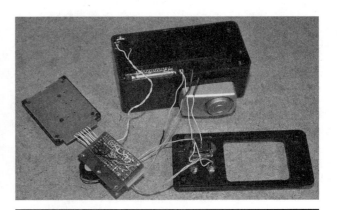

Figure 20-23 Completing all of the hardware wiring connections.

Figure 20-24 Adding the master power switch to the system.

There is also plenty of room in the enclosure in case later hardware upgrades are needed.

The master power switch will cut power between the battery pack and the circuit, and it is also the only way to shut off the alarm once it is triggered by an intruder. Figure 20-24 shows the rocker switch we chose to use as the master power switch. The square switch requires the notching tool to be used again to cut an accurate square hole, but since we had a bunch of these switches salvaged from old computer power supplies, it made sense to use one.

To secure the keypad to the inside of the enclosure lid, we used a bead of hot glue around the edges as shown in Figure 20-25. The keypad also had four small mounting holes, but the hot

glue was just as effective in hold the panel securely to the plastic lid. We also ran a bead of glue around the master switch just to ensure it would not come loose from the hole. As a travel alarm, you will want to make your hardware as durable as possible, especially if it will travel in a suitcase. Be warned, though, that airport security might not allow you to bring the device in your carryon or checked luggage—for obvious reasons.

In portable mode, the alarm trigger switch is connected to the alarm enclosure so that the entire system is one piece. This allows the alarm to protect such things as luggage, dresser drawers, filing cabinets, doors, and any other object that will move when an intrusion occurs. By installing the mercury or ball switch on a short piece of stiff wire on the connecting jack as shown in Figure 20-26, it is easily aligned in any position, allowing the alarm box to be "set" at any angle and armed. Other types of sensors such as microswitches, magnetic switches, or trip wires could also be used in portable mode, allowing the alarm to be set up to protect just about any area or object from intrusion. In this case, you will also want the alarm activation countdown timer to be set to a very short duration—or none at all—so that the instant the alarm is disturbed, the siren begins to wail.

With a roll of double-sided tape handy (Figure 20-27), the portable alarm system can easily be installed on doors, drawers, or even inside

Figure 20-25 Sealing the hardware using a hot glue gun.

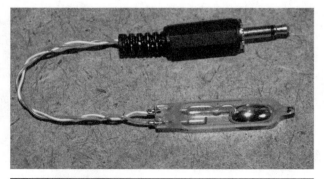

Figure 20-26 Creating the motion trigger switch for portable mode.

Figure 20-27   Double-sided tape will help install the alarm in portable mode.

a vehicle to alert you and scare away potential intruders. The motion switch is shown sitting on the top of the alarm cabinet in Figure 20-27, waiting for any movement of the box to set off the extremely loud siren. By adjusting the angle of the motion sensor on the stiff wire, you can fine tune the alarm sensitivity the way you want it.

Figure 20-28 shows how easy it is to adjust the motion sensor on the end of the wire so that the alarm can be installed at any angle or position. Here it is set to go off if our highly secret electronic prototype drawer is infiltrated by corporate spies! Using the ball switch or mercury switch, it is nearly

impossible to open the drawer without setting off the alarm, even if you know it is there.

One of the reasons we made this alarm system was so that we could bring it on travels and protect our rented room or luggage from thieves. It's not fun when a thief slips into your room at night and "borrows" your laptop, so this simple alarm will scare would-be thieves away before they get to your valuables. Sure, this may not be the most high-tech security system available, but one thing we have learned from the school of hard knocks is that you don't need the best security system in the world, just one better than your neighbors'! Seriously, thieves normally want an easy target, which means a room without any security system at all. Figure 20-29 shows the alarm system working to protect the interior of an unlocked room by leaning it up against the door in motion-activation mode.

We also used this alarm system to protect several storage buildings on our property that did not have power available. By installing magnetic door and window switches via wires to the alarm system, it was able to act much like a typical home alarm system, allowing time to set or deactivate the alarm using the secret code during entry and exit. In this mode, the alarm unit needs to be inside the building in a place that is out of the way, just in case the intruder is brave enough to hunt around for the alarm unit as it is screeching away. The intruder will likely

Figure 20-28   Protecting our top secret prototype storage cabinet from spies.

Figure 20-29   A simple way to scare away burglars.

Figure 20-30 A semipermanent installation using a magnetic door switch.

flee the scene right away, and even if he/she does try to find the alarm box, it will be very difficult as the high pitch of the siren is not at all easy to track. Figure 20-30 shows the alarm mounted in one room while the magnetic door sensor is in another room. Multiple sensors can be chained in parallel this way as long as they are all either "normally open" or all "normally closed" circuit.

This project leaves plenty of room for improvements and modification, and you can use just about any sensor imaginable to trigger the alarm system. Infrared motions sensors, gas sensors, light sensors, sound sensors and even radiation sensors can be used as long as you power them externally. If you build this project using surface mounted components, it could be made very small and easily concealed inside another object, making it very stealthy. You will never have to be without some kind of security system with this simple project!

PART SEVEN
Digital Camera Spy Projects

Camera Trigger Hack

THERE ARE TIMES when you need to acquire a very high-resolution image, triggered by some external event such as movement, time, or computer control. Video security cameras are very limited in resolution, often to less than 640 × 480 pixels, which in digital camera terms is less than half of 1 megapixel. Nowadays, a small digital camera that can take an image with a resolution of 4000 × 3000 pixels can be purchased for mere pocket change, so even if your subject is at a long distance from the camera, the details will still be present in the image when zoomed on a computer screen. This simple project demonstrates how to hack into the camera's shutter release button to add some kind of external control to allow automated picture taking.

Figure 21-0 **This external controller allows any device to control the shutter release.**

this hack is fairly easy to do as long as you can find a way to open the cover on your donor camera. Once completed, the resulting relay controller will allow any external electronic device to focus and then take a photo, essentially duplicating the operation of the two-position shutter release trigger on your camera. Also, note that your camera will not be usable for regular photography after this hack as the original shutter release switch will be removed.

The sacrificial camera shown in Figure 21-1 is an HP Photosmart M547 digital camera with an 8-megapixel imaging system. This camera has been around the world and dropped in an ocean, but despite some dents and scratches it still functions perfectly, so it will begin a new life as a covert spy gadget. To open one of these small digital cameras, you will need a set of tiny screwdrivers, a small knife, and a whole lot of patience. Since the goal of manufacturing is to keep costs to a minimum, the cases on these cameras are often snapped together,

PARTS LIST

- Resistors: R1 = 1 K
- Diodes: D1 = 1N4001 or similar
- Transistors: Q1 = 2N3904 or 2N2222A or similar NPN
- Relay: Single pole relay rated for 5 to 12 V
- Battery: 6- to 12-V battery or pack

Because this project is a hardware hack, you should not try this with a good camera, or one that you are worried about breaking. There is always a possibility of destruction when cracking the case open on such a small electronic device that is jammed full of tiny components. Of course, if you are good with small tools and a soldering iron, then

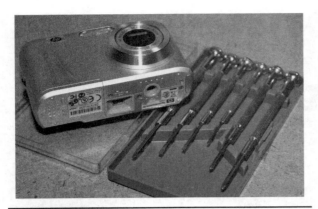

Figure 21-1 This camera will be converted for external shutter control.

Figure 21-3 Often, the screws will be hidden behind doors or under labels.

which will require some careful prying to open them up.

The tiny screws shown in Figure 21-2 are about the same size as the head of a pin, so don't expect to recover one that falls on the floor! A magnetic screwdriver really helps out when working with these tiny parts, and you can improvise by simply placing a small magnet on the shaft of the screwdriver to make it temporarily magnetic. Also, some camera manufacturers have contests between their engineers to see who can use the most different sizes of screws in a single camera in order to make putting one back together the most difficult, so if you start to see multiple screw sizes and lengths, make a diagram as you go. Luckily,

Figure 21-2 This camera will be converted for external shutter control.

this camera had only two different-sized screws—silver ones on the outside and black ones on the inside. We have seen much worse!

To make your hacking work even more tedious, screws may be placed in hard-to-find locations, behind trap doors (Figure 21-3) or even under logo stickers and plates, so if you can't seem to get the case open, look around for those secret screws. Remember, these small consumer devices were not made to be taken apart, repaired, or hacked by us evil geniuses! Work your way around the entire camera and try to get all of the obvious screws removed from the casing.

Taking out all of the visible screws is the easy part, and now comes the real fun—prying the plastic casing apart to reveal the internals. Most likely, the two or more plastic casing parts will be snapped together, held by friction on the inside, so you will need to carefully pry open the parts using a small knife blade or flat screwdriver blade. If you are afraid of damaging your camera then it's not too late to chicken out and put back the screws. If you decide to continue, then expect a few scrapes along the edges as you pry away with your screwdriver.

To begin to pry apart the casing, carefully wedge a small flat screwdriver blade into the edge as shown in Figure 21-4 so you can start to force the two plastic halves apart. Don't dig in too deep or you run the risk of damaging the electronics.

Figure 21-4   The plastic casing will usually be snapped together from the inside.

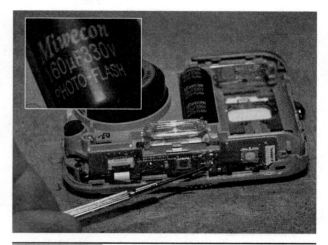

Figure 21-5   There is a danger lurking inside every camera that includes a flash.

Since these cameras are full to the brim with small components, expect that there will be no space between the inside of the casing and the circuit board. Once you dig in enough to gain some leverage, pry the screwdriver until you get the first plastic latch to pop open. Once you have one gap, the rest will be fairly easy, but do take your time and try to keep the screwdriver from reaching into the cabinet as you go along.

Now is a good time for a warning. Inside any camera that includes a flash is a 350-volt (V) capacitor that will hold a charge for months at a time. We guarantee that once you get your fingers across the charged terminal, you will take notice to never do it again! How can something so small be so dangerous you ask? Well, the capacitor has to hold a voltage that is more than three times the voltage that you will find in your AC outlets in order to fire the xenon flash tube in the camera, and although the current is kept fairly low, the resulting shock from the charged capacitor will wake you up in a way you will never forget. Trust us—we have zapped ourselves more times than we care to remember.

Figure 21-5 (inset) shows the beasty you are looking to avoid—the small half-inch (in) black cylinder with two leads coming out of one end. Once you have identified the capacitor, short a screwdriver across the leads, if you can reach them, so that you don't end up taking the voltage with

your fingertips. If you can't see the capacitor, then you "may" be safe from the high voltage, but do work carefully.

This camera has not been in use for more than a month, but after we removed the flash capacitor just for fun, it still measured almost 300 V as shown by our meter in Figure 21-6! Getting this kind of DC voltage across your finger feels like touching a red hot stove element, and getting a shock across both hands feels like swimming in a pool full of toasters that are plugged into live AC current! In other words, it hurts. Don't do it!

Figure 21-6   After more than 30 days, the capacitor still holds almost 300 V!

Figure 21-7   The camera shutter release switch will be mounted to a small circuit board.

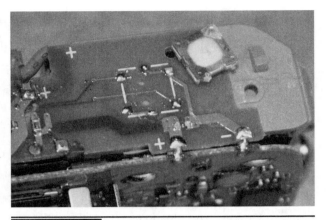

Figure 21-8   The traces are revealed on the circuit board after removal of the small switch.

Once you have managed to pry apart your camera casing, you should be able to identify the tiny shutter release switch, which will look like a small square with a round center button as shown in Figure 21-7. This tiny switch is actually a two-position switch that first sends the focus command to the camera and then the shutter release command. If you remember from using the camera, you can focus by holding the switch half-way down and then take a photo by pressing it all the way. For this reason, there will be three or four contact points that need to be hacked into in order to bring control out to the external relays. From here, there are two ways you can continue with this project: removal of the switch to add wires or by tapping into the circuit to add wires while leaving the switch installed. The easiest method is to simply unsolder the switch, as you may not be able to identify or even tap into the traces leading to the switch. On this camera, you can clearly see the three connecting traces in Figure 21-7, but we just decided to remove the switch since this camera was no longer in normal service.

The tiny shutter release switch will probably be top-mounted to the small circuit board, so you will need to either heat up the contacts and pry one edge at a time, or use a solder wick to remove the solder on one side at a time. Don't try to pry up the small switch, even if you don't plan to use it again as

you may also rip up the thin copper traces, making it impossible to install the new wires. Figure 21-8 shows the four small solder pads left over after removing the pushbutton switch from our camera. We also added some solder to the pads so it would be easier to install the new wiring.

Although we have not seen one for many years, older digital cameras may have larger pushbutton switches that are soldered as through-hole parts, requiring the removal of the small circuit board to get at the soldered leads. The camera shown in Figure 21-9 is a year 2008 digital camera, and had a through-hole pushbutton switch. Even a camera this old will acquire an image with a resolution

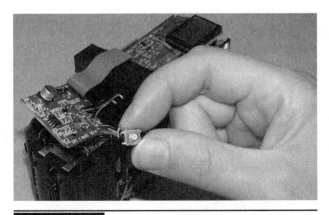

Figure 21-9   This older camera required removal of the small circuit board.

many times greater than any security camera, so don't throw out your old digital camera—hack it into something better.

Depending on the design of your camera, there will be three or four contact points. Most likely, you will only need three wires to control both functions of the switch (focus and shoot) as there will usually be a common ground. If you can still get the batteries into the camera with the casing apart, it is easy to test which contact points control which functions by placing a wire over the points to see how the camera responds. In our camera, there were four points under the small switch, but two of them were a common ground so we soldered the needed three wires to the circuit board as shown in Figure 21-10. Use the finest wire you can find, and leave enough wire to reach whatever connector or control box you intend to add to the camera later.

Before you embark on the incredible journey of putting the camera all back together, try to feed in the batteries and test your new switch wiring by placing the wires into a breadboard as shown in Figure 21-11. You may have to hold the battery door in place just to make the connection, but it should be possible to test the camera before putting all of the tiny screws back in place. We used a pair of small pushbutton switches—one for the focus function, and one for the shoot function. Both switches share a common ground point.

Figure 21-11 Testing the connections using a pair of pushbutton switches.

Once we had the battery door in place, the camera lens zoomed out, and the liquid crystal display (LCD) screen showed that the camera was indeed functioning properly. We could now figure out which button was for focusing and which button was used to take a photo. Actually, you could get away with only a single button as most cameras will manage to focus and shoot if you just pressed the shutter release button all the way down. The exception is when there is a fast-moving target or when lighting conditions are suboptimal. All of the projects we intend to base this camera interface on will assume that both the focus and shoot buttons are connected, but it would be easy to modify them for just the shoot button alone (Figure 21-12).

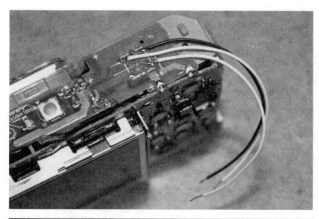

Figure 21-10 Adding the wires that will control the pushbutton functions.

Figure 21-12 The camera powers up and takes a photo when the buttons are pressed.

Figure 21-13   The wiring will be placed through a slot in the plastic casing.

Figure 21-15   A three-conductor 1/8-in stereo jack will connect the camera to the device.

The last puzzle to solve before putting back all of the tiny screws is how to get the wiring out of the camera casing. The easiest way we found was to cut a small notch into the edge of one half of the casing, using a pair of wire cutters so that the wires can be placed through the casing as shown in Figure 21-13. Since there is no possible way you will have room to include a connector in the camera case, running the wires through a small opening is the only option. Avoid bending the wires on an extreme angle from the solder point or you may end up lifting the copper traces from the very small circuit board.

The assembled camera shown in Figure 21-14 is retested using the breadboard and switches just to make sure that nothing was damaged in the

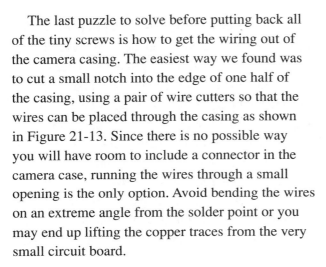

Figure 21-14   Once reassembled, the camera is tested using the breadboard and switches.

reassembly process. The camera powered up and tested just fine once the batteries were installed. Oddly, we only had one screw left over after putting the camera back together. Normally we have more than one! How did you do?

In order to attach the wiring to the external triggering device(s), some type of connector will be necessary. If your system uses three wires, then a 1/8-in stereo-jack pair like the ones shown in Figure 21-15 will be perfect. For systems with four wires, a telephone-type connector or computer network connector would work. You could actually avoid using any connectors if you intend to build a circuit and have it attached directly to the camera, but this would make it difficult to try out multiple triggering devices with your system. Since we intended to make several triggering devices, we opted for the removable jack installation.

Since there is barely enough space inside the camera casing for just the wiring, there would be no chance of installing the connector inside the camera. As shown in Figure 21-16, we just affixed the female 1/8-in connector to the camera body using a bit of double-sided tape. This position was convenient for installation and would not block any of the camera controls or sensors. Having the connector in this position also meant that the camera could function normally if we soldered a pair of tiny switches to a male connector to be used manually just like the original functionality of the camera.

Figure 21-16 The female connector is affixed to the body of the camera.

The male connector for the 1/8 stereo jack is shown in Figure 21-17, having three conductive rings. The larger base area is ground (or common), and the two rings at the end are for each camera function. Since there are no high-frequency signals in this project, you don't have to use shielded cable or worry about how you connect the terminals as long as you know which pins control each function.

At this point, the camera trigger hack is complete, and you can connect the switch wires to any compatible voltage source or digital switch in order to control the camera. The only problem is that you have no idea what polarity or voltage the camera expects to see between the common wire and the one or two control wires. Chances are, any

voltage from +3 to +5 V into the control wires will trigger an event, but we certainly cannot guarantee this. The next few steps will show you how to make what we consider a universally safe interface for your camera trigger using relays.

Because there is a chance you may damage the camera electronics with an overvoltage from say a microcontroller, it would be safer to emulate the function of the original switch, rather than try to trigger the functions using a voltage output. To do this, all you have to do is install a relay at the end of the signal wires so that when the relay closes, it acts just like the original switch, closing the circuit. Your external controller is now free to drive the relay driver transistor(s) with any positive voltage, making it easy to connect your hacked camera to any project with zero risk of damage to the internal electronics.

Any small voltage relay can be used, and you can often salvage relays like the ones shown in Figure 21-18 from old modem cards, telephones, and answering machines. A relay rated for 5 or 12 V will be a good choice, and since you are not switching any current, the smaller the better. The round relays (reed relays) shown in Figure 21-18 need such low current to activate that they can be driven directly from the pins on most microcontrollers, so you don't even need the driver transistors.

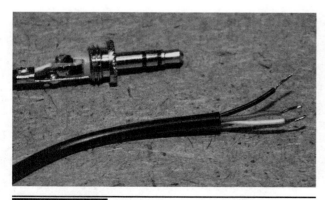

Figure 21-17 Adding the male connector to plug in the external controller.

Figure 21-18 Low-voltage relays can be used to make a universal interface.

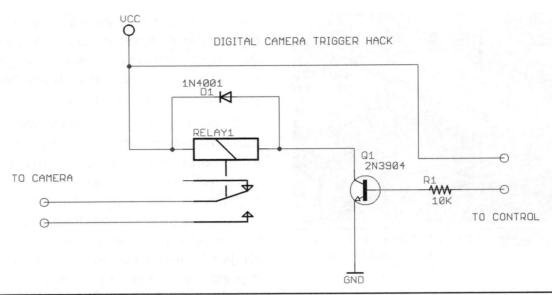

Figure 21-19   This simple transistor circuit will drive any small DC relay.

To drive a relay from any microcontroller or small voltage signal, you will need only a small NPN transistor, a diode, and a resistor. The simple schematic shown in Figure 21-19 will drive relays from 5 to 12 V with an input of only a few volts. The transistor is any common small-signal NPN type such as a 2N2222 or 2N3904, and the diode is any small-current diode such as a 1N4001 type. The value of resistor R1 can be anything from 1 to 10 K. The circuit is very simple—a voltage passed through resistor R1 reaches the base of the transistor, switching the current from VCC to GND through the relay coil, causing the contacts to close. The diode protects the transistor from back voltages that can be caused from inductance as the relay coil turns off. Since your camera trigger is only connected to the relay contacts, the camera circuitry is completely isolated and safe from your external controller's electronics. So now you could even trigger your camera from a 120-V relay running from an AC line powered security device such as a motion-activated light.

It's always a good idea to first build a circuit on a solderless breadboard so you can test or modify it before needing a soldering iron. Figure 21-20 shows the dual relay controller being built and

Figure 21-20   Building the relay controller on a solderless breadboard.

tested on a solderless breadboard. We are using 2N3904 transistors, 1N4001 diodes, 1-K resistors, and two 5-V relays pulled from old computer modem cards. This controller worked fine on voltages from 5 V up to 12 V. An input signal as low as 1.5 V was able to engage the relays.

To test the relays using various input voltages, we connected the hacked digital camera to the solderless bread board as shown in Figure 21-21 and then used pushbutton switches that would switch the positive supply into the resistors driving

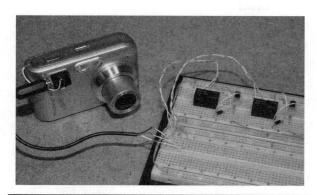

Figure 21-21 The hacked digital camera is connected to the relay controller.

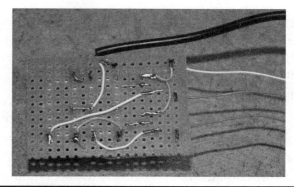

Figure 21-23 The circuit is made using wires on the underside of the board.

the base of the transistors. The circuit ran fine from a wide range of voltages thanks to the general-purpose 2N3904 or 2N2222 NPN transistors. The circuit was now ready for a more permanent home.

Whenever we need to make a proper circuit board for a simple circuit like this, we just use perforated board since there is very little wiring to be done. This relay driver circuit is so simple that you could actually just solder the leads of the components right to the relay pins and not use any circuit board at all. Figure 21-22 shows the small 2-in square piece of bare perforated board and the eight components that will be installed onto it.

To build a simple circuit like this on some plain perforated board, insert the components in a logical position and then complete the circuit by soldering wires on the underside of the board to form traces. Figure 21-23 shows the underside of our board

after completing most of the wiring. This method of prototyping creates a reliable circuit that is very easy to modify later. Of course, this method isn't nearly as nice as a real circuit board, but it is hardly worth waiting 2 weeks and paying $150 to have one made for such a simple project. The total time it took to build this circuit was measured in minutes.

The completed relay controller circuit board was once again connected to the camera and powered by a 6-V battery pack. The small solderless breadboard shown in Figure 21-24 has two pushbutton switches that are connected between the relay driver inputs and the positive supply voltage so that both relays can be tested with the camera. The circuit worked perfectly on the first attempt, and it was also discovered that this camera did an

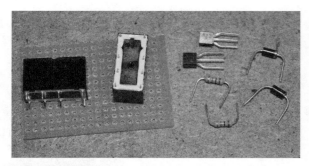

Figure 21-22 Perforated board is a good choice for making simple prototypes.

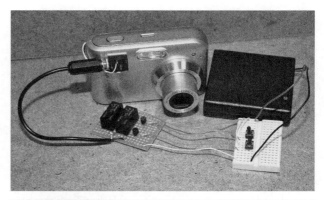

Figure 21-24 Testing the completed relay controller using manual pushbutton switches.

autofocus if only the "shoot" button was pressed without using the "focus" function first. That means we could have got away with only a single relay and switch, but it can't hurt to have a proper timed focus control as well.

This universal camera trigger hack can now be controlled with any microcontroller or electronic circuit using a low-current, low-voltage signal line. Because the relays simulate the operation of the original dual-stage pushbutton switch that was pulled from the camera, the system can be interfaced to any external device without risk of damage to the camera's internal electronics. This simple hack will allow a multitude of spy gadgets to be built, using the high-resolution imaging system in the camera to acquire your information automatically, triggered by some external event.

PROJECT 22

Repeating Camera Timer

THIS PROJECT WILL EXTEND the camera trigger hack project, allowing a timer to control both the focus and shutter release functions on a digital camera at an adjustable rate. This method of repeating time-delayed image taking is also referred to as "time-lapse photography," and can be used to speed up time by piecing together hundreds of photos taken over the span of hours or even days. By first focusing the camera before the shot, the camera will be able to acquire moving targets with far fewer missed or blurry exposures. In this project, a timer feeds a 10-stage counter, allowing up to 10 individual control points, although only two are needed in order to control the camera relay interface.

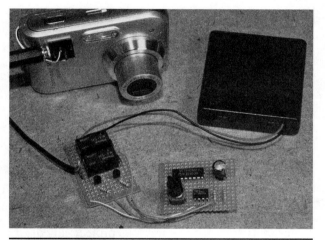

Figure 22-0 This system will focus and shoot a photo at some repeating interval.

PARTS LIST

- IC1: LM555 analog timer
- IC2: 74HC4017B decade counter
- C1: 1 to 470 µF to control timer
- Resistors: R1 = 1 K, R2 = 1 K, R3 = 1 K, R4 = 1 K, R5 = 1 K
- VR1: 50-K variable resistor
- Diodes: D1 = 1N40001 or similar
- Transistors: 2N3904 or 2N2222A or similar NPN
- Relays: Single-pole relay rated for 5 to 12 V
- Battery: 3- to 6-V battery or pack

By using the other eight digital-output pins on the decade counter, several more cameras can be controlled, or more relays can be added to allow the controlling of various other electrical devices such

as solenoids, alarms, lights, or even AC-operated appliances. The rate of photo taking can be controlled by a variable resistor, and by altering the value of the timer capacitor, rates of several photos per second, all the way down to single photos every hour, can be set. This project assumes that you have previously built the hacked camera trigger project, although you could certainly interface it to some other hardware as well.

The small board shown in Figure 22-1 is a previous project called camera trigger hack, and it allows any electronic device to issue a focus-and-shoot command to the camera. We call this a hack because it requires removal of the original switch from the camera in order to hack into the two functions that control the focus and shoot signals on the cameras circuit board. You "may" be able to build this project without the previous project as long as your camera board will accept

Figure 22-1 **This is the relay interface that controls the camera shutter switch.**

the 5-volt (V) digital signals from the 4017 decade counter into the camera's board, but to be safe, this previous project adds a level of safety to ensure your camera will not be damaged by any external device or voltages.

Looking at the schematic shown in Figure 22-2, you can see that it also includes the two relays and driver transistors from a previous project called

the camera trigger hack. The new components include the 555 timer-oscillator and the 4017 digital-decade counter, which will allow up to 10 digital devices to be actuated one after the other. The operation of this circuit is very simple as it is just a variable-rate pulse timer driving the clock input on the decade counter. To give the camera some time to focus before shooting, the focus relay is set off by count 0, and the shoot relay is set off by count 5. The counter always counts from 0 to 9 and then repeats, so this spacing offers the most time between functions.

Transistors Q1 and Q2 are relay drivers, which are actuated by the digital signal coming from the currently active pin on the 4017 decade counter. Potentiometer VR1 allows the pulse rate to be controlled to a large extent, but the capacitor C1 is responsible for setting the overall range of time, which could be as fast as several pulses per second to only a few pulses per hour. A value of 1 microfarad (μF) for C1 will allow VR1 to set the

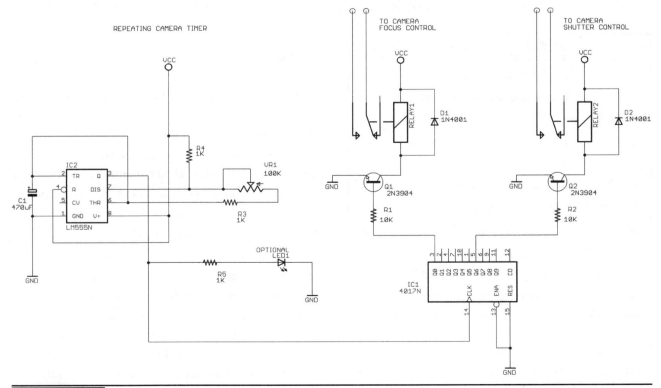

Figure 22-2 **This is the schematic for the repeating camera timer project.**

rate from several shots per second and a value of 1000 μF for C1 will slow down the pulse rate to 10 minutes or longer. You will have to experiment with larger capacitor values if you want to really slow down your frame rate. A value of 10 μF is probably a good start.

Resistor R5 and the LED are optional, as it is just a heartbeat that will show you the rate of pulses coming from the timer. It is nice to have the light-emitting diode (LED) installed during the testing phase as it gives some indication of the circuit operation. During our build, we also added LEDs to the control lines, so we had three LEDs in total.

The first capacitor we tested in the circuit shown breadboarded in Figure 22-3 was only 1 μF, so it gave a fairly quick pulse rate, allowing adjustment from about one photo per second to speeds that were too fast to control the camera. This rate was too fast to be of much use, but made it easy to take videos of the circuit in action. Our intention was to use this system to perform time-lapse photography, so we ended up installing a large 470-μF capacitor after we finished testing. You will have to experiment with various values for capacitor C1 to find the range you are looking for.

The small capacitor shown in the bottom left of Figure 22-4 is only 1 μF. This gave a frame rate of between 1 and 100 frames per second. The camera could only handle taking photos at a speed of about

Figure 22-4 Various capacitor values we tried in order to control the frame rate.

two per second, so for fast time-lapse photography or with a very fast camera, this might be a decent value to use. With the flash off, the camera was actually much more capable of taking photos quickly, but at several per second, the memory card would be filled in a hurry. The large capacitor on the lower right of Figure 22-4 is 1000 μF, and we ended up using that one to set the rate of photo taking to about 1 per hour. At this slow rate, the 4017 taps can be changed from 0 and 5 to 0 and 1 to keep the focus and photo functions closer together.

Figure 22-5 shows the completed relay driver board jacked into the solderless breadboard where

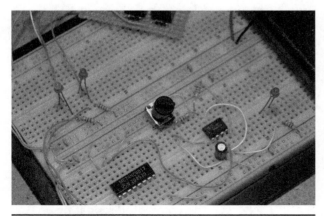

Figure 22-3 Breadboarding the circuit to test various capacitor values.

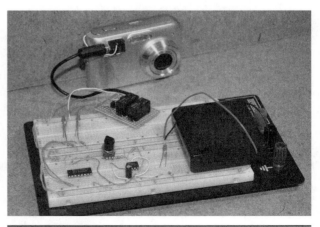

Figure 22-5 The complete system is being tested with the timer on a breadboard.

the timer and counter portion of the circuit are being tested. While you are prototyping this project, some interesting options might include using a beeper on one of the free counter pins to alert you before or after a photo has been taken, or possibly controlling more relays or multiple cameras. There is certainly a lot of room for expansion with only 2 of the 10 counter outputs in use.

The relay drive board is shown in Figure 22-6. This is a previous project called the camera trigger hack. The idea behind the relays is that no voltage is sent to the camera, only the closing of a circuit, which emulates exactly the function of the original shutter release switch that has been hacked out of the camera. Since most digital cameras have a two-position shutter release switch (focus and shoot), there are two relays used to copy this operation. You could get away with only one connection to the shoot function, but at the expense of having fast-moving targets possibly out of focus.

Once we had the timer working at a rate we were happy with, we added the six components on a small piece of perforated board and made the connections on the underside using some small wire. A circuit of this size hardly merits a real printed circuit board, and it only took a few minutes to move the parts from the solderless

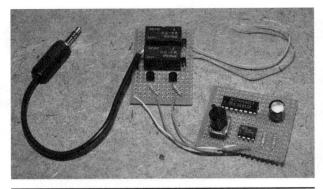

Figure 22-7   The timer circuit is built on another small perforated board.

breadboard onto the perf board and then solder the wires to form traces (Figure 22-7). We decided to keep the relay board separate as it may end up being connected to some other activation circuit in a future project.

The completed project ran fine from any voltage between 5 and 12 V, so we decided on a 6-V battery pack made up of AA batteries as the power supply. The completed rig, including the relay board is shown in Figure 22-8, connected to the hacked digital camera form a previous project. This system works very well for time-lapse photography or as a high-resolution security system.

With the camera connected to an external power supply, and a large memory card, an

Figure 22-6   This is the relay driver board that safely controls the camera functions.

Figure 22-8   The completed repeating camera timer project with the relay board.

entire day of high-resolution image capture can be done. Having the ability to take images at 8 megapixels means that details such as faces and license plates are highly visible, even when the subject is far away from the camera. Compared to a regular video security camera, this system has more than 10 times the clarity, but this is at the trade of frames per second. If high-detail long-term surveillance is your goal, then this rig will deliver.

PROJECT 23

Sound-Activated Camera

THIS SIMPLE PROJECT WILL add sound-activated control to any digital camera by adding a time-controlled pulse into a relay board that is connected to the dual-stage camera shutter switch. The sound is picked up by a sensitive microphone and then fed into an operational amplifier that is set up as an adjustable comparator so that the sensitivity can be controlled. The level of sound activation can be adjusted to respond to very faint sounds such as voices or footsteps and also adjusted to respond only to loud sounds such as music or hand claps. As a security device, this project will allow a high-resolution image to be captured in response to some type of nearby sound.

Figure 23-0 This circuit will add sound-activated control to a camera.

PARTS LIST

- IC1: 74HC121 one shot
- IC2: LM358 dual operational amplifier
- Resistors: R1 = 1 K, R2 = 10 K, R3 = 10 K, VR1 = 10-K potentiometer
- Capacitors: C1 = 10 µF
- LED 1: Low current red LED
- Microphone: Dual-pin electret microphone
- Battery: 3- to 9-V battery or pack

This project makes use of a circuit from a previous project called Camera Trigger Hack, which is a pair of relays and drivers fed into the shutter release switch on a digital camera in order to mimic the original functionality of the switch. You could probably feed the output from this circuit directly

into the camera shutter switch, but, to be safe, the relay adds a level of isolation from the camera circuit board. Some cameras also have an external remote control jack. This could also be used.

The small board shown in the left of Figure 23-1 is a previous project called Camera Trigger Hack, and it allows any electronic device to issue a focus-and-shoot command to the camera. We call this a hack because it requires removal of the original switch from the camera in order to hack into the two functions that control the focus and shoot signals on the cameras circuit board. You "may" be able to build this project without the previous project as long as your camera board will accept the 5-volt (V) digital signals from the 74121 one-shot into the camera's board, but to be safe, this previous project adds a level of safety to ensure your camera will not be damaged by any external device or voltages.

Figure 23-1 This is the relay interface that controls the camera shutter switch.

The small circuit board shown in the right of Figure 23-1 is the board that will be presented here, and it consists of a small microphone, an op-amp preamplifier (LM358), and a digital one-shot switch (74121), which controls the pulse time to the relay board that will trigger the shutter release switch. The one-shot is needed, as the camera expects the shutter release to be pressed down by a human, and the pulses sent from the microphone preamplifier will be much too short

to be taken seriously by the camera. The one-shot takes this millisecond input pulse and then sends out a digital pulse for a duration that is long enough for the camera to respond.

Looking at the schematic shown in Figure 23-2, you can see that the output from a small microphone is sent into the LM358 op-amp, which acts like a comparator, responding to changes in voltage from the microphone. Because the microphone is of the electret type, it includes its own high-gain amplifier right inside the tiny metal can, further increasing the sensitivity of the comparator circuit. The variable resistor VR1 controls the sensitivity of the comparator so sounds as faint as a whisper or as loud as a clap can be used to activate the camera switch relays.

The output from the op-amp comparator is sent into a signal-conditioning IC called a one-shot. The 74121 one-shot will take any input voltage, no matter how short of a duration and send an output signal of some calculated time based on the values of resistor R3 and capacitor C1. For the camera trigger, a time of about 1 second is fine, so the

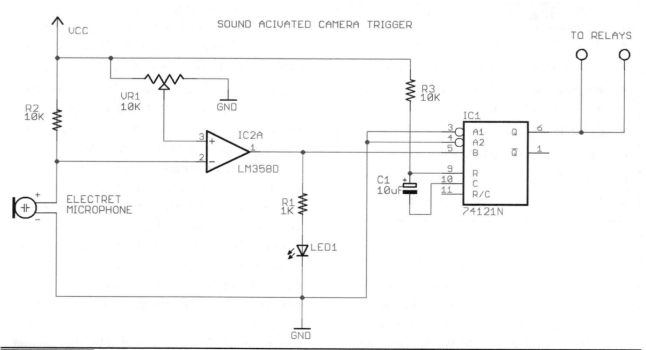

Figure 23-2   This is the schematic for the sound-activated camera trigger.

values of 100 (microfarad) µF and 10 K set this time. The reason it is necessary to have the one-shot in the circuit is because the signals from the comparator will be modulated by the sound, and will likely be too fast to allow the relay to close long enough to trigger the camera shutter release. The one-shot simply lengthens the pulse, which also gets rid of multiple pulses from the same quick sound. Since the camera needs to respond quickly in this application, both the focus and shoot relays are tied together, allowing the image to be taken as fast as the camera can respond to the sound. A very fast camera would be able to perform stop-motion photography, like the capture of a balloon popping or water drop.

The 74121 is an interesting and very useful integrated circuit (IC), as it can create a pulse varying in length from nanoseconds to 30 seconds by varying the values of the capacitor and resistor used for timing. By having such precise control over pulse widths, very accurate timing systems

can be made, although in this application, the 74121 basically acts like a switch debouncer to simulate the operation of a human button press. The datasheet (Figure 23-3) shows that the switch is controlled by a pair of timing pins and has multiple inputs to allow different pulse edges to create the output. In this circuit, the one-shot is set up to send out a 1-second-long digital pulse every time the input goes low. The input is the AC voltage signal from the op-amp comparator that is responding to changes in sound picked up by the microphone.

Electret microphones are extremely common in just about any electronic device that needs to hear sound. The small round can microphones are extremely sensitive to any sound, and they include a built-in electronic audio preamplifier right in the tiny metal can. What's interesting about these microphones is that they are powered by 3 to 12 V, and use the same pin for power as for output, so there are only two connections to the can. Figure 23-4 shows several electret microphones

DM74121
One-Shot with Clear and Complementary Outputs

Order Number	Package Number	Package Description
DM74121N	N14A	14-Lead Plastic Dual-In-Line Package (PDIP), JEDEC MS-001, 0.300 Wide

Connection Diagram

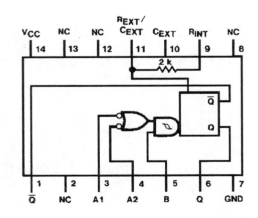

Function Table

Inputs			Outputs	
A1	A2	B	Q	Q̄
L	X	H	L	H
X	L	H	L	H
X	X	L	L	H
H	H	X	L	H
H	↓	H	⊓	⊔
↓	H	H	⊓	⊔
↓	↓	H	⊓	⊔
L	X	↑	⊓	⊔
X	L	↑	⊓	⊔

H = HIGH Logic Level
L = LOW Logic Level
X = Can Be Either LOW or HIGH
⊓ = A Positive Pulse
⊔ = A Negative Pulse
↑ = Positive Going Transition
↓ = Negative Going Transition

Figure 23-3   Part of the datasheet for the 74121 one-shot IC.

Figure 23-4 An electret microphone is both a microphone and high-gain amplifier.

that we salvaged from various audio appliances. You can see that there are two pins or solder points on the bottom of the can. Looking carefully at the pins, you will be able to see that one pin is also connected to the outer can by a small metal strip. This will always be the negative terminal.

The sound-activated triggering system is shown in Figure 23-5 built on a solderless breadboard for initial prototyping and testing. There are not many wires or components to this circuit, so it is very easy to create. You can actually omit the light-emitting diode (LED) and resistor R1 if you want because it is there only to help set the level of the sound activation as VR1 is adjusted. When you are adjusting VR1, the LED will turn off right at the point where the comparator is set to its maximum

response, so this helps when setting the maximum sensitivity of the system.

When you are testing the circuit, either have the output from the 74121 one-shot feeding the relay board or another LED through a 1-K resistor so you can see or hear the response from the system as you adjust VR1 and tap on the microphone with your finger. You will know the circuit is working when the relay closes or when the LED flashes for 1 second in response to some sound. The circuit will run fine from 3 to 6 V and will respond to sounds as subtle as a finger snap when adjusted properly. If you do not see any activation from sounds when adjusting VR1 through the entire range, check the polarity of the microphone, as this does matter.

The relay drive board is shown in Figure 23-6. This is a previous project called the Camera Trigger Hack. The idea behind the relays is that no voltage is sent to the camera, only the closing of a circuit, which emulates exactly the function of the original shutter release switch that has been hacked out of the camera. Since most digital cameras have a two-position shutter-release switch (focus and shoot), there are two relays used to copy this operation. In this project, however, both functions are tied together as the camera will need to respond quickly to the sound activation. The downside to this is that the camera may not have time to spot focus on extremely fast-moving targets.

Figure 23-5 The sound-activation circuit on a solderless breadboard.

Figure 23-6 This is the relay driver board that safely controls the camera functions.

Figure 23-7 The sound activation circuit connected to the relay driver board.

Figure 23-8 Our solderless breadboard is now spying on us!

Figure 23-7 shows the complete rig, ready to take a photo every time it hears any sounds in the room. We have the sensitivity adjusted so that we can clap or snap anywhere in the room and get the camera to take a photo. If there is some ambient background noise, then you will have to adjust VR1 in order to reduce the sensitivity and only respond to louder sounds, or your memory card will be filled in a hurry. Being an audio circuit, you could also include some type of frequency filter to help deal with background noises, and as luck would have it, the LM358 has one free op-amp left over for such a function.

Besides having the polarity of your electret microphone reversed, there is not much else that can go wrong with this simple circuit besides a wiring error. If you hear your relay clicking in response from the sounds but your camera fails to respond, then you can try using a larger capacitor for C1 in order to lengthen the pulse time from the 74121 on-shot. A simple way to figure out the approximate time needed to have your camera respond is to manually drive the relays by shorting the base of the transistor to VCC to activate the relays. Most cameras should be fine with a 1-second switching, but you may have to experiment with your camera (Figure 23-8).

There are so few components and wires in this simple circuit that it only takes a few minutes to

move from the solderless breadboard onto a bit of perforated board as shown in Figure 23-9. The components are placed through the holes in the perf board, and then a circuit board is formed on the underside using wires and solder. You could certainly include the relay driver board and the sound activation board on the same circuit board, but having the ability to mix and match various projects is nice as well.

Our sound-activated camera was going to be installed on location for some security job, so we installed the two circuit boards into a small project box as shown in Figure 23-10. The plastic box will also have enough room to hold a 9-V battery and a power switch to turn it on and off.

Figure 23-9 The completed sound activation system on small circuit board.

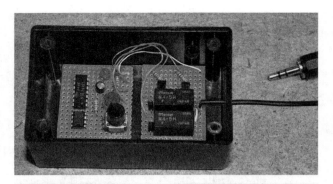

Figure 23-10   Adding the two circuit boards into a small project box.

Figure 23-11   The completed sound-activated camera, waiting to spy on trespassers.

To run a 5-V circuit from a 9-V battery, you only need to include one other component—a 7805 regulator to step the voltage down to 5 V. If you alter the voltage on this circuit, you will also have to readjust the sensitivity, but this is made easy by using a chassis-mounted variable resistor so that it can be set anytime.

Figure 23-11 shows how our completed sound-activated camera looked after installing the two small circuit boards and battery into a plastic project box. The knob is attached to the variable resistor VR1 to allow easy adjustment of the sound sensitivity, and the electret microphone is installed on the side of the box so that sound can enter through a small hole in the plastic.

We also included an external 1/8-in audio connector so that we could activate the camera from an external microphone or some other voltage signal. Since the LM358 op-amp comparator can be set to respond to just about any changing voltage, other devices such as light-operated circuits, or motion-activated circuits, could be used to trigger the camera. Having a high-resolution camera responding to some external event makes for a very useful security device, able to pick up the smallest details in a scene, even when placed at a long distance from the subject.

Motion-Activated Camera

THIS PROJECT USES a heat-sensing motion detector to trigger the shutter-release button on a hacked digital camera so that high-resolution images can be captured anytime a person or animal crosses in front of the motion-sensing zone. By hacking into an old motion-activated floodlight, the cost is kept to a minimum and based on a pre-existing system that is known to work well. This project converts the motion sensor for DC battery operation, allowing it to become portable and safe from high voltages.

This project also makes use of a circuit from a previous project called Camera Trigger Hack, which is a pair of relays and drivers fed into the shutter-release switch on a digital camera in order to mimic the original functionality of the switch. You could probably feed the output from this circuit directly into the camera shutter switch, but to be safe, the relay adds a level of isolation from the camera circuit board. Some cameras also have an external remote-control jack. This could also be used.

Motion-sensing lights like the one shown in Figure 24-1 use a special heat sensor to detect movement of warm-bodied creatures. This sensor detects the body heat of the subject as it passes across a pair of small sensing elements inside the heat sensor. This heat sensor is also known as a Passive InfraRed (PIR) sensor and is the magic behind every one of those inexpensive outdoor security lights as well as most indoor motion-sensing units. Because these devices are mass manufactured, they are easy to find as surplus or even new for very low cost. Many security lights are tossed out when the plastic degrades or when the relay that controls the AC lights fails, so you can probably salvage the needed parts even from one that is deemed to be nonfunctional.

| Figure 24-0 | This rig will take a photo anytime it senses motion or heat changes. |

| Figure 24-1 | Any common outdoor motion-sensor light can be used for this project. |

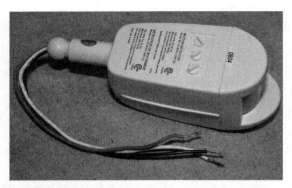

Figure 24-2   Only the motion-sensor part of the security light is needed for this project.

Figure 24-3   Motion sensors are also available that run directly from DC current.

Most motion-sensing security lights will have an AC cover plate that includes one or more light sockets and a plastic sensor unit that can be removed from the base plate. Since you will not need any of the AC electric parts, tear down the sensor light until you have isolated only the motion-sensor box as shown in Figure 24-2. This part of the system is a standalone device that uses 120-volts (V) AC to switch an AC load through a mechanical relay. You could actually run this system from the AC line by isolating the relay contacts to switch your control electronics, but to play it safe, it is best to convert the motion sensor to run from low-voltage DC so you can power it from a battery pack.

If you are not interested in hacking up an AC appliance for battery operation, you could just purchase a ready-to-use motion-sensor board or kit from an electronics supplier. The two motion sensors shown in Figure 24-3 are from Sparkfun.com and are ready to run directly from 5-V DC. If you do plan to order a PIR sensor kit from an electronics supplier, ensure that you also receive the small Fresnel lens needed to spread out the sensing area over a larger area. The PIR sensor by itself is only capable of sensing movement a few inches away from the tiny window on the device, so the Fresnel lens is necessary. This lens will look like a small bubble or plastic sheet with tiny grooves on its surface.

To open these motion-sensor boxes, you will need some brute force and ugly hacking since they are glued tight to keep out water and never intended for repair or modification. Of course, that will not stop the determined evil genius with a knife in hand! The easiest way to split open the plastic box is to run a sharp knife along the seal as shown in Figure 24-4, breaking the glue so that the parts can be pried apart. Once you have one side free, you can insert a thin screwdriver blade and split the two halves apart. Try to avoid damaging the half with the lens as you will be using it again later.

Once you have the front part of the motion-sensor casing open, you will be able to get at the circuit board inside. You can see the Fresnel lens in Figure 24-5, glued to the front half of the plastic

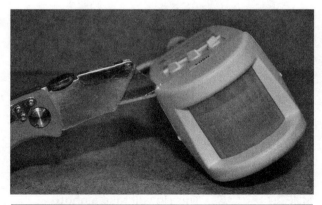

Figure 24-4   Sometimes the term *hacking* must be taken literally!

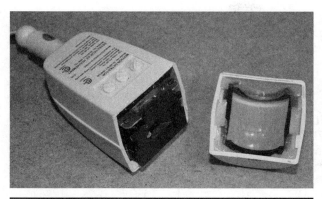

Figure 24-5 The two halves open up to reveal the circuit board inside.

Figure 24-7 This is a typical example of a motion-sensor circuit board.

casing. When you are done hacking the sensor, the goal will be to have the lens placed back in the same spot ahead of the sensor that it was originally in order to keep the field of view the same. Usually, you can glue the plastic cabinet back together or just fasten the front half over the circuit board or onto another project box that holds both parts together.

There are usually one or more adjustments on the sensor to control time on, sensitivity, and delay, and these controls will be placed on the outside of the plastic so that they can be set. You may have to pop out the small plastic knobs as shown in Figure 24-6 to remove the circuit board from the casing as there will often be a connecting rod between the plastic knobs and the small timers soldered to the circuit board. Remember that these things are designed

to be cheap, not to be easily taken apart so expect some parts to be glued in place or held with friction. You may have to do a little more hacking in order to free the circuit board.

The circuit board will often be made of two parts as shown in Figure 24-7—a main circuit board that contains some safety electronics and a relay to control the AC lights, and a smaller vertically mounted circuit board that will contain the PIR sensor and its control electronics. Most of the time, you can reduce the circuit down to only the small sensor board, as the AC board is not going to be needed in this project. Of course, this requires some reverse engineering of the circuit board, but being such a simple circuit, it does not take much effort at all.

Chances are you will not be able to track down a schematic for your security sensor, but most sensors work the same way and use similar components. The schematic shown in Figure 24-8 is a fairly common example of a motion-sensor schematic, using a few op-amps to control the amplification and sensitivity from the PIR sensor. The current trend is to use a single motion-sensor duty integrated circuit (IC) on the board, so if your electronics look a lot more simple than the board shown in Figure 24-7, it is probably one of the newer cost-reduced types. Either way, reverse engineering the system for DC operation is fairly easy.

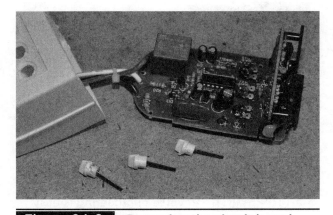

Figure 24-6 Removing the circuit board from the plastic housing.

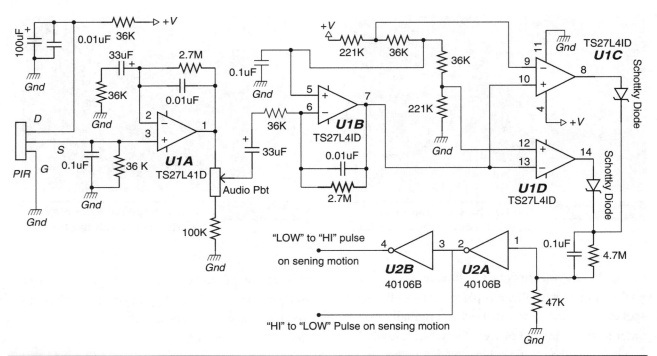

Figure 24-8 An example schematic from a common security-sensor system.

If your board has an easily identifiable IC such as the CD4011B NAND gate shown in Figure 24-9, then your reverse-engineering work is already half completed. Looking at the datasheet for the 4011B, it can be seen that like most logic DIP parts, the top left pin is VCC (positive) and the bottom right pin is GND (negative). So knowing that, you could feed the required DC voltage right into these pins and most likely have the entire board working from

a battery pack. Of course, this is an ugly hack, and it is better to trace back to the regulation system to input your power there instead.

A single-sided printed circuit board (PCB) is very easy to trace and reverse engineer because you have all of your traces on the bottom side and can see through the board by holding it in front of a bright light as shown in Figure 24-10. Seeing the components connected by the traces makes

Figure 24-9 Identifying one of the board ICs to trace back the power and ground points.

Figure 24-10 Reverse engineering is easy on a board like this since you can see through it.

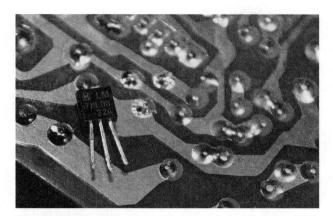

Figure 24-11 This is the voltage regulator that powered the sensor board.

it very easy to follow the ground or power lines around the board in order to figure out how the circuit is powered. In this photo, you can see that there has already been a 3-pin part removed from the board—that was the voltage regulator that we found by following the lines back from the 4011B IC on the board.

The security lighting system runs from a 120-V AC power line, but internally needs only 5 or 9 V to operate. The goal is to reverse-engineer the board so you can locate and bypass the onboard voltage regulator in order to power the sensor form a much safer DC-battery source. The voltage regulator will probably look like a common transistor, having three pins like the one shown that was removed from the board in Figure 24-11. Common values

for the voltage regulators will be 7805, LM7805, LM78L05, 7809, LM7809, LM7808, and many other variations of the 78xx series of regulators.

The 78L08 regulator we pulled from the board is a common 8-V positive voltage regulator that will change voltages from 12 to 20 V back to 8 V, which is the DC voltage that the electronic circuit was designed to run from. We actually found that our circuit ran fine on voltages as low as 5 V, and tested it as high as 9 V without any problems. To identify your voltage regulator, just look at the numbers printed on the components and enter them in your favorite Internet search engine to see if you can bring up a datasheet. If there is a silkscreen on the circuit board, then the regulator will be marked as IC1 whereas a transistor would be marked as Q1 or T1.

Once the voltage regulator has been identified, you can look at the data sheet in order to find the pinout diagram as shown in Figure 24-12. On the regulator, pin 1 was the output, pin 2 was ground, and pin 3 was the input—and this corresponded to what we found by following those traces to the various components and ICs on the board. Capacitors are marked on the negative side of the can, and so are diodes, so this can also help when trying to figure out which pins or traces are positive or negative.

If you can identify the voltage regulator as well as a datasheet to acquire the pinout information,

3-Terminal Positive Regulators
General Description

The LM78LXX series of three terminal positive regulators is available with several fixed output voltages making them useful in a wide range of applications. When used as a zener diode/resistor combination replacement, the LM78LXX usually results in an effective output impedance improvement of two orders of magnitude, and lower quiescent current. These regulators can provide local on card regulation, eliminating the distribution problems associated with single point regulation. The voltages available allow the LM78LXX to be used in logic systems, instrumentation, HiFi, and other solid state electronic equipment.

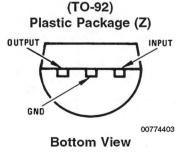

Figure 24-12 The datasheet will show the pin diagram for the regulator.

then unsolder the regulator from the board, taking note of its orientation. You can install a pair of wires into the regulator's place, adding one wire for positive and one for negative. Your new power supply will feed DC power directly to the board, so its positive wire will go into the hole where the output of the regulars was, and your new negative wire will live in the hole that is marked *ground* or *common* on the regulator pinout. Basically, you are removing the original regulator to replace it with a set of wires that will feed DC power right into the board.

The board we modified for direct DC power is shown in Figure 24-13, with the new DC power supply wires soldered into the holes that once held the onboard regulator. Now, the only other lines to trace are the ones that come from the relay. There are a few different ways you can take this project from here. It will depend on the design of your motion-sensor board and how you hacked into the digital camera trigger. If you hear the relay click when you power up your sensor board under your new DC power supply, then that is good news because you have enough voltage to engage the relay coil. If you do not hear a relay click, but the motion indicator LED seems to be working, then your relay may require more voltage or have been supplied from another current-limiting power supply on the board. In the later case, you will

need to remove the relay and figure out which trace feeds the relay driver to activate the coil.

Our board had some odd AC supply powering the relay through an SCR, so we had to pull the relay and then trace back to figure out which point on the board fed the relay driver transistor or SCR. This job is easy to do once you know that the motion sensor is working because you can just probe the obvious pins with a voltmeter. When the test LED is on, the voltage will be somewhere between 1 and 5 volts as compared to zero volts when the test LED is off. This signal will likely come directly from the motion-sensor IC or the op-amp on the main board. Another sneaky trick is to just take this signal from the test LED, as it is always on when the relay should be on. It can be fed into your own relay driver board. We found a point on our board that went into the gate on the relay driver SCR and just soldered a wire right from there.

With the power running to the board, as well a line connected to the relay driver circuit, our hacked motion sensor was now self-contained and running from a 6-V battery pack. Figure 24-14 shows the hacked sensor board with an LED connected to the line from the relay driver. Each time we ran our hand over the PIS sensor, the LED in our small breadboard lights for some amount of time depending on the settings of the various adjustment

Figure 24-13   The hacked board now runs from 9-V DC rather than 120-V AC.

Figure 24-14   Testing the hacked sensor circuit board using an LED to signal motion.

resistors on the board. From here, the sensor board is ready to connect up to our hacked digital camera trigger board.

If you enjoy the reverse-engineering aspect of this hobby, then you can further reduce your motion-sensor board by removing all of the AC regulation circuitry and the safety electronics. Since only DC is used now the safety circuit is pointless as well as the AC regulation components. On many of these sensors, the only board you need will be vertically mounted to the AC board, and it can be reduced to run on its own by doing a little creative reverse engineering.

Figure 24-15 shows a pair of sensor boards we hacked down to their minimal state by removing them from the AC board to run directly from DC power. If you take the reverse engineering this far, then you will also be losing the relay and driver, but the sensor system will still be fully functional. Instead of driving a relay, though, the sensor will output a 5-V signal whenever motion is detected. This will certainly work well with the hacked camera trigger relay board built in a previous project.

When you are done hacking and reverse engineering the motion-sensor board, it will be ready for your other projects, but will require the Fresnel lens to be attached again. If your sensor casing split so that the front half included the

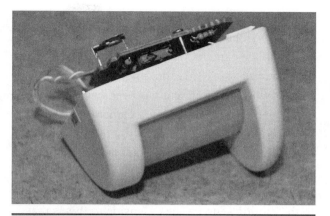

Figure 24-16 Attaching the Fresnel lens to the reverse-engineered sensor board.

Fresnel lens like the one shown in Figure 24-16, then you can just use a little hot glue to fasten the circuit board in the same place it was before you hacked the system. The Fresnel lens expands the field of view, allowing the sensor to see an entire room or yard rather than just a few inches in front of its tiny glass window. Of course, feel free to test the sensor response with no lens as it may work perfectly in your own application, if short range is all that is needed.

The small relay driver board shown in Figure 24-17 will create a bridge between your newly hacked motion sensor and the hacked digital camera. This

Figure 24-15   Reducing the sensor board to their absolute minimal configuration.

Figure 24-17   This is the relay driver board that safely controls the camera functions.

is a previous project called the Camera Trigger Hack. The idea behind the relays is that no voltage is sent to the camera, only the closing of a circuit, which emulates exactly the function of the original shutter release switch that has been hacked out of the camera. Since most digital cameras have a two-position shutter release switch (focus and shoot), there are two relays used to copy this operation. In this project however, both functions are tied together as the camera will need to respond quickly to the motion sensor. The downside to this is that the camera may not have time to spot focus on extremely fast-moving targets.

If you kept the original relay working on the motion-sensor board, then you do not need to use a separate relay board as you will already have the required isolation offered by the relay on the motion-sensor board. As long as you are only connecting to the contact points on the relay, then no voltages will be sent from your motion-sensor system, only an open or closed circuit, just like the original shutter release switch in the digital camera.

With all of the components working together perfectly from the 6-V battery pack, we found a suitable black box to create a self-contained unit that could be sent into the battle zone. The motion sensor was glued to the top of the enclosure as shown in Figure 24-18, with a hold drilled out

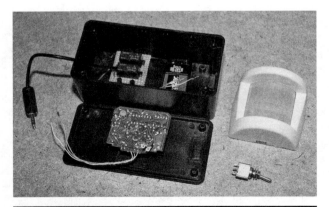

Figure 24-18 Creating a self-contained motion-sensing camera system.

Figure 24-19 The motion-activated camera waiting for the next victim to set it off.

so that the PIR window could see the outside world. The Fresnel lens part would then be glued over the box lid to expand the field of view. The battery and relay board were also mounted into the small plastic box so that the unit would be a fully functioning motion trigger system that could be plugged into the hacked digital camera or any other device we felt like operating under motion-detected control.

The completed motion-activated camera is shown in Figure 24-19, waiting to capture in high detail the next person who dares to wander into our living room. As a security system, this project works very well, especially when a high level of detail is necessary. On a fresh charge, the camera can run for hours, snapping photos until the batteries are dead or until the memory card is full.

This system also works well in nature photography, where a stealthy setup is needed in order to capture some elusive creature or one that you don't want to be anywhere near due to danger. We also found this project useful when we wanted several group photos that we were supposed to be included in and did not feel like fiddling around with the single self-timer on the camera. With this system, you just turn it on and the camera will keep on acting like your personal paparazzi as long as something is moving in its field of view!

Camera Zoom Extender

WHEN YOUR JOB is to collect information covertly, there are times when you will need to be both far from the subject, but at the same time collect as much high detail as possible. A typical 10-megapixel camera will often be able to capture a face or license plate from a block away, but there are times when you need even great imaging power, and the built-in optics will not cut it. Some cameras offer a digital zoom function, but this really only expands the pixels of the image so no extra detail is captured using this function. To collect data from a greater distance, your camera needs more optical zoom power.

There are basically two ways you can bring your digital camera into the extreme zoom zone: using a proper interchangeable zoom lens for a single-lens reflex (SLR) camera, or by retro-fitting your camera to use a binocular or telescope or gun sight like the ones shown in Figure 25-1. The first option will require you to own a digital camera that allow lenses to be changed. These become expensive, especially the telephoto zoom lenses—very expensive! The second option will work with any cheap digital camera, allowing it to see further than any costly SLR zoom lens, but at the expense of a little quality. By adapting a $100 digital camera to one side of a binocular, you will have a camera capable of taking images from a great distance, and although this setup will have some blurring around the edges, it will have similar zoom capabilities to those $10,000 lenses used by paparazzi to invade the privacy of many entertainment personalities.

If you own a digital SLR (single lens reflex) camera or some other quality digital camera that

---

PARTS LIST

- Camera: Any digital SLR or pocket camera

- Optics: Gun site, monocular or binoculars

- Base: 2-in wide ¼-in thick plywood or 1/8-in steel

Figure 25-0 A digital camera can be adapted for ultra long range photography.

Figure 25-1 Digital zoom is never as good as true optical magnification.

Figure 25-2   SLR digital cameras offer the ability to change the lenses.

Figure 25-3   This telephoto lens will fit onto any Nikon SLR camera body.

allows the changing of lenses, then you can simply purchase a telephoto lens that will enhance the optical powers of your camera. Of course, these quality cameras are not cheap; the lenses are often more expensive than the camera itself. The Nikon D60 shown in Figure 25-2 is a good-quality 12-megapixel camera that can be purchased for under $500 from many camera stores. This camera has many lens options available, including wide angle and telephoto lenses. The standard 18 to 55 mm lens that came with the camera is shown removed from the camera body.

A quality camera that can image over 10 megapixels will often reveal extremely high details at long distances as long as you set the image quality at the highest possible setting. The lower settings take smaller images or use higher compression, often blurring extremely fine details in distant scenes. Using only the 18 to 55 mm lens, this Nikon can easily capture a face or license plate a block away as long as we take photos at the highest quality and then show them nonscaled on a computer program. If you plan to do a lot of surveillance work and want a robust camera that will bring in the details, then make no mistake— a good SLR camera with a 10-megapixel imager and a standard 15 to 55 mm lens will blow away any cheap 20-megapixel camera with an optical zoom rated for 16x. There is no comparison at all really. The pocket cams can't even come close to capturing the clarity of that a digital SLR can.

The huge lens shown attached to the Nikon SLR digital camera (Figure 25-3) is a 55 to 200 mm telephoto lens. It basically takes over where the standard 18 to 55 mm leaves off. This lens could pull the time off your watch from across the street on a clear day. Although it was almost as expensive as the actual camera, the results are amazing. When we use this lens on Nikon D90 for security work, it is amazing how much detail is shown on the computer when viewing the images without any scaling. Anyone serious about surveillance should have an SLR camera with both the standard 18 to 55 mm lens as well as a telephoto lens like this one. It only takes 10 seconds to swap the lenses, and the camera is ready to go from imaging a block away to seeing as far as you can with your naked eyes.

Now that we have explored the expensive way to shoot long-range photography, it's time to move into the hacking zone. At the sacrifice of a little quality, any cheap pocket digital camera can be made to see as far away or much further than those equipped with expensive telephoto lenses. The idea is to have the camera look through the eyepieces on a binocular, gun sight, or telescope in order to see a distant scene just as you would while looking through the optics. The interesting thing about retro-fitting a camera to an optical device is that the camera can actually see more detail than you could, so once the images are viewed full size on a computer screen, details you might not have

Figure 25-4 This 3× optical zoom gun sight can be used as a makeshift camera lens.

Figure 25-5 Figuring out the optimum focal point by trial and error.

been able to see will be visible. A hundred-dollar pocket camera connected to the gun sight shown in Figure 25-4 will actually see further than the 55 to 200 mm lens shown in Figure 25-3, but some of the edges around the image will be blurred.

We will be using our Nikon D60 as the base camera for the remainder of these experiments, but any small pocket digital camera would work. The smaller the diameter of the lens, the easier it will be to adapt to the eyepiece on the binocular or telescope. So a cheap digital camera will actually be better for these experiments than the SLR camera we are using here. The resulting system will certainly not offer the kind of quality you need to send your photos to a gossip mag for print, but you will certainly be able to compete for imaging distance with many of the ultra-expensive telephoto lenses used by the pros. Any optical system that has been designed for you to look through can be used as an optical magnifier for a digital camera. Binoculars work very well for this project.

To adapt an imaging system that was designed for human eyes to the cameras lens, you will have to mess around with both the zoom settings on the camera and the distance from the cameras lens to the eyepiece on the optical device. Pocket cameras with smaller lenses seem to like a distance about the same as your eyes would be from the lens on the optical device, so that is a good starting point. Larger lenses like the one on the Nikon shown in

Figure 25-5 will probably need to be further away from the eyepiece to acquire a larger imaging area.

Start by setting your camera zoom to the midpoint and then simply look through the viewfinder or LCD while holding the lens close to the eyepiece on the optical device. Aim the optical device out of your window and see how the camera responds at various distances and zoom levels as you watch the live image on the viewfinder. It will be fairly easy to get an image, but you will likely have a large black border around the image or see blurring to one side. These errors are usually corrected by proper alignment of the camera lens with the optical device, but at this point, just mess around and try to find that best mix of distance and zoom to get the largest possible image with the least amount of black border. The more powerful the optics, the more border you can expect with the gun sight having almost no border as compared to at least 50 percent border with a telescope.

The gun sight we tested worked well when held at about 2 inches (in) in front of the camera lens, but the binoculars shown in Figure 25-6 needed to be less than half an inch from the lens for optimal focal range. Avoid touching the cameras lens with any object, even the rubber eyepieces from the binoculars or you will have to clean the lens. There will be no reason to ever have anything touch the camera lens surface, with ¼ in being about as close as you will need to get, even when adapting

Figure 25-6 The binocular eyepieces need to be closer to the lens than the gun sight.

Figure 25-7 Making a solid base to align the optical device to the camera lens.

for use with a telescope. Don't worry about the side-to-side alignment right now as you will not be able to get it right on by simply holding the two parts together. Alignment will be corrected later.

Once you have figured out the optimal zoom setting and physical distance from the optical device that works best for your camera, the next step will be to align the two so that the image is not blurred on one side or off center. You may not be able to completely eliminate the black borders around the image. But, as long as you are getting sharp imaging in the center, there will be incredible detail available as you take photos using the cameras with highest-quality setting.

Alignment of the camera with the optical device is easy to do using a sturdy base that will hold both components in alignment with each other. All cameras have a tripod mounting bolt hole in the bottom, so this will be used to fasten the camera to the base, but as for the optical device, you will have to find some way to secure it to the base as well. Fastening the gun sight shown in Figure 25-7 was easy because the sight already had a pair of mounting brackets. The sight was too low while just sitting on the base, but that will be easy to correct with a few washers once the holes are drilled in order to fasten it in place with a pair of bolts. If your optical device has no mounting

apparatus, then you will need to make your own brackets using wood or metal in order to fasten it to the base. The base is made from a piece of solid wood or metal, and needs to be as straight as possible.

Since our gun sight had a pair of mounting brackets that would take a bolt, we just placed it on the base and then marked the drill hole positions as shown in Figure 25-8. It is important to have the optical device in line with the base so that it ends up aligned with the center of the camera lens. As you have probably discovered when holding the optical device in front of the camera lens by hand, any slight misalignment will result in a blurred or shifted image.

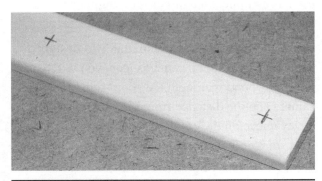

Figure 25-8 Marking the holes on the base to mount the gun sight.

Figure 25-9 Using the mounting bolt from a tripod base to fasten the camera.

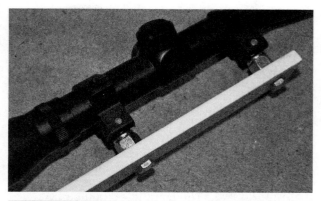

Figure 25-10 These bolts will lift the gun sight to the level of the camera lens.

If you can't find the exact size bolt needed to fasten your camera to the base through the hole drilled in it, just borrow one from a tripod mounting base as shown in Figure 25-9. The standard tripod bolt is a ¼-in diameter, 20 threads per inch bolt, and will only screw into the camera body to a depth of about ¼ in. If you want to make your own tripod bolt, find one with the same threads and saw it off to the length needed to pass through your mounting base material and into the camera body.

The alignment between the optical device and the camera lens has to happen in two dimensions: left to right, and up and down. Left-to-right alignment is taken care of by the base as long as your mounting holes are in the center, but the up-and-down alignment may require some work. Our gun sight was too low for the camera lens, so we added a pair of bolts as shown in Figure 25-10 that just happened to be the correct thickness in order to align the gun sight perfectly with the camera lens. You will know when you have found the perfect alignment as the image will be centered in the viewfinder, with an equal amount of black border around the edges and no corner blurring.

If you intend to mount the completed system to a tripod, then you will have to drill another hole in the base for a tripod bolt. To find the optimal placement for the main tripod bolt, balance the base

with both the camera and optical device installed so that you can find the center of gravity. Figure 25-11 shows the base balanced on its center over a roll of tape while both the camera and gun sight are installed.

Once the center of gravity has been found, drill the appropriate size hole in the base material and then find a bolt and nut that will allow your tripod platform to be fastened to the base material as shown in Figure 25-12. The completed unit should balance on the tripod platform, which will keep the stress away from the mounting bolt and help keep your tripod balanced for easy adjusting.

Figure 25-11 Finding the center of balance with both devices mounted to the base.

Figure 25-12   The completed gun-sight adapter sitting on the tripod mounting platform.

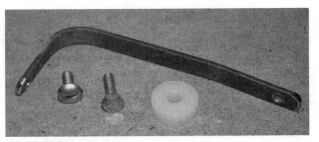

Figure 25-13   A piece of bent steel flatbar will be made into a binocular adapter.

Binoculars are particularly good for this project as they offer a long range or adjustable zoom factor and adapt very well to smaller digital cameras with lens diameters of less than 1 in. The other benefit is that you can look through the unobstructed eyepiece to find your target while pressing the shutter release button to take an image of anything you are looking at. This allows the camera viewfinder to be turned off, saving battery life and helping to keep your viewing area dark, which is definitely a good thing in a stealth operation.

Since binoculars have no center tripod mounting bolt hole, they can be a challenge to adapt to a camera, so you will have to do a little creative thinking in order to create a system that holds the camera lens in perfect alignment just behind the binocular eyepiece. Many binoculars have a bolt hole in the front where the two pieces hinge together. This can usually be found by popping off the plastic cap that is fit over the open end. If your binoculars have this bolt hole, then you can find a small length of steel flatbar like the piece shown in Figure 25-13 and then just bend it into a shape that will allow it to be mounted to the binocular bolt hole at one end while the other end holds the camera in place using the tripod mounting hole on the camera. It takes a bit of fiddling to get the metal into the right shape, but this does work well once you find that perfect alignment.

The bent length of steel flatbar is shown mounted to both the camera and binoculars in Figure 25-14 after spending some time fine-tuning the alignment. Another thing to consider is which eyepiece to use with the camera. Since this digital camera had most of the body on the right side of the lens, we opted for the right eyepiece so that the camera would not be in our way when looking through the left eyepiece. Figure 25-14 shows the underside of the binoculars, and the camera is installed over the right eyepiece.

The completed binocular-cam is shown in Figure 25-15, easily grabbing all of the detailed text from a scene that was about 500 feet away. We were able to zoom the camera in to get rid of the entire black border around the image, and the

Figure 25-14   The metal flat bar holds the camera over the right eyepiece in perfect alignment.

Figure 25-15 Testing the binocular-cam with a scene that has some detailed text.

Figure 25-16 The camera and monocular mounted to a box for automatic operation.

system worked extremely well, with only a little blurring around the outer edge of the image. This system does not have the color quality offered by the expensive SLR zoom lens shown in Figure 25-3, but it can see almost 10× as far and costs 10× less, even when you consider the price of the camera and binoculars! Not bad at all for a 2-hour hack.

These binoculars also had an adjustable zoom with an optical power of 16 × 50, so they were great for both medium-distance and very-long-distance surveillance. You will, of course, be giving up the ability to use the camera's built-in flash, so set your camera for no flash but also short exposure if you can. Fast-moving scenes need fast exposure, but if you are just hunting for license plates or small details in a static scene, long exposure will also work well as long as you have your system mounted some way so it is held steady. If you affix your system to a tripod and set the camera for night shots or long exposure, you can almost see in the dark, bringing in high details in dark scenes as long as there is nothing moving. We found that the camera could see much better than we could at night, just under the ambient light from streetlights.

The version of the zoom extended shown in Figure 25-16 is one that we built for time-lapse photography using the hacked camera trigger project and trigger timer project shown earlier. The

binocular was made into a monocular by removing one of the hinged sides and then both the camera and monocular were fastened to a steel box that would also contain the timer electronics and a battery pack. The monocular made for a smaller configuration, but focusing had to be done live on the viewfinder when setting up the scene for the first time.

The version of the camera zoom extender shown in Figure 25-17 can detail a license plate from miles away. It does take some effort to set up the scene and get the camera to focus, but once

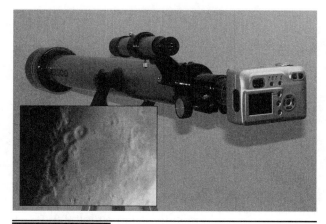

Figure 25-17 For ultra-long-distance photography, a telescope works well.

working, this system can see as far away as the telescope can, bringing in details that are far out of reach from most digital camera lenses. The only drawback to using such a large amount of magnification is that the camera has to be triggered either by the self-timer or remote control as the small movements cause by touching the camera would create massive shifts in the scene. With a steady tripod and a remotely- or self-timed camera, the images are fairly crisp, and the system can see at such magnification levels that moon crater photographs are even possible!

To adapt a small digital camera to a telescope, use the standard eyepieces that you would need to view the image yourself and then affix the camera so that the lens is about ¼-in away from the lens on the eyepiece. The system shown in Figure 25-17 uses a small plastic tube with set screws to hold the camera to the eyepiece directly, although this option only works well if your camera lens sticks out of the camera body. The flatbar system shown in Figures 25-13 and 25-14 would also work for a telescope mounting.

For ultra-high-quality long-distance imaging, we use our Nikon SLR along with the gun sight mounted to the tripod as shown in Figure 25-18.

This system allows the acquisition of images with ultra-high details from a distance of a block away when using the camera at full-quality resolution. There is only a slight darkening around the edges of the image, but other than that the quality is very good considering how easy to build and inexpensive this setup was.

To test our gun-sight version of the zoom extender, we set the camera for no flash, quick exposure, and then shot a scene over 500-feet away on a dreary snowy day. The image was bright, crisp, and full of detail as shown in Figure 25-19. The license plate was huge when we viewed the image on the computer screen with no scaling. We could have probably seen the color of a person's eyes at this distance using the zoom extender. The crosshairs were a bit annoying, so we later removed them from the gun sight.

The binocular version of the zoom extender did not have the ultra crisp clarity of the gun sight system, but this is to be expected since it can see a scene at 4× more magnification. When you look further, you will always be trading light and clarity for distance, but with a 10-megapixel camera, there is still a lot more detail captured than you can see with your naked eyes. Figure 25-20 shows both of

Figure 25-18   The completed gun-sight zoom extender using a digital SLR camera.

Figure 25-19   At over 500 feet away, a license plate is no problem at all for this system.

Figure 25-20 Both long range zoom systems being tested on the same scene.

the long-range zoom cameras being tested on the same winter scene. Even with the snow falling and lack of bright sunlight, the images captured were very high in detail.

This simple project proves that you don't have to spend thousands of dollars on a high-end camera and lens to spy on distant scenes. The quality of the images captured with the $50 pocket digital camera adapted to the $100 binoculars are very good, and the ability to see distant scenes with this setup rivals what can be done with a $5,000 telephoto lens.

PROJECT 26

Nikon Clap Snap

Have you ever wanted to remotely activate your Nikon camera? Maybe you already own the optional keychain remote but find it to be lacking in features? How about complete hands-free remote triggering? Yes, our clap snap project will not only allow hands-free operation, but it has a continuous 10-second time-lapse feature built in as well so you will never have to mess around with the camera menus or be caught pointing that keychain at the lens in your photos! This project feeds the output from an audio amplifier into the analog-to-digital converter on an inexpensive microcontroller that decodes three sharp sounds and then issues the infrared signal to the camera to take a photo. This project works on all of the Nikon SLR cameras we have tested, and may work on others as well. Modifications to the design to support different features and makes of cameras would certainly be easy.

Figure 26-0   Build the nikon clap snap remote control.

fit into just about any low-end 8-bit microprocessor with a little code reworking. The source code is presented in AVR assembly, but would be easy enough to convert to C.

It took some time to decode the pulse train that came from the original Nikon keychain remote control, but in the end it was found to be very similar to most of the VCR or TV remote controls; using a series of on and off intervals of a 40-kilohertz (KHz) carrier signal. This old and time-tested protocol is actually called "RC5" and although it varies from one manufacturer to another, it is basically a modulated 38- to 40-KHz carrier wave. The Nikon was found to use a 40-KHz carrier wave as per the pulse train measured on the output of the optional Nikon keychain remote control system.

We use our Nikon cameras a *lot*, and often we find ourselves pecking at the menus to get that

PARTS LIST

- IC1: ATMEL ATMega88P or comparable microcontroller
- IC2: LM386 1-W single IC audio amplifier
- Resistors: R1 = 1 to 10 K to bias electret microphone, R2 = 1 K, R3 = 1 K, R4 = 1 K, R5 = 1 K, VR1 = 10 K potentiometer
- Capacitors: C1 = 0.1 µF, C2 = 0.1 µF
- Microphone: Electret 2-wire style microphone
- Buzzer: 2-wire piezo buzzer element
- Battery: 3- to 9-V battery or pack

The clap snap program is designed for an AVR ATMega88 microprocessor, but is simple enough to

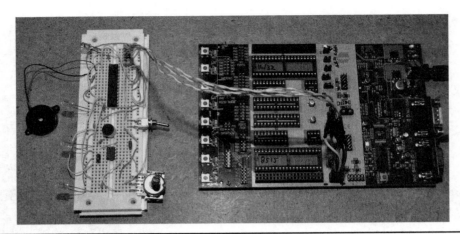

Figure 26-1   The original prototype was built on a solderless breadboard.

self-timer activated. The issue we have with the self-timer is that not all of Nikon cameras allow more than a single shot to be taken, and by the time we get back into place, the timer seems to already be taking the photo! We also have the keychain remote, but again, it forces us to point at the camera and then attempt to set up the shot with the remote in our hands. We wanted hands-free operation as well as a continuous 10-second time-lapse system that would offer a hint of warning before each shot so the clap snap project was born. Our goals were to have three claps, snaps, or sharp sounds trigger a shutter release while ignoring most background sounds. We also wanted to have a nonstop 10-second timer mode. A low-cost ATMega88 was chosen as it had more than enough power for the job and would run from its internal oscillator.

Once we had the timing parameters for the pulse train that the Nikon cameras needed, we created a very simple amplifier system that would send a voltage spike to the analog-to-digital input on the microcontroller (Figure 26-1). The amplifier responds very well to a short and sharp sound, but does only a fair job of ignoring any ambient background noise. Further processing done in the microcontroller will allow some "intelligent" listening to happen, making the camera only respond to three sharp claps or snaps in a row rather than false triggering every time the background noise level is high. To allow some feedback on operation, three visible light-emitting

diodes (LEDs) are added to the circuit. A red LED shows the sound spikes on the input of the ADC, a blue LED indicates that claps have been heard and that the window of opportunity is open, and a green LED flashes when the system has heard three claps and is about to take a photo. Control of the Nikon camera is through an invisible infrared LED, just like the optional factory-made remote control.

The schematic is very simple, only using the ATMega88, an LM386 audio amplifier, and a few other semiconductors (Figure 26-2). Any high-gain audio amplifier that can boost the output from the electret microphone would work, but the LM386 seemed to be very good at the task and is a common component. An electret microphone is a tiny microphone in a can that also includes a sensitive amplifier built in, and these can be found in just about any consumer device that requires a sound input. Old telephones, answering machines, tape recorders, and many toys have an electret microphone so check your scrap pile before looking to purchase a new one. An electret microphone will usually be about the size of a pencil eraser and have two wires on the underside, or two points to solder wires. One point will always be connected to the can, and this is the negative or ground side. This polarity is very important since the electret requires its own power source as shown in the schematic.

The 10-K variable resistor biases the output from the LM386 amplifier to VCC so that the level

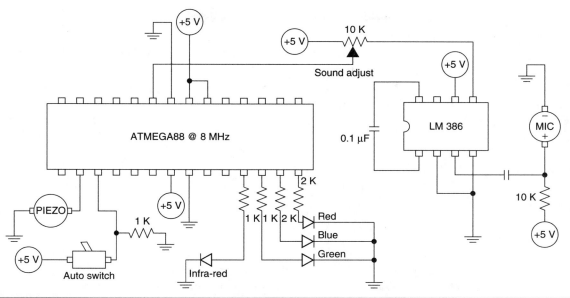

Figure 26-2 This is the basic clap snap schematic using an ATMega88.

of sound can be adjusted to help minimize false triggering if there is a lot of ambient background noise. The program code does a great job of detecting and counting three claps, but if the noise is overwhelming the ADC, there will be false triggering. The adjustment makes it easy to set the unit for proper operation in just about any environment. We added our variable resistor, the printed circuit board (PCB), and only set it once, but we have not yet needed to make any changes. Adding it to the cabinet with a knob would allow fine tuning at any time.

```
;   **********************************************************************************
;   ********** NIKON CLAP & SNAP : (C) 2008 BY RAD BRAD FOR LUCIDSCIENCE.COM
;   ********** TARGET : ATMEGA-88P WITH INTERNAL OSCILLATOR AT 8 MHZ
;   **********************************************************************************

; ATMEGA88 DEFINITION
.include "m88def.inc"


;   **********************************************************************************
;   ********** REGISTER DEFINITIONS
;   **********************************************************************************

.def t1 = r16
.def t2 = r17
.def t3 = r18
.def t4 = r19
.def clap = r20
.def reps = r21
.def lctr1 = r22
.def lctr2 = r23
.def wctr1 = r24
.def wctr2 = r25
```

```
; ***********************************************************************************
; ********** RESET AND INTERRUPT VECTORS
; ***********************************************************************************


; RESET VECTOR
reset:
rjmp startup
startup:


; STACK POINTER
ldi t1 ,low(ramend)
out spl,t1
ldi t1 ,high(ramend)
out sph,t1



; ***********************************************************************************
; ********** ADC AND IO PORT SETUP
; ***********************************************************************************


; MICROPHONE ANALOG INPUT
cbi ddrc,0
cbi portc,0

; RED STATUS LED
sbi ddrb,0

; BLUE STATUS LED
sbi ddrd,7

; GREEN STATUS LED
sbi ddrd,6

; INFRARED OUTPUT LED
sbi ddrd,5

; PIEZO BUZZER OUTPUT
sbi ddrd,0

; AUTO SWITCH INPUT
cbi ddrd,1

; SETUP ADC FOR FREE RUNNING
lds t1,ADMUX
cbr t1,(1<<REFS1)
cbr t1,(1<<REFS0)
sbr t1,(1<<ADLAR)
cbr t1,(1<<MUX3)
cbr t1,(1<<MUX2)
cbr t1,(1<<MUX1)
```

```
cbr t1,(1<<MUX0)
sts ADMUX,t1
lds t1,ADCSRA
sbr t1,(1<<ADEN)
cbr t1,(1<<ADATE)
cbr t1,(1<<ADIE)
sbr t1,(1<<ADPS2)
cbr t1,(1<<ADPS1)
cbr t1,(1<<ADPS0)
sts ADCSRA,t1


; *****************************************************************************
; ********* STARTUP AND INITIALIZATION
; *****************************************************************************

; FLASH LEDS AND BEEP ON STARTUP
rcall delay
sbi portb,0
sbi portd,7
sbi portd,6
rcall beeper
rcall delay
cbi portb,0
cbi portd,7
cbi portd,6

; RESET REGISTERS
clr t1
clr t2
clr t3
clr clap
clr reps
clr lctr1
clr lctr2
clr wctr1
clr wctr2
clr xl
clr xh



; *****************************************************************************
; ********* MAIN LOOP
; *****************************************************************************
main:

; RUN AUTO MODE IF AUTO SWITCH IS ON
sbic pind,1
rjmp aut1
```

```
; 10 SECOND AUTO TIMER DELAY FLASH
ldi t1,4
adl:
sbi portb,0
sbi portd,7
rcall delay
cbi portb,0
cbi portd,7
rcall delay
dec t1
brne adl

; COMPLETE DELAY AND TAKE PHOTO
sbi portd,6
rcall beeper
rcall delay
rcall photosnap
cbi portd,6
rjmp main
aut1:

; READ ADC INPUT VALUE AND SET CLAP REGISTER
lds t1,ADCSRA
sbr t1,(1<<ADSC)
sts ADCSRA,t1
adcloop:
lds t1,ADCSRA
sbrc t1,ADSC
rjmp adcloop
lds clap,ADCH

; CLAP DETECTION LED CONTROL & COUNT REPS
cpi clap,150
brne cdl1
sbi portb,0
ldi lctr1,20
ldi lctr2,255
cdl1:
cpi lctr1,0
breq cdl2
dec lctr2
brne cdl2
dec lctr1
cdl2:
cpi lctr1,0
brne cdl3
sbic pinb,0
inc reps
cbi portb,0
cdl3:
```

```
; TRIPLE CLAP DETECT WINDOW LED CONTROL
cpi clap,150
brne dwl1
sbi portd,7
ldi wctr1,255
ldi wctr2,255
dwl1:
cpi wctr1,0
breq dwl2
dec wctr2
brne dwl2
dec wctr1
dwl2:
cpi wctr1,0
brne dwl3
clr reps
cbi portd,7
dwl3:

; DETECT THIRD CLAP AND TAKE PHOTO
cpi reps,3
brne tcv1
clr reps
clr lctr1
clr lctr2
clr wctr1
clr wctr2
cbi portd,6
cbi portd,7
sbi portd,6
rcall beeper
rcall delay
rcall photosnap
rcall delay
cbi portd,6
tcv1:

; CONTINUE WITH MAIN LOOP
rjmp main



; *****************************************************************************
; ********* 1 SECOND DELAY ROUTINE
; *****************************************************************************
delay:
ldi yl,40
dly1:
ldi xh,high(60000)
ldi xl,low(60000)
```

```
dly2:
sbiw xl,1
brne dly2
dec yl
brne dly1
ret



; ****************************************************************************
; ********** PIEZO BEEPER
; ****************************************************************************
beeper:

; BEEP TONE 1
ldi t1,100
bp1:
sbi portd,0
ldi xh,high(1500)
ldi xl,low(1500)
bp2:
sbiw xl,1
brne bp2
cbi portd,0
ldi xh,high(1500)
ldi xl,low(1500)
bp3:
sbiw xl,1
brne bp3
dec t1
brne bp1

; BEEP TONE 2
ldi t1,100
bp4:
sbi portd,0
ldi xh,high(1200)
ldi xl,low(1200)
bp5:
sbiw xl,1
brne bp5
cbi portd,0
ldi xh,high(1200)
ldi xl,low(1200)
bp6:
sbiw xl,1
brne bp6
dec t1
brne bp4
```

```
; BEEP TONE 3
ldi t1,100
bp7:
sbi portd,0
ldi xh,high(1000)
ldi xl,low(1000)
bp8:
sbiw xl,1
brne bp8
cbi portd,0
ldi xh,high(1000)
ldi xl,low(1000)
bp9:
sbiw xl,1
brne bp9
dec t1
brne bp7
ret


; ****************************************************************************
; ********** NIKON REMOTE CONTROL SHUTTER SIGNAL
; ****************************************************************************
photosnap:
ldi t3,2
snaploop:

; CYCLE 1 = 16000 CLK MOD
ldi t1,77
c1:
sbi portd,5
ldi t2,33
m1:
dec t2
brne m1
nop
nop
nop
cbi portd,5
ldi t2,33
m2:
dec t2
brne m2
nop
nop
nop
dec t1
brne c1
nop
```

```
; CYCLE 2 = 222640 CLK PAUSE
ldi xh,high(22264)
ldi xl,low(22264)
dl1:
nop
nop
nop
nop
nop
nop
sbiw xl,1
brne dl1

; CYCLE 3 = 3120 CLK MOD
ldi t1,15
c2:
sbi portd,5
ldi t2,33
m3:
dec t2
brne m3
nop
nop
nop
cbi portd,5
ldi t2,33
m4:
dec t2
brne m4
nop
nop
nop
dec t1
brne c2
nop

; CYCLE 4 = 12640 CLK PAUSE
ldi xh,high(1264)
ldi xl,low(1264)
dl3:
nop
nop
nop
nop
nop
nop
sbiw xl,1
brne dl3
```

```
; CYCLE 5 = 3360 CLK MOD
ldi t1,16
c3:
sbi portd,5
ldi t2,33
m5:
dec t2
brne m5
nop
nop
nop
cbi portd,5
ldi t2,33
m6:
dec t2
brne m6
nop
nop
nop
dec t1
brne c3
nop

; CYCLE 6 = 28640 CLK PAUSE
ldi xh,high(2864)
ldi xl,low(2864)
dl4:
nop
nop
nop
nop
nop
nop
sbiw xl,1
brne dl4

; CYCLE 7 = 3200 CLK MOD
ldi t1,15
c4:
sbi portd,5
ldi t2,33
m7:
dec t2
brne m7
nop
nop
nop
cbi portd,5
```

```
ldi t2,33
m8:
dec t2
brne m8
nop
nop
nop
dec t1
brne c4
nop

; CYCLE 8 = 505600 CLK PAUSE
ldi xh,high(50560)
ldi xl,low(50560)
dl5:
nop
nop
nop
nop
nop
nop
sbiw xl,1
brne dl5

; CYCLE 8 = 505600 CLK PAUSE
ldi xh,high(50560)
ldi xl,low(50560)
dl6:
nop
nop
nop
nop
nop
nop
sbiw xl,1
brne dl6

; REPEAT SEQUENCE TWICE
dec t3
sbrs t3,0
rjmp snaploop
ret
```

The other components in the schematic make up the visible LED status lights and the infrared output that the Nikon camera will respond to. The visible LEDs are actually optional, but it does make it easier to see if the unit is operating and how it is responding to the sound. We also added a beeper that would alert us as to when the photo was being taken just in case we had to get ready for the shot. The toggle switch allows the system to work in sound-activation mode or nonstop 10-second timer mode when you need to do some time-lapse photography or just want a lot more than one image.

Even if you plan to build your version of the clap snap exactly as shown, it is highly recommended that you first build it on a solderless breadboard before attempting to hard wire the circuit (Figure 26-3). This makes it easy to program the microcontroller, test the operation of the unit, and of course make any necessary modifications to suit your needs. There are several free pins on the microcontroller, and the program hardly uses any of the flash memory, so there is plenty of room for improvements and modifications to this project. Our next version is going to have a full liquid-crystal display (LCD) and menuing system, but maybe you will beat us to it by expanding this project?

The values of the resistors on each LED shown in the schematic are based on the operating characteristics of the LEDs we have, so you will probably have to change them to get optimal brightness from your own LEDs. Try the 1-K resistor with the infrared LED at first, and if you find that the clap snap does not work within 10 feet of the Nikon remote control eye, you can either study your LED datasheet to calculate the optimal value or drive the infrared LED with a transistor. Range is not a goal here since the clap snap is designed to be within a few feet of the camera, so the 1-K value should be enough to drive

the infrared LED. Again, experimentation is the key, so mess around with the values and change the design as needed.

To ensure the unit is working properly before heating up the soldering iron, real tests should be done using your Nikon remote-enabled camera (Figure 26-4). If you have never used your camera with a remote control, then a little digging in the manual will be necessary as the camera needs to be set for "remote-control shutter release" or something similar. Also, identify the little infrared sensor window in your camera body; it will be the quarter-inch black plastic circle facing the same direction as the lens. If you are not sure where the remote window is on your camera body, there is always that dusty owners' manual!

Place the camera near the infrared LED so that the output is only a few inches away from the Nikon infrared-sensor window. Since you can't see the infrared output, this is the only way to verify that the circuit is actually functioning.

If your camera fails to respond to the signal to take a photo, then either you do not have the camera set up to receive the infrared signal or your circuit is not put together properly. You can try to capture the output from the infrared LED on an oscilloscope to ensure that there is a pulse train or just move the piezo buzzer to the infrared output

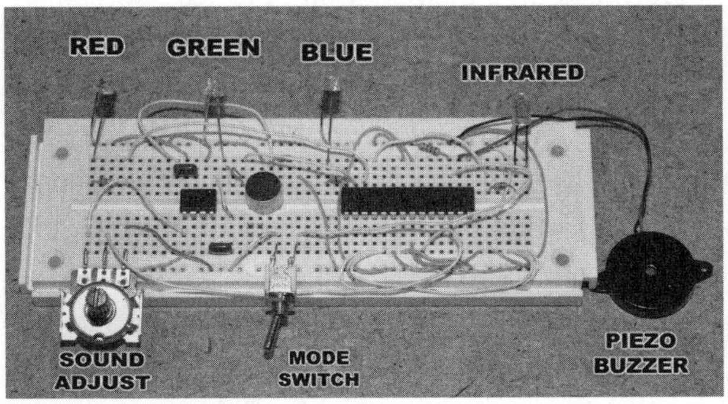

Figure 26-3   Breadboarding a home-built device makes debugging very easy.

Figure 26-4   Testing the prototype with an actual Nikon camera.

Figure 26-5   Adding the components into a small box.

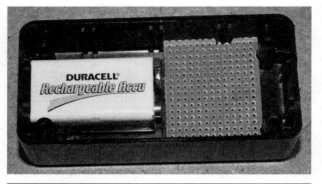

Figure 26-6   Fitting the perfboard and battery into the project box.

pin on the microcontroller and "listen" for the signal. When the microcontroller outputs the pulse train, you will hear a faint *buzz-buzz-buzz* sound on the piezo buzzer. If you are certain that your camera is ready to respond to the remote control signal and you cannot get it to trigger, then you have something wrong in your circuit, so backtrack and fix it!

Once you have verified the operation of the circuit using a camera, the parts can be soldered to a bit of perforated board or a PCB and mounted into some kind of project box (Figure 26-5). We decided to go with a 9-volt (V) battery as the power supply, so we used an LM78C05 regulator to step the voltage down to 5 V. The size of the battery will dictate the overall size of your project box, so start there.

Perforated board (perfboard) is the perfect method of taking your project from the solderless breadboard into the real world. Because the circuit is fairly simple and contains only a few parts, the hand wiring that has to be done will only take a few hours. Making a real circuit board for such a small project is more work than creating the entire project, so unless mass production is your goal, the perfboard is certainly the best option. We use perfboards that are just a wafer with holes, but you can also find perfboards with solder pads and even strips to match what you have on the solderless breadboard. If you are really in a hurry and on a budget, just pop the pins through some

cardboard and solder the wires in the underside of the cardboard!

We wanted to keep the unit as compact as possible, so we found a 2 × 4 inch (in) plastic box that had just enough room to fit the battery, small perfboard, LEDs, and switch inside (Figure 26-6). Always plan ahead, taking into account the space needed for things like wiring, battery clip, screws, switch leads, and even the box lid when you are choosing a project box.

To ensure that the parts will not bounce around inside the plastic box, the perfboard was clipped on the sides to fit snugly in the plastic box and the battery just happened to squeeze in between the two plastic screw standoffs. Hot glue or even Velcro strips are also a great way to hold down loose parts in a project box, if necessary.

Before adding the perfboard, the accessories such as LEDs and control switches are added to the box by drilling the appropriate holes. To save space, all leads are cut short or bent out of the way to allow more space for the circuit board and battery. Space will be tight, but it looks as though all of the parts will fit into this small plastic box. At this point, all common ground connections and the wires coming from the LEDs and switches can be attached (Figure 26-7).

The microcontroller and amplifier integrated circuit (IC) are the first components to be placed on

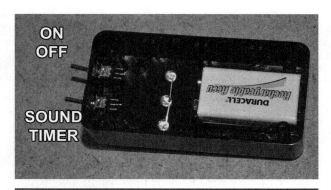

Figure 26-7 Adding the visible LEDs and control switches.

Figure 26-9 Adding the other components on the small perfboard.

the small perforated board (Figure 26-8). Placing the large ICs first towards the center of the board allows the smaller semiconductors to find room around the ICs as close to the necessary pins as possible. The idea is to cram all of the parts on the small board so that all wiring can be kept as short as possible. By placing parts close together, it may be possible to simply bend the pins and solder them together.

It is a good idea to use a socket for all microcontrollers when building a project like this since you may want to remove it later for reprogramming and upgrades (Figure 26-9). A socket makes it very easy to swap out a dead part that was fried by making a wiring mistake—not that we ever make those! We did not have the odd 28-pin socket needed for the ATMega88, so we just used a standard 20-pin socket and then cut off the

extra 8 pins from another to make a single socket. Ugly, we know, but it worked just fine!

The resistors (including the variable resistor) were also added to the board at a position that allowed one end to be soldered directly to the corresponding pin in the ICs, keeping the wiring that had to be done clean and minimal. New capacitors also come with extra-long leads, so you can also bend them around to create makeshift traces when using a perfboard to make your final circuit.

The input/output (IO) wires can also live next to their corresponding pins on the perfboard to help keep the final wiring job as minimal as possible. We added all of the wires needed to connect the LEDs, switches, battery, and buzzer to the board using wires that were longer than necessary so they could be trimmer later (Figure 26-10). Also

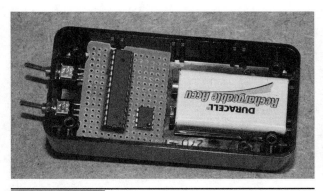

Figure 26-8 Laying out the components on the small perfboard.

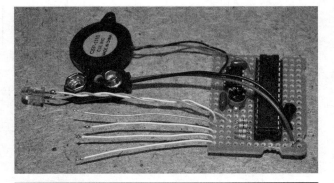

Figure 26-10 Adding the power, LED, and switch wires to the board.

shown in the perfboard is the small 5-V regulator that will cut down the voltage from the 9-V battery. As a testament to the robustness of the AVR microprocessor, we had the regulator connected wrong, and our project was actually functioning fine for over a week seeing a full 9 V! We only realized the error because the timer was running too fast. To this day, we are still using the same microcontroller.

Originally, we considered adding a digital countdown display using either an LCD or 7 segment LED display, but simplicity seemed better for this project, not to mention there was almost no room left in the small plastic project box. The final schematic was drawn up based on the original

prototype, but does leave plenty of program space and free IO pins to make a later upgrade (Figure 26-11). The general rule of DIY is that a project is never completed, and the monster will always need to grow!

Once all of the components have been placed onto the perforated board, the leads that are close enough to each other are soldered together, and then the rest of the circuit is made up of small bits of scrap wiring (Figure 26-12). We like to start with the power and ground connections first, as these always make up the bulk of the wiring and will be the source of any "dangerous" wiring errors, you know, the kind of mistakes that let out the magic smoke?

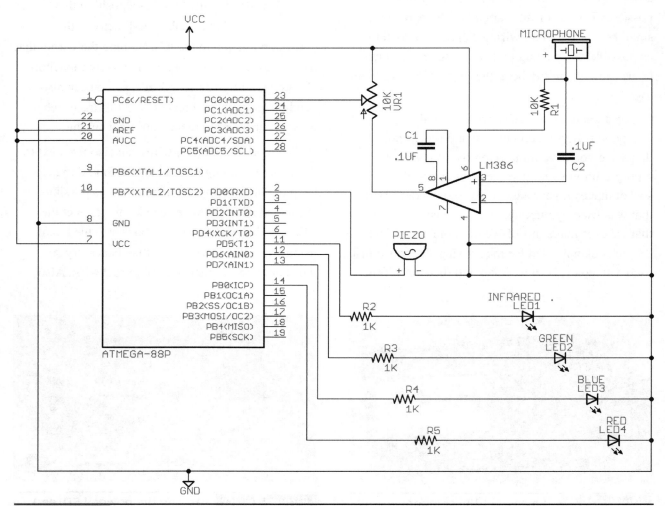

Figure 26-11 The clap snap final schematic diagram.

Figure 26-12 Creating a circuit board from a mess of wiring.

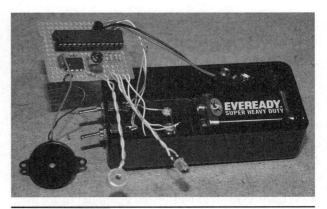

Figure 26-13 Final wiring and initial power test.

Once power and ground paths have all been checked for errors, the other semiconductors are soldered together as per the original schematic diagram. Always triple-check your work, and when it comes time to power up your project, use a current-limited power supply if you have one just to make sure you don't create a smoke show due to a crossed power connection. The clap snap will flash its LEDs and send a welcome beep when it starts up, so that will give an indication of what is working or not working. We also place our finder on the important ICs when power is first applied to ensure that nothing is getting warm, a clear indicator that power is reversed. It sometimes amazes us what extreme abuse an IC can take and still work!

With all of the wiring on the underside of the perfboard completed as well as the external component connections, it will be time for the moment of truth—will the project work, fail, or monumentally fail? We always shoot for instant success, but most of the time there are at least a few wires either left out or crossed (Figure 26-13). The only time a crossed wire is an issue is when it deals with VCC or ground, so we always take a great deal of time with those connections right from the start.

Using a current-limited power supply set to clip around 50 milliamps (mA) is always a safe way to initially power up a hand-wired board because

the needle with pin or the short circuit LED will come on before any of the components overheat. We prefer the excitement, so we just completed the wiring and threw the 9-V battery across the plug to see what would happen. We played our hand, and won—no smoke, and the buzzer greeted us with the expected blip-blop sound from the microcontroller. Okay, we had to fix the LED connections, though, as the colors were completely reversed from what we wanted. Not bad!

With everything working as expected, we carefully pushed all of the extra wiring into the box and fit the lid into place (Figure 26-14). The battery

Figure 26-14 Fitting all of the parts in the box one last time.

was on top so it would not be a huge hassle to change, and the wires coming from the perfboard were still long enough to allow the board to be removed for debugging without having to unsolder any of the wires. The clap snap was now complete, although it still needed some method of affixing it to the Nikon camera body when we used our tripod.

Once the last screw was installed on the box cover, the clap snap was tested again to ensure that none of the wires became crossed as they went into the box (Figure 26-15). Success again! The unit performed perfectly, and could trigger the Nikon infrared sensor from a distance of over 10 feet, although we never designed the output to have any real range. The goal was always to have the clap snap near the camera and allow the sound system to work from a very long distance from the source of the claps. The unit beeps as expected when the *on* switch is flipped, and both sound-activated mode as well as time lapse photography mode worked as expected.

One of the features we found by accident when testing various Nikon cameras is that the clap snap can trigger any Nikon camera that is within range of the infrared output. This means that multiple

Figure 26-16 The clap snap can trigger more than one camera at once.

camera angles can be used to acquire the same image, although we are not sure if one flash is good enough for more than one camera yet (Figure 26-16). We had our D90 and D60 firing after three claps, which was something that we will probably find useful when wanting more than one angle of the subject to be photographed.

It was also found that besides claps, snaps, or sharp bangs, spoken words could trigger the system, making it seem like it was actually voice controlled! We would yell "One, two, three" or even "Take my photo," and the system would recognize the burst of sound as three distinct voltage spikes into the analog to digital converter. It was fun to demonstrate the unit and tell people that it was full voice-activated! Of course, that just gave us new ideas for later, more advanced versions of this project.

One thing that was missing was the ability to easily affix the clap snap in an appropriate position while using the camera on a tripod. Since the clap snap has limited infrared range, it really needs to have a line of sight between the infrared LED and the Nikon photo sensor, which is located on the front of the camera body. To solve this problem, we added a small metal band that would just clip onto the strap clip, which was perfect since we never use a camera strap (Figure 26-17).

Figure 26-15 The completed and functional clap snap ready for use.

Figure 26-17 Adding a clip to allow tripod operation.

Figure 26-19 The mounting hook just drops into the strap clip.

We also thought about an adapter that would connect to the tripod, but the ability to hook it directly to the camera meant that we would never be messing around when it came time to use the clap snap. The metal band is just a 1/16-in thick strip of sheet metal cut and filed out to sit in the camera strap connector. There are many other ways that a mounting device could be made, including Velcro or some kind of wire body.

To affix the mounting hook to the project box, we had to open the cover, pull out all of the components one more time and then drill the appropriate bolt holes (Figure 26-18). The steel part

had to be bent in such a way that when the clap snap was hooked in place, it would not interfere with any of the camera controls or focusing hardware.

We also set the project box at an angle so that the visible LEDs would be clearly visible across a room. This simple mounting system has been working perfectly, although it may not be the optimal solution for all of the Nikon camera models (Figure 26-19). Of course, the clap snap can still be placed standing up anywhere near the camera as long as the line of site between the infrared sensor and LED is not more than 10 feet.

Overall we are very pleased with the way this project turned out, and it certainly does the job it was designed to do. When we decide to build version # 2, we will probably add real voice activation, multiple timer modes, a full backlit LCD screen, a keypad menu system, and external jacks for long-range triggering, and special FX modes for time lapse or tricky stop-motion shots. For now, we are enjoying the current version, which is being used for taking the photos that you see on our website.

Figure 26-18 The mounting hook is bolted to the project box.

We hope you enjoyed this project and find your new camera toy to be extremely useful in those situations where you need hands-free operation or nonstop time-lapse photography (Figure 26-20). If you make any cool modifications, please stop by our forum or send us a photo for our gallery so others can enjoy your work! Ideas and suggestions are always welcome, and since we are already working on the details for Clap Snap II, we are open to suggestions.

Figure 26-20   The clap snap lets you drive your Nikon with no hands!

Index

Note: Page numbers referencing figures are followed by an *"f."*